Basic of Veterinary Pharmacy

Basic of Veterinary Pharmacy

Jhansi Rani Mallam

RANDOM
RANDOM PUBLICATIONS
NEW DELHI (INDIA)

Basic of Veterinary Pharmacy

ISBN 978-93-5111-883-1

Published in 2016 in India by

RANDOM PUBLICATIONS

4376-A/4B, Gali Murari Lal, Ansari Road
New Delhi-110 002
Phone : +9111-43580356, 011-23289044, 011-43142548
e-mail: sales@randompublications.com,
info@randompublications.com, randomexports@gmail.com

Reprinted 2026

Type Setting by : Friends Media, Delhi-110089

Preface

Veterinary pharmacy is a specialized area of practice within the field of pharmacy as a whole. It is in the veterinary academic setting that pharmacists have established themselves as an integral and important part of the veterinary health care team in that veterinary hospital pharmacists are engaged in many different activities involving drug distribution, clinical services, teaching, and research.

Veterinary pharmacists are also often involved in clinical research with veterinary hospital clinicians and, less often, in areas of basic research with other faculty members of the veterinary school. Veterinary pharmacy is predicted to continue to grow, expand, and evolve in those areas in which it has already become established, namely, the veterinary schools and their associated teaching hospitals. There is also ample opportunity and need for pharmacists to become involved in other areas, such as the veterinary pharmaceutical industry, veterinary regulatory agencies, and agricultural and livestock production, which affect not only veterinary medicine but also public health as a whole.

Veterinary Pharmacy is a practice that has slowly evolved from the role of a sole Veterinarian selecting, dispensing and sometimes manufacturing his/her own medications to a highly regulated, research-based pharmacy profession that requires a vast background of knowledge about pharmacology, therapeutics, law and compounding of drugs.

It is hoped that this book would provide to a wide cross-section of people an understanding of some of the issues on significant aspects of areas of concern.overing the latest syllabus prescribed by most of the Indian universities.

– Author

Contents

1

Livestock Management and Production

ANIMAL CARE PERSONNEL

The principal scientist or animal management supervisor should make all animal care personnel aware of their responsibilities during both normal work hours and emergencies.

A programme of special husbandry procedures in case of an emergency should be developed. It is the reserach facility management's responsibility to ensure that personnel caring for agricultural animals used for research or teaching are appropriately qualified or trained. This responsibility may be delegated to an IACUC. Qualification by experience and training must be documented.

The animal facility manager must ensure that all animal care personnel are aware of their responsibilities during and outside normal work hours. Protocols for emergency care must be developed and made available to all personnel.

OBSERVATION

Animals in intensive accommodations should be observed and cared for daily by trained and experienced caretakers. Illumination must be adequate to facilitate inspection. In some circumstances, more frequent observation or care may be needed. Under extensive conditions, such as range or pasture, observations should be frequent enough to detect illness or injury in a timely fashion, recognize the need for emergency action, and ensure adequate availability of feed and water. A disaster plan must be developed for observing animals and providing care during emergency weather or health situations. Regardless of accommodations, animal observations should be documented and husbandry or health concerns reported to the animal facility manager or attending veterinarian as appropriate.

EMERGENCY, WEEKEND, AND HOLIDAY CARE

There must be a means for rapid communication in case of an emergency.

In emergencies, facility security and fire personnel must be able to contact staff members responsible for the care of agricultural animals. Names and contact information for those individuals should be posted prominently in the animal facility and provided to the security department or telephone center. If posting names and contact information poses privacy or security issues, a contact number for a security or command center should be used instead.

The institution must ensure that emergency services can be contacted at any time by staff members. The institution must assure continuity of daily animal care, to encompass weekends, holidays, unexpected absences of assigned personnel, and emergency situations. Staff assigned to weekends and holidays must be qualified to perform assigned duties. Cross-training of staff and establishment of standard operating procedures is encouraged to assure consistent, high-quality care.

Emergency veterinary care must be readily available after daily work hours, on weekends, and on holidays. In the event that weather conditions or natural disasters make feeding temporarily impossible, every attempt should be made to provide animals with a continuous supply of water. Absence of feed for up to 48 h should not seriously endanger the health of normal, well-nourished juvenile or adult cattle, sheep, goats, horses, poultry, or swine. Feed should be provided within 24 h to very young animals that are not nursing their dams.

EMERGENCY PLANS

A site-specific emergency plan must be developed to care for agricultural animals that are used for research and teaching. The goal for a plan should be to provide proper management and care for the animals regardless of the conditions. However, some conditions may be so unusual and extreme that it will not be possible to provide immediate care for the animals and to simultaneously ensure employee safety.

Thus, emergency plans should define proper animal management and care and parameters to ensure employee safety. Emergency plans should name employees or positions that are considered essential for providing proper animal management and care.

Those employees should a priori understand that responding to emergencies is a condition of employment and that they will be held accountable should they fail to care properly for the animals.

Plans should focus on emergencies that are most likely to occur in the specific geographic area or the research or teaching facility. Emergency plans should include animal evacuation plans specific to the research or teaching facility and actions to be taken if transportation is interrupted.

ANIMAL IDENTIFICATION AND RECORDS

Animals should be permanently identified by a method that can be easily read. Identification of individual animals is desirable, but, in some circumstances, it is acceptable to identify animals by group, cage, or pen. Individual birds may be wing-banded or leg-banded. Ear-notching, ear tattooing, electronic transponders, and branding may be used for individual identification of other species, and each has its advantages and disadvantages. Ear notches and tattoos are permanent and effective, but notching constitutes elective surgery and tattoos generally cannot be read without restraining animals.

Electronic transponders require special sensor units or stations, but should be considered when possible. Cattle and horses are most consistently identified using freeze-branding on the hip, shoulder, rear leg, or side. In addition, when freeze branding is used on more than one breed of horse, branding is performed under the mane. Some states require that cattle be permanently identified by branding with a hot iron; however, this procedure is more stressful than freeze-branding.

Ear and neck chain tags, although readable at some distance, can become lost and are therefore not necessarily permanent. In addition, neck chains and straps should be avoided in situations in which the animal could become entangled in a fence, rock outcropping, or other feature of the environment. Any associated pain and distress should be considered when determining the method of identification.

In some cases, it may be necessary to identify animals in multiple ways. Individual records are needed for most animals. These records should include information about the animal, its source and location, its productivity, its reproductive performance, protocols the animal is assigned to, and its ultimate disposition. Records for individual animals or groups should also include dates of vaccination, parasite control measures used, blood testing dates and results, and notations as to whether castration, spaying, or other elective procedures have been performed.

Applicable veterinary data to be recorded include dates of examination/treatment, clinical information/diagnosis, names of medications and amounts and routes of administration, descriptions of surgical procedures, and resolution of surgical procedures or illnesses. Principal scientists or animal facility managers may wish to record nutritional information. Research protocols often dictate that additional information be recorded.

VERMIN CONTROL

Programmes should be instituted to control infestation of animal facilities by vermin. The most effective control in facilities prevents entry of vermin into the facility by screening openings and ceilings; sealing cracks; eliminating vermin breeding, roosting, and refuge sites; and limiting access of vermin to feed supplies and water sources. Building openings should be screened with 1.3-cm mesh, and ceilings with ridge vents should be screened with 1.9-cm mesh to minimize rodent and bird entry. Smaller mesh sizes are recommended where they will not interfere with airflow. Mesh may need to be installed along foundations below ground level, especially with wood foundations.

Pesticides should be used only as approved. Particular caution should be exercised with respect to residues in feedstuffs, which could injure animals and eventually pass into the meat, milk, or eggs. Pesticides should be used in or around animal facilities only when necessary, only with the approval of the scientist whose animals will be exposed to them, and with special care. A pesticide applicator or a commercial service may be used.

In some regions, wildlife and stray cats and dogs may spread zoonotic diseases, including rabies, to agricultural animals. In highrisk locations, institutions should implement an educational programme that includes training scientific and animal care personnel to recognize the signs of rabies in both wildlife and agricultural species and to handle and report potentially rabid animals. Inoculation may be advisable for humans who may come into contact with animals in regions where rabies is endemic. Many agricultural institutions keep cats for pest-control purposes. Although the use of free-roaming cats is a traditional form of pest control for agricultural facilities, cats may limit the ability for baiting and may present hygiene or accident risks or serve as disease vectors. However, when cats are present, proper veterinary care and oversight should be provided to these animals. Veterinary care should include vaccinations, parasite control, and neutering.

MANAGEMENT OF LIVESTOCK

When we study about improving our livestock we learn how they must be sheltered, fed, and mated, what kind of drinking water should be given to them and how the sick and diseased animals ought to be treated? This way we learn to manage our livestock for better production and utilization.

FEEDING OF ANIMALS

All animals must be fed properly. The food should contain the requisite nutrients *i.e.* carbohydrates, proteins, fats, minerals, vitamins and water.

The food which is given to cattle can be divided into two categories:

1. Concentrates like cotton seeds, oilcakes, cereal grains, bran etc. They are very rich in most of the nutrients.
2. Roughage includes fibrous and rough food like straw and stems of cereal crops. Generally roughage has a low nutrient content.

An average Indian cow eats about 15-20 kg of green fodder and 4 to 5 kg drygrass, which is mixed with a sufficient amount of grain. A cow drinks about 32 litres of water. Goat and sheep eat grass, herbs and waste products from the farms. Pigs are usually given cereals and their products to eat. Poultry birds are given a mixed feed consisting of cereals, bone meal, minerals and vitamins.

Housing of Animals

We must protect our animals from too much heat, rain and cold. We must, therefore, be careful where we house them. Their houses should have proper sanitation and ventilation. Too many animals should never be kept in a small space. Different animals require different types of houses. Hens and fowls are kept in cages while sheep and goats stay in open yard, which is partially covered with roof made of straw. This open yard should have a hedge of iron wires all around to prevent the animals from running away.

Water and its Supply

To keep these animals healthy they should be given clean water to drink and in sufficient quantities. For example, on an average a cow consumes about 27-36L of water, pigs require 5-23L, camel 8-90L and poultry birds require about 240mL of water. Besides this we must also bathe the cattle with clean water.

Some Common Diseases of Animals and Vaccination

Sometimes domestic animals may be afflicted by diseases.

Some of the common diseases of animals are listed below:

- *Viral Diseases*: Pox in cattle, goats, sheeps, and fowls; dermatitis in goats and sheep; foot and mouth disease in cattle.
- *Bacterial Diseases*: Tuberculosis in cattle and poultry birds; cholera in fowls; diphtheria in calf; diarrhea in chicks; foot rot in sheep.

Most of these diseases can be prevented by proper sanitation, a controlled diet, proper housing and also by vaccinating the animals against these diseases at the proper time and age.

Animal Breeding

Breeding means to reproduce. In case of animals, breeding is done to obtain animals with desired characters. The two individuals of desirable characters can be selected as parents. These are then crossed to obtain new breeds of animals, *e.g.* by cross breeding a cow of low milk yielding breed, we can get breeds of cow which produces more milk.

Artificial insemination is an important and effective method of breeding. The process involves injecting the semen obtained from desired bull belonging to high milk yielding breed into the reproductive tract of female during heat period. It generally gives important breeds, and is widely used to improve the qualities of cow, buffaloes, poultry, horse and goat etc.

ANIMAL PROCUREMENT

When an institution acquires new animals, attention must be paid to applicable international, federal, and state regulations and institutional procedures, particularly those dealing with transportation and animal health. All animals must be obtained and transported legally. The attending veterinarian, in conjunction with the principal scientist, should formulate written procedures to assess the health status of a herd or flock obtained from a vendor before acquiring animals. The institution should develop a mechanism and process of control for animal acquisition that ensures coordination of resources that will preclude the arrival of animals in advance of preparation of adequate housing and appropriate veterinary quarantine procedures. Quality control for vendors and knowledge of the history of purchased animals is part of an adequate institutional veterinary care programme.

Animals of unknown origin or from stockyards should only be used if necessary; such animals may pose significant unknown health risks compared with animals of known origin and therefore should be handled appropriately. Newly acquired animals should undergo a quarantine and acclimation period, including preventive and clinical treatments as appropriate for their health status.

ACCLIMATION AND STABILIZATION

Newly arrived animals require a period of acclimation. Acclimation refers to a stabilization period, before animal use, which permits physiological and behavioural adaptation to the new environment. The attending veterinarian should establish general acclimation guidelines for each species. Any modifications to the general programme should be discussed with the attending veterinarian before animals are shipped. In some cases, animals may require an extended acclimation period because of their history or health status. On the other hand, some studies, such as comparisons of metaphylactic treatments for shipping fever, need to begin as soon as animals arrive. Such exemptions from the acclimation period must be scientifically justified and approved by the Institutional Animal Care and Use Committee (IACUC).

QUARANTINE

Quarantine is the separation of newly received animals from those already in the facility or on the premises until the health of the new animals has been evaluated and found to be acceptable. The attending veterinarian should ensure that quarantine facilities or locations are appropriate and that quarantine procedures are consistent with current veterinary practices and applicable regulations. The quarantine period should be long enough to observe signs of infectious disease or obtain diagnostic evidence of infection status.

Quarantine and testing of animals before introduction is especially important for herds or flocks that have attained specific-pathogen-free status, but these additions should be discouraged. If the health history of newly received animals is unknown, the quarantine programme should be more comprehensive and sufficiently long to allow expression or detection of diseases present in the early incubation stage. Exceptions to quarantine practices should be approved by the attending veterinarian in advance of shipment of the animals. The attending veterinarian, or skilled personnel under the direction of the attending veterinarian, should perform an initial examination and subsequent daily observations for newly arrived animals. Animals should be observed in quarantine until they are cleared for introduction into a herd or facility. During the quarantine period, animals should be vaccinated and treated for diseases and parasites as appropriate to protect their health and maintain the health of animals in the home facility. In addition to having adequate quarantine procedures, research facilities and animal use protocols should be designed to minimize the risk of introducing or transmitting disease agents.

SAFETY AND ANIMAL HANDLING

Stockmen working in animal housing and associated handling facilities are exposed to a number of potential risks and health hazards. Many stockmen work alone, handling big, sometimes dangerous animals as well as large groups of animals.

Several work tasks involve daily lifting, carrying of heavy loads and awkward working postures, which may cause overload injuries on the worker's musculo-skeletal system. Other potential risk factors contributing to accidents as well as to physical health problems and physiological stress for the worker is the fact that they operate in an environment where they are exposed to airborne particles, gases, noise.

The work environment is partly the same environment, which is provided for the animals, and it should be noted that there is a strong relationship between the animal welfare issues and the handling of the work environment problems in cattle houses.

HEALTH AND INJURIES AMONG LIVESTOCK FARMERS

Accident Agents

In an accident situation there is a complex interaction between man and injury agents. In animal husbandry the human is subjected to a high degree of exposure to a large number of factors that singly or in combination can be accident provoking, such as animals, equipment and building structures. Risky situations are often due to wet and slippery floors as well as systems for material and animal handling.

Prevention of Accidents

With reference to accident statistics some accident preventing facilities and measures can be listed:

- Good cattle handling and handling facilities

- Pens from which the herdsman can easily escape if necessary in case of for instance frightened cattle and aggressive bulls
- Personnel passages through yard fences, properly placed
- Wide feeding passage or possibility to shut out the cattle from the feeding area
- Non slip flooring
- Satisfactory lighting considering quality and quantity
- Well designed ladders and stairways
- Protective railings around the fodder hay-chute in two storey houses
- Handles and bolts that do not stand out in alleys and pathways
- Protectors for the knuckles on barrows
- Emergency stops on machinery

Table. Recommended Illumination Level for Different Work Areas in Buildings Used for Beef Production. Lux Values Refer to the Work Plane. In Addition to the Quantity, the Quality of Light must be Considered

Building or Work Area/Task	Intensity of Illumination, Lux
Animal housing area	150
Animal handling area	250
Veterinary treatment	1,000
Feed processing, feed alley	250
Silo, feed storage	100
Ladders and stairs	250

Ergonomics

Any normally fit adult person, regardless of sex, physical strength and body dimensions should be able to work comfortably in livestock buildings. Despite an increasing degree of mechanisation and automation many jobs in animal husbandry are still associated with lifting heavy loads, moving and carrying equipment, feed and other materials with manual methods.

An important step towards improved human welfare within cattle housing is to eliminate unnecessary loads on the human musculoskeletal system.

Health Aspects Related to Air Quality

It has been known for decades that farm workers suffer health risks due to inhalation of gases and airborne particles commonly referred to as dust. The particles may be liquid droplets or solid. Several respiratory disorders have been reported, which are associated with dust and work in confined livestock buildings such as hypersensitivity, pneumonia, acute inflammation, chronic bronchitis, occupational asthma and toxin fever.

The respiratory problems connected to gases range from mild irritations of the respiratory tract to lethal effects. For the most frequent components in the air environment in livestock buildings there are statutory threshold limit values, which differ from country to country. It should be noted that threshold limit values in connection with animal welfare in many countries often are lower than the work environment legislation.

Dust and Dust Reducing

Dust in animal houses is generated inside the building and arises primarily from the animals and bedding and feed material. The airborne particles are often carriers of biologically active material like endotoxins, pathogens and allergens that can highly be disease provoking.

Of special interest is the concentration (*i.e.* number of particles per air volume) of respirable particles with a diameter less than 5 μm that can be inhaled and deposited in the lung.

The smaller particle the higher possibility that the particle reaches the lower respiratory tract and the deeper the particle is deposited the higher is the risk for illness. Even if "common dust" can be seen by human eyes (larger diameter than 100 ìm), the very small respirable particles cannot be, so we are indeed dealing with a danger our senses are not able to detect.

There are basically four approaches that can be used for airborne dust reducing:

1. Prevent particle formation
2. Prevent particle release
3. Remove suspended particles from enclosed workspaces
4. Isolate workers from dust clouds in workspace.

Of course the best approach in this matter is to have the cattle and the working sites outdoors or in open barns. In the cattle barn the most effective way to obtain a low dust concentration level is to prevent formation by always using hygienically perfect forage and bedding material. Providing an effective ventilation system in a barn is a good practice to remove suspended respirable particles as well as gases.

Gases and Reduction of Gas Levels

In animal buildings over 150 different gases and volatile compounds have been identified. Most of the volatile compounds originate from biodegradation of animal excreta or are produced by the animals themselves such as carbon dioxide (by the respiration of the animal). The gases that are found in the highest elevated concen-trations are ammonia, carbon dioxide, hydrogen sulphide and methane.

It must be emphasised that it is dangerous to enter any manure tank without either using a self-contained air supply or combining several measures: testing the air, constant and adequate ventilation of fresh air and using an harness and lifeline on the person entering the tank. As well, human and animals positioned over a slatted floor with manure storage underneath are at risk during agitation, mixing or pumping of manure due to the gases, especially hydrogen sulphide.

A good strategy is to evacuate dangerous zones during these manure-handling actions.

Ammonia

Ammonia is the most common polluting gas in the atmosphere of the cattle barn. Animal manure is the main source of ammonia. The gas together with hydrogen sulphide a major component in what is termed odour, noticeable at ammonia concentrations of 5 ppm or more. Higher concentrations than that cause irritation to the respiratory organs and can aggravate the negative health effects of high dust concentrations.

Many factors affect the emission and concentration of ammonia such as air flow rate, manure and air temperature, manure surface area, density of animals, the degree of mixing of urine and faeces, time intervals between manuring, the pH-value, carbon/nitrogen ratio, mois-ture content and type of bedding. Consequently, many measures can be taken such as: proper ventilation and sufficient number of air changes, low air temperature in the barn and good urine drainage characteristics of the floor.

Hydrogen Sulphide

Hydrogen sulphide is the most toxic gas in animal confinement houses and 200 ppm and more are lethal for humans. The presence of the gas is of concern in buildings with liquid manure systems particularly. During slurry clearing operations concentrations in the range of 1,500 to 2,000 ppm have been reported. Since hydrogen sulphide is lethal the aim should be to keep it below detectable concentration at all working sites. Special care has to be taken when pumping or agitating the manure. During emptying a slurry cellar the workers and the animal should be evacuated from the barn. In mechanical ventilated, closed animal houses continuously evacuation of the gases in an indoor slurry pit through a perforated duct under the floor and along the pit can reduce the gas concentrations above the slatted floor. By removing the slurry daily from the cattle barn to an outdoor storage it is possible to hold the concentration at an acceptable level.

Methane

Methane is generated when manure is stored under anaerobic conditions. This occurs in all non-aerated manure storages. Primarily, methane is dangerous because of its flammability. If methane is mixed in a proper proportion with oxygen a spark will be able to set off an explosion.

Carbon Dioxide

Carbon dioxide is present in all air, indoor as well as outdoors. Only extremely high concentrations (70,000 ppm or 7 per cent and over) could have severe health effects. Mostly, incidents and accidents involving carbon dioxide are connected with asphyxiation due to oxygen deficiency rather than the direct effect of carbon dioxide itself. For instance, it is dangerous to enter a haylage silo during storing. In animal buildings the gas is more an indicator of ventilation efficiency than a health risk.

Noise

Noise can be defined as "unwanted sound, rapid, annoying pressure vibration in the surrounding air". Ventilation fans, grinders, vacuum pumps and mechanical feeding and manure systems generate noise. High sound levels are a real stress factor. Exposure to noise will be able to cause temporary deafness and permanent hearing loss after a variable period if the level is over 85 dB(A). The unit, dB(A), is exponential and the human sense of hearing will perceive the sound level as halving if a reduction of 10 dB(A) occurs. It is fairly easy to reduce the sound level of ventilation fans and other mechanical noise generators. Maximum noise level recommendations and regulations differ from country to country.

Furthermore, it is important to reduce noise when handling cattle in order to improve animal movement and to make cattle handling safer. Cattle are more sensitive to high frequency noise than human. Animals will be calmer and easier to handle, if noise level is reduced. Clanging and banging metal parts should be silenced with rubber pads. Equipment operated with hydraulics should be engineered for quietness.

CATTLE HANDLING AND CATTLE HANDLING FACILITIES

Handling facilities are an essential part of a safe, easy and rapid handling of cattle. Appropriate handling and handling facilities remove much of the stress and frustration of the workmen, which inevitably occurs with excited, stubborn or aggressive animals. Properly constructed facilities confine cattle safely and efficiently with minimal animal stress and risk of injury to both cattle and workers.

Animal Behaviour and Improving Animal Handling

Understanding cattle behaviour can help farm workers to avoid dangerous situations and minimise accidents to handlers, as well as to design the handling facilities appropriately. Animals have natural boundary called the Flight Zone.

Deep penetration into the Flight Zone can cause panic and escape attempts. Handlers should have the possibility to remove, and should remove themselves from the Flight Zone, if the animal becomes aggressive. To move an animal forward: stand in the shaded area position B, behind the Point of Balance at the shoulder. Keep out of the Blind Spot at the rear of the animal. To stop movement: back off to position A. To make an animal back up: stand in front of the Point of Balance. To make an animal turn left or right: approach the animal head on. Factors reducing an animal's Flight Zone size: frequent contact with people; history of gentle handling; calm environment. Factors enlarging the Flight zone: infrequent contact with people; history of rough or abusive handling; excitement.

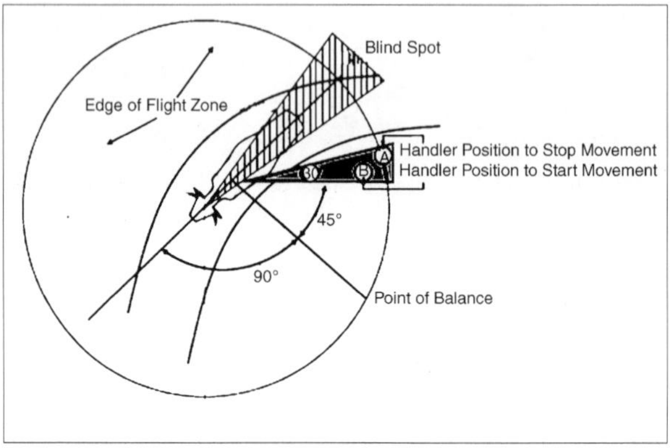

Fig. The Flight Zone of Animals

To improve animal handling and animal movement, many technical measures can be taken such as:

- Appropriate handling facility design.
- Animals tend to move from a dark to a more brightly lighted area. The light should illuminate the chute up ahead. Eliminate shadows and patches of light and dark, which may confuse animals. An approach is to illuminate the entire working area. Lamps should not shine into the eyes of approaching animals because glaring and blinding light impedes movement. Illumination should be uniform and diffuse.
- Prevent distractions, such as a chain hanging down in an entrance. Avoid sparkling reflection in a puddle, a moving reflection on a sheet of metal or bars of shadow across an otherwise sunlit alleyway. Dark colours can create shadow effects. Bright colours such as white and light yellow have been proven satisfactory.
- Reduce noise
- Animals might refuse to move, if they can see people ahead. Install shields to prevent animals from seeing farther ahead. Gates can be rigged with motor controls so a handler standing behind the cattle can open them.
- Solid sides that prevent the cattle from seeing outside the fences should be provided in the races and crowding pens. The crowding gate on the crowding pen should be solid as well, preventing animals from attempting to turn back to were they come from.
- Cattle are sensitive to changes in type and texture of floors and fences.

Changes in type of flooring can cause balking. Use the same type of flooring throughout a facility, if possible. Use non-slip flooring. Drains should be located outside main drive alleys, chutes and crowd pens.

FUNDAMENTAL CHANGE IN THE NATURE OF THE HUMAN-ANIMAL RELATIONSHIP

ANIMAL DOMESTICATION

Domestication is most commonly portrayed as a fundamental change in the nature of the human-animal relationship. Some authors have envisaged domestication more as a continuation of existing human-animal relations; in other words, as an extension or elaboration of the hunter-prey relationship, rather than as a complete break from it. Domestication can in this way be viewed as the end product of a series of gradually intensifying relationships between humans and other animals. With respect to this, Jarman and Wilkinson (deem it inappropriate to focus merely on the dichotomy between the wild and the domestic.

They suggest that this represents only one aspect of a wide range of close relationships between humans and other animals. Marginal cases, such as the reindeer economies and game-cropping of the present day, they argue, imply that this dichotomy may not have been so clear cut in the distant past. Similarly, Hecker has proposed a broad spectrum of human-animal interactions, ranging from the cooperative driving of animals to specialised hunting and culling to finally the selective breeding of domestic livestock. Although this view of domestication highlights varying degrees of cultural manipulation and control which evolved over time to obtain animal protein, it fails to fully account for the radical changes in human social life and behaviour that animal domestication brought with it in its wake. As Bokonyi points out, although human activities have—in some way or another—always interfered with the lives of other animals, domestication is 'an interference of a quite different kind'.

The profound social significance of the transition from hunting to herding has, however, been incorporated into several recent definitions of domestication. Meadow, for example, describes animal domestication as being a 'selective diachronic process of change in human-animal relationships involving, at the very least, a change in focus on the part of humans from the dead to the living animal and, more particularly, from the dead animal to the principal product of the living animal—its progeny'.

This process, he argues, manifests itself in two respects: first, in terms of the transformations in the social and economic structure of the human societies which associate with the animals; and secondly, in the behavioural, morphological and physiological changes which the animal undergoes as a consequence of domestication.

Similarly, Clutton-Brock defines a domesticated animal as 'one that has been bred in captivity for purposes of economic profit to a human community that maintains complete mastery over its breeding, organization of territory, and food supply'. She goes on to argue that domestication is both a cultural and biological process which 'can only take place when tamed animals are incorporated into the social structure of the human group and become objects of ownership'. The morphological transformation of the animal occurs subsequent to its initial integration into human society. The successful domestication of animals was most likely the product of a long-term process of trial and error. It has been suggested that early human efforts to tame other animals were a product of the highly social nature of humankind. Clutton-Brock, for instance, contends that the enfoldment of other species into human society was an extension of the practices of 'sharing, nurturing and protecting weaker members of the human group'. The assumption that our ancestors would have been prepared to tolerate or support weaker persons, other than perhaps infants and young children, let alone members of other species, is somewhat suspect. Historical and anthropological accounts, however, confirm that in some societies, women suckled young mammals along with their human offspring.

This suggests that within early hunting and gathering societies, juvenile animals were perhaps trapped, nurtured and raised alongside humans and were granted a certain level of protection by their human captors. The European explorers of the eighteenth and nineteenth centuries recorded many instances of pet-keeping and affection for small animals amongst the indigenous peoples they encountered. On the basis of their accounts, during the late nineteenth century, Francis Galton postulated that the 'savage' penchant for taming and caring for small animals as pets provided the basis for the development of livestock keeping.

This notion has in fact provided a cornerstone for many explanations of the origins of domestication, most particularly with regard to the early assimilation of dogs into human society. Archaeological findings suggest that *Canis familiaris* was probably the first animal species to undergo domestication. Unlike later domesticates, dogs were not—as far as we are aware—much eaten. Instead, they were most likely used as an aid for obtaining meat. It is widely assumed that the domestic dog descends directly from wolves. Wolves exhibit complex social behavioural patterns similar to humans. They are efficient group predators with a social structure based on a dominance hierarchy. It is thought that our ancestors began to develop a close association with wolves by rearing young cubs that they had caught or found.

The hypothesis follows that some of the more placid of these animals reached maturity and accepted human beings as pack members in their adulthood. These tamed animals began to breed in human captivity and, over several generations, eventually developed behavioural characteristics distinct from their wild relatives. This process ultimately culminated in the evolution of a separate kind of animal: the dog. Humans and wolves were probably close competitors for food. They shared the same prey and it is likely that they came into close contact, wolves possibly learning to scavenge on the leftovers from human game drives and the parts of animals which humans preferred not to eat in times of plenty. Although one can speculate on a mutual interest in proximity, it is perhaps more probable that humans saw a way of surpassing their lupine competitors in predation and securing food for themselves by using tamed adult wolves—and eventually dogs—to help in the hunt by detecting and tracking game, and later to help herd other animals rather than prey on them.

The dog is unique amongst early domesticates, given that it was probably not domesticated specifically for food. Most species that underwent domestication were probably intended as 'walking larders'. Sheep, goats and cattle, for example, were most likely exploited as transportable sources of meat and other animal by-products. At a later stage, species such as the horse, donkey and camel were domesticated to provide muscle power for transport and traction, although their meat and milk were also consumed long before they began to perform these roles in human society.

Although humankind was effectively able to secure a constant supply of animal protein for itself through the enfoldment of animals into human society, the fact remains that only a handful of species were ever domesticated successfully. There are comparatively few domesticated species commonly found in Europe today. The majority of these species originated in the Near East; others descend from the Americas and were only introduced to the European continent following the discovery of the New World. Whilst the appearance and behaviour of these species have changed dramatically since their incorporation into human society, they have flourished under human protection and have, in some instances, outlived their wild or ancestral counterparts. However, as Clutton-Brock points out, the benefits of living under the umbrella of human protection are somewhat dubious. Despite the massive population size and geographic distribution of domesticated animal species in contrast to their wild progenitors, these species have suffered 'irretrievable loss of genetic diversity and evolutionary autonomy'.

DOMESTIC ANIMAL DIVERSITY

Domestic animal diversity (DAD) is rightly labeled one of the most threatened aspects of biodiversity by the Food and Agriculture Organization (FAO), the UN entity charged with global oversight of DAD documentation and conservation. Much of today's remaining diversity in domestic animal breeds survives in traditional farming and herding communities in the South, where it was generated by local/indigenous knowledge and social organization. Yet FAO and other international organizations have made little effort to integrate such knowledge and practice into their global strategies for understanding and maintaining DAD.

Their rationale for saving local/indigenous breeds from impending extinction seems to lie mainly in these animals' possession of valuable genetic material that may be of potential benefit to the North or to humanity at large. (Ironically, many indigenous breeds are now at risk due to cross-breeding policies previously promoted by the same formal-sector institutions now seeking to save them.) Scant attention has been paid to the fact that endangered breeds are frequently associated with marginalized social groups whose economic and cultural survival depends directly upon these animals - and who thus have an even stronger and more immediate interest in their conservation.

Agro-biodiversity is composed of crop and livestock diversity at the levels of genes, species and habitats. In general parlance, however, farm-animal diversity refers to the recognized breeds of ruminants, swine, equines and poultry that humans have developed across the millennia from less than a score of wild animal species. This DAD is now at grave risk. According to the Food and Agriculture Organization, only some 5,000 breeds or strains of all species of farm-animal domesticates still exist, from among untold thousands developed historically. Europe alone has lost nearly half the breeds of farm animals found there at the beginning of the twentieth century. Worldwide, about a third of the remaining 5,000 are in danger of extinction, with breeds disappearing at the rate of more than one per week.

In the North, most livestock is raised under intensive or even industrialized production systems that make for impressive outputs of meat, milk, eggs and fibre. These outputs have been achieved by using such high-tech reproductive operations as artificial insemination and embryo transplantation, selecting solely for productivity without regard to other traits such as fitness. In the process, the genetic base of the species so treated has been greatly narrowed and their natural survival traits have been compromised.

The resulting food-producing 'machines' are very vulnerable to disease and generally less hardy. They therefore require high veterinary and other inputs, such as special feeds, expensive housing and sophisticated husbandry. Further, these industrial breeds' fertility and reproductive performance have been undermined to such an extent that they often have difficulty mating, giving birth and mothering their young. It is questionable whether they could even survive outside the managed environments and factory farms in which they are raised.

To ensure at least a modicum of fitness and vitality in future populations of food-producing animals, and to keep genetic options open, access to fresh genetic material will always be required. But most of the wild relatives of today's animal domesticates are extinct. This means that a major source of such material lies in the more rustic livestock kept by farmers and herders under extensive, subsistence-oriented production systems in the South.

Variously termed local/indigenous or unimproved (that is, by Northerners), these Southern landraces of animals still harbour such invaluable characteristics as natural disease-resistance, strength, hardiness, good libido and fertility, and other survival-oriented traits. Indeed, in the absence of livestock with these traits, the majority of the world's people could not survive. Most of the world population is still rural, and all rural groups rely on domesticated animals for a key part of their livelihood; but few can afford or access the high-tech inputs and breeds typical in the North. Fortunately for people everywhere, however, Southerners' landraces often embody special traits that are of present or future economic importance. Their genetic makeup may offer possibilities for tastier, healthier or more specialized foods than the bland, generic products from factory animals.

But if, like their wild relatives, indigenous livestock breeds also disappear, then future human generations will have no genetic manoeuvring room in which to adapt their food-animal production systems to emerging disease threats or changing environmental and economic conditions. Thus serious DAD conservation efforts by FAO and other formal-sector institutions like the International Livestock Research Institute (ILRI) are vital to future food security.

Article 8 of the UN Convention on Biological Diversity states that genetic resources should be conserved in the 'surroundings where they have developed their distinct properties'. For livestock, this means the pastoral and farming communities that nurture most of the world's existing DAD. The diversification of a livestock species into many different strains or breeds is the outcome of different ethnic and social groups' managing that species in a wide variety of habitats and manipulating its genetic composition according to localized biophysical (climate, disease and predator threats, availability of forage, water, minerals, etc.) and socio-cultural (*e.g.* knowledge, beliefs, labour, social organization) conditions in the context of these human groups' varying needs and preferences for livestock goods and services.

In other words, albeit without the North's high tech, people in the South and in traditional societies everywhere have also consciously shaped breeds to their own ends - contrary to ethnocentric beliefs among many animal scientists that livestock in developing countries evolved without human intervention. Although targeted studies of animal breeding in traditional cultures are rare, recent overviews of local/indigenous knowledge of and practices in animal reproduction, breeding and other husbandry arenas attest to a wide range of astute interventions in livestock genetics. Given their direct dependence on livestock for daily survival, nomadic pastoralists in particular have accumulated a rich store of knowledge and expertise in animal breeding.

KNOWLEDGE OF BREEDING

All long-time stock raising peoples have a practical, working knowledge of genetics, at least at a phenotypic level. Many pastoral and agro-pastoral peoples also keep detailed mental or oral stock records. Indeed, even children can often identify the pedigrees of all the animals in their own and nearby families' herds. This is because an animal's ancestry is often encoded in its name, and names are not changed when animals are sold or exchanged. Usually, records follow female bloodlines.

Furthermore, many pastoral peoples restrict or even taboo the sale of (especially female) breeding stock outside their home community or ethnic segment (tribe, caste). They may do so because they consider such animals their prime capital, as do practically all camel pastoralists or ideological explanations may be offered. For example, Andean agro-pastoralists' believe that the souls of animals slaughtered far from their home community will be unable to make their way back to be re-born, thus prejudicing continued herd reproduction. Whatever the emic reason, etically such social restrictions isolate animal populations, and thus encourage and stabilize the development of unique, localized breeds.

Breeding Goals

Nearly every traditionally stock raising society has developed one or more distinctive livestock breeds to suit its particular environment and animal-product needs and wants. People often have multiple breeding goals for a given species. Still, the first consideration is sheer survivability. Thus stock are selected and bred for: resistance to dangerous endemic diseases; general hardiness, as in an ability to survive even when forage, water and minerals are scarce and to trek long distances in search of them; and adaptability to local climatic extremes, whether in temperature, precipitation or altitude.

Also often figuring in breeding choices is animal behaviour and temperament as these relate to livestock survival and human management thereof, for example, good mothering instincts and aggressiveness towards predators, but also herd ability, tameness, and loyalty, the latter reflected in stock finding their own way home and resisting contact with humans other than their masters/mistresses.

Of course, in the absence of high-tech operations, good libido and natural fertility are general desiderata. Other breeding goals may involve the particular goods and services sought from a given species or animal, beyond just maximum product yields. To take a few examples: strength or fleetness in draught, pack or riding animals; intelligence in lead or work animals; and particular fibre Colour or qualities, milk-fat content or eggshell hardness, according to market demands and conditions. Finally, various aesthetic and cultural considerations may also be at work, as when a people prize certain phenotypic characteristics (*e.g.* a particular horn or body conformation, feather configuration or physiognomic feature) for special social or ceremonial purposes.

Breeding Practices

These consist of the techniques and social institutions by which breeding goals are achieved. The most common is simple selection of animals allowed to mate. As a rule of thumb, traditional stock raisers focus selection on male animals; although typically they also take into account the male's female forebears and relatives.

Selection is negatively implemented by such techniques as: castration - probably the most common way of preventing unsuitable male animals from breeding; culling, slaughtering or selling animals considered unfit as breeding stock; subdividing herds by sex and age; tying aprons or sheaths over the genitalia; and in one culture, inserting stones into the vagina as a sort of IUD. It is positively implemented by, for example, corralling together or manually coupling prime mating pairs; purchasing, hiring or borrowing preferred studs; 'stealth' techniques such as grazing or losing ones' female animals in another's herd that has desirable studs; and other social mechanisms such as multi-household stud ownership.

The practice of offspring testing is reported for camel breeders in Somalia and India. Also widely used in scientific breeding, offspring testing refers to restricting a male animal's mating until his first crop of progeny can be evaluated. Only if his first offspring live up to expectations will the male then be used more widely.

Some stock raising societies fastidiously avoid inbreeding, while others see no harm in it. Attitudes about animal inbreeding sometimes mirror local marriage practices. For instance Arabian Bedouin are endogamous and so permit inbreeding, whereas the exogamous pastoralists of Rajasthan never allow it. Indeed, breeding practices are often intertwined with cultural beliefs and local social organization. To take another example, the historically superior quality of cattle in the Marwar region of Rajasthan has been attributed to the fact that all sub-standard bulls were systematically castrated. This practice hinged on the availability of low-caste people to perform this socially debasing task. Nowadays, however, few such people are willing to do this work, and cattle quality has thus suffered.

Breeds

Many local/indigenous livestock species and breeds have a close association with particular ethnic groups with whom they have co-evolved. Indeed, these groups may couch their very identity in terms of the human/animal bond. The survival of the two - animals and humans - is interlinked. Without special interventions, animals will seldom retain their genetic uniqueness outside the traditional production systems that gave them rise. Conversely, the ethnic group in question cannot continue their production system and thus their cultural (or even physical) survival without these animals. A few examples follow, drawn just from among stock raising peoples of a single country, India.

- One of the best indigenous cattle breeds is the Sahiwal, which combines exceptional hardiness with high milk yields. Muslim pastoralists are the hereditary keepers of Sahiwal, which they carefully maintain as purebreds by vigourously resisting all attempts by government and development agencies to introduce cross-breeding.
- Rebari pastoralists developed the Gir and Kankrej cattle breeds, which became the founding stock of famous beef cattle breeds in Brazil and the southern US But their original creators are being pushed off their pasturelands, thereby imperilling these breeds' continued existence in their original surroundings.
- A hereditary camel-breeding caste, the Raika can no longer earn a living from their breeding work because alienation of pastures has reduced the sizes and reproductive rates of their herds. But regarding themselves as guardians of the camel, some of them continue to keep camels anyway.
- Bastar tribals in Madhya Pradesh are the only group to have protected the rare Aseel chicken, which is extinct everywhere else.

Such examples could be multiplied many times over across the globe. The point is that the human groups who have nurtured and developed special animal genetic resources should be regarded as their rightful owners. As such, these peoples can play, and should be accorded, a vital key role in initiatives to conserve these invaluable resources.

Conserving Livestock Diversity

FAO has a global mandate to study, advice and set guidelines on conserving livestock genetic resources for present and future food security. FAO's Initiative for Domestic Animal Diversity began by establishing a database known as the DAD Information System. The system is designed to inventory and monitor DAD worldwide. The Initiative further seeks to conserve those breeds classified in the database as endangered and critical, and to promote exchange of these precious genetic resources. To these ends, the Initiative has established: an intergovernmental mechanism, a technical programme of management support for UN member nations, a cadre of experts, and a country-based global infrastructure of national coordinators.

On paper, FAO recognizes the importance of stock raisers' local/indigenous breeding knowledge. 'The indigenous knowledge that has helped to produce and maintain domestic animal diversity is largely unexplored and yet this knowledge is essential in order to understand and continue developing these animal genetic resources'. This recognition has not been translated into any specific activities, however.

Data Basing

So far, no efforts have been made to recover local/indigenous knowledge about breeds and then integrate it into the DAD-IS. Database information is provided only by designated national coordinators in FAO member countries. Stock raisers are not directly consulted, even though they may be the only ones who know of many local breeds and strains. To illustrate, only recently did scientists discover a new camel breed with high milk-production potential - although naturally the pastoralists from whom scientists learned of the breed had long known about it.

Neither does the DAD-IS taking account of stock raisers' breed classification systems. These typically differ from scientists' and often are more refined than scientific classifications. To take just one example, scientists opine that India's donkey population is comprised of a single breed. But local donkey-breeders distinguish at least three types that are phenotypic ally distinct and hail from three different areas. By scientific criteria, these features make them, in all probability, three breeds.

The DAD-IS characterizes breeds according to production features and population size. The former include: milk yield, lactation length, milk fat, litter size, birth weight, adult weight and adult wither height. But stock raisers evaluate breeds differently, often using non-productivity-related characteristics. Many such traits are manifest only under traditional management; they may be invisible in stabled animals or on government research farms. For example, transhumant Gujjars of the Himalayan foothills keep a buffalo breed that instinctively knows when to migrate to higher or lower pastures. But this behaviour would be unobservable in Gujjar buffalo penned on a farm or research station.

Population data recorded in DAD-IS include, for example, total population size, total number of females bred, total number of males used for breeding, and so forth. Again, this information does not systematically draw on stock-raiser knowledge, breed classifications and terminologies. Consider the case of Tharparkar cattle. Indigenous to the Thar Desert along the Indo-Pakistani border, Tharparkar are famous for their good yields of high-fat milk even under arid conditions.

Thus they are accorded a high conservation priority. Yet the word Tharparkar means nothing to local people, who instead say Sindhan. Scientists themselves cannot agree on how to define and hence enumerate this breed. Some count the entire cattle population in the two districts of India where it occurs; others consider only the couple of hundred animals kept on state breeding farms to be 'true' Tharparkar.

SELECTION PRESSURE MODEL OF TECHNOLOGICAL CHANGE

The aim of this section is to draw the threads of this chapter together. In the technical sphere, the BCM mode is dependent principally on the interplay of innovations in the physical, mechanical, chemical and biological sciences. Each of these three contributing 'disciplines' shapes the heuristics which guides research in the others. At the institutional level, professional bodies, career structures, and education and training programmes anchor the mode.

The innovations emerging as a result of these heuristics aim either:

- To create new wants in consumers; or
- To meet an 'obvious and compelling need', removing bottlenecks in the sequence of technological change; or
- To exploit new opportunities arising as a result of the use of new technologies.

An economy becomes, 'a web of transformations of products and services among economic agents. Over time, "technological evolution" generates new products and services which must mesh together "coherently" to jointly fulfil a set of "needed" tasks' The mesh becomes finer and the web increases in complexity as economic interests grow and diversify, as institutions are transformed, and as industries mature, increasing the degree of specialisation in production.

The heuristics shaped by the key disciplines of the BCM mode are the source of adaptations which increase the 'fitness' of either a machine, a seed variety, or a chemical input within the economic web. The organisation and development of science, and especially, its splintering into disciplines, have favoured a reductionist view of agriculture. Furthermore, the chronological order in which scientific disciplines have matured has focused attention first on pieces of the physical, then the biological world.

A new technology or an adaptation of an old one would deform the fitness landscape to which other technologies are coupled. Thus, the ability to breed for uniformity alters the fitness of mechanical innovations which function best where plants are uniform. Evidently, not all technologies are coupled to all others. The way in which the coupling process itself evolves is a characteristic of the trajectory (of the BCM mode).

Thus, the economic web encompasses arrays of coupled, co-evolving entities, the path of whose development is traced out as combinations drive the evolution of the web through deforming the fitness landscape, and challenging coupled technologies to evolve accordingly. In the BCM mode, agricultural input supply, engineering, and food processing industries have co-evolved with the ideal plant concept. The demands of these industries increasingly shape, and are shaped by, the appearance of the ideal plant. Institutional changes also deform the fitness landscape through altering incentives for investment in particular activities. Agricultural policies, and property rights regimes, especially regarding germplasm, are important in this regard.

The fitness of some erstwhile 'unfit' technologies can be improved by new technologies. In such instances, it might be more appropriate to speak of co-mutation than co-evolution, as the very existence of one technology in the web is predicated upon the existence of the other. The tomato harvester and tomatoes with skins able to withstand that process were developed in tandem. Similarly, some technologies which could improve the fitness of technologies used in other techno-economic modes never emerge, being rejected by the dominant mode.

The heuristics applied in each of the disciplines mentioned may vary to some extent over time as the state of scientific knowledge changes, or as new demands, or potential markets, are recognised. However, such changes are confined within boundaries. It is important to understand that the nature and extent of this confinement changes with time. In the early development of a techno-economic mode, the confinement may be relatively loose owing to the fact that little coupling of technologies exists. A range of possibilities may co-exist (interpretative flexibility).

The web is immature and the shape of the fitness landscape is independent of inter-related technologies owing to their absence. As regards the ways of meeting a given end or purpose, the fitness landscape might be represented by a relatively stable topography with large numbers of low peaks, or small mounds. Each technology used is part of the way up one of the peaks, though there are also peaks which signify relatively under-used technologies.

As technologies mature in market economies, so the web increases in complexity. Specialisation of tasks, and the new opportunities opened up by the use of new technologies, causes the web to increase in the number of coupled technologies. These processes – the emergence of interrelated technologies, the evolution of institutions, the effects of learning processes, financial inducements, political decision-making, historical contingencies, and others – exert their influence in this model of technological change through deforming the fitness landscape, and selecting out one, or sometimes more than one, way of doing something, and effectively deselecting alternatives.

These events and the emergence of one or other technology represent a co-evolutionary process where increasing returns operate. The greater the support for one or other technology, the more powerful is the interests which effectively lobby in favour of its continuing. On the fitness landscape, the peaks become fewer, farther apart, and more sharply defined. In these circumstances, it becomes likely that adaptive hill-climbing on one of the dispersed peaks will be the easiest path towards improvement, rather than seeking another of the widely dispersed peaks. Institutions support the development of this peak, lowering others in the process. It is important to recognise that the landscape's topography is shaped, in part; by the way people see the world. If one were to remove this influence upon the landscape, the topography might well be substantially altered by virtue of the fact that the number of 'ways of doing something' exceeds the number which we conceive of at any one time, a reflection of the fact that our thinking is path-dependent. Kauffman writes:

The actual web is a subset of the possible; and the actual evolution of economic webs must be a path through the ever expanding set of next possible economic webs... Since the actual evolution of the web is caught on the edge of addition and disappearance, under endogenous non-economic and economic forces, an eventual dynamical theory of web evolution must meld all these.

Similarly, a theory of lock-in seeks to explain why what disappears is rarely allowed to re-appear, and why these disappearances influence what can and cannot be added in the future. Options that are available may disappear due to seemingly insignificant events triggering the building of a new web on one peak as opposed to another, as the work of Arthur and David suggests. Since, as the web increases in complexity, each new good or service generates possibilities for other new innovations which are coherent with the web structure, so the web's evolution takes on something close to an autocatalytic character.

Note also that it is not impossible for more than one technology to develop in fulfilment of a particular role. The way in which the fitness landscape is deformed does not preclude the simultaneous development of two or more technologies on different peaks which effectively compete in the marketplace as long as they both remain competitive. The rate of incremental innovation, the institutional framework (especially issues concerning the structuring of markets) as well as political factors, will affect the way in which the market is shared.

Where changes occur as a result of changes which appear only indirectly related to goings on in the area under examination, it is useful to talk in terms of historical contingency. This term captures some of the historical events described by Arthur and David which may be 'small' and appear, at first glance, of limited relevance to the systems whose subsequent evolution they may determine (although they can equally be 'big', like war).

Whilst this idea has some general implications, it has some quite particular ones in the context of agriculture. As farming has become more commercialised and more specialised, partly due to technological changes 'outside' agriculture, so the fitness of one or other innovation has come to be determined by quite different selection pressures to those that might have operated previously. Indeed, because of the nature of the BCM mode itself, in particular, its desire to control the production process, the selection pressures themselves are increasingly derived from anthropogenic institutions, in particular, markets, rather than from 'natural' forces. Some of the risks associated with the 'natural' characteristics of agriculture are reduced, only to be replaced by risks emanating from exchange rate and interest rate policies, and commodity price fluctuations.

The fitness landscape representation allows one to visualise the historical process by which the BCM mode of doing agriculture has evolved. As will be shown in the case study material, the evolution of a way of doing agriculture based on genetic uniformity has proceeded in the context of evolving ways of interpreting the world, changing institutions, new opportunities for making money, and political decisions concerning the role of the state in agricultural research. In addition, genetic uniformity has enabled the sorts of specialisation of agricultural tasks, in terms of mechanisation of harvesting and processing, that have enabled an increasingly complex technological web to emerge. Moreover, it necessitates the use of synthetic pesticides (sooner or later) and as such, is both the glue which both binds the BCM mode, and the heart of the process of locking-in. Through elaboration and use of various concepts, a way of understanding technological change in agriculture within an evolutionary framework has been proposed. Having hinted that technological change generally is possessed of an evolutionary character, and that certain paradigmatic forms for such change exist, I have suggested that technological change in agriculture has taken place within a mode in which mechanisation, agrochemicals, and breeding are closely integrated, a more or less tangible illustration of which has been the elaboration of an ideal plant concept whose architecture enables and demands such integration. Lastly, I have proposed a way of viewing technological change in terms of the fitness landscape concept first used by Sewall Wright in his work on evolutionary biology.

The fitness landscape conceptualisation allows for a visual representation of the 'vestedness' of vested interests. It also makes clear how, as the techno-economic trajectory is carved out, and as an increasingly complex web arises to support it, it becomes increasingly difficult to do things in ways other than the way they are being done. Unless the mode reaches such a stage of maturity that exhaustion of the possibilities for adaptive uphill walks on an increasingly tightly defined landscape seems to be close at hand, then the only changes which are likely to be encouraged are those which would be selected by the prevailing selection environment, or something very close to it, in which case, the extent to which such a change really constitutes a change at all is likely to be subject to some debate. In the absence of some external intervention which forces a change in mode, the situation of lock-in is likely to persist.

POULTRY

FEEDING SCHEDULES FOR POULTRY

In PB-1 females, ME restriction was successful in achieving the desired body weight (83 g/week and 5. 75 per cent growth rate) at 20 weeks of age.

Karanj cake in the diet adversely affected Krishibro chicks, but feeding of detoxified cake improved the performance.

The production performance of PB-2 line was better with organic Zn supplementation at 60 ppm and hatchability at 80 ppm. Trace mineral retention in tissues of broiler chicks was linearly related to their level in diet. Herbal antifungal compounds, *viz. Piper longum*, *Ferula assafoetida* and *Cymbopogon citratus* oil were identified as potent antifungal herbal compounds against *Aspergillus parasiticus* in poultry feed under *in vitro* conditions.

QUALITY PROTEIN MAIZE IN DIET

Dietary replacement (50 to 100 per cent w/w) of normal maize (higher in lysin and tryptophan by 44 and 33 per cent, respectively, over normal maize) resulted in 6.93 per cent higher body weight gain, 7.69 per cent improvement in feed utilization efficiency and 5.15 per cent higher breast meat yield in broiler chicken.

AMELIORATION OF HEAT STRESS IN BROILERS

Amelioration of heat stress in broiler chickens was achieved through supplementation of vitamin C and potassium chloride in diet. Heat shock protein 70 (HSP70) was detected in muscle and blood lymphocyte in birds as potential marker of heat stress.

ARECA SHEATH

Areca (*Areca catechu*) is a commercial crop and the fallen areca sheath is a good source of dry roughage to animals. The nutritional value of areca sheath is similar to paddy straw (CP, NDF, ADF) with an advantage of having less lignin, silica and more calcium, sulphur and copper. Feeding of areca sheath to sheep and dairy animals replacing paddy straw could be an alternate approach.

LIGNIN DEGRADATION

Activities of lignolytic enzymes produced by immobilization on areca sheath recorded temperature and pH stabilities over a wide range, thus making it desirable for supplementation in animals.

FEED ADDITIVES

Out of the 44 plant products, *viz.* seed pulp (5), leaves (20), spices (13), citrus fruit peel (3), oils (3), and 23 plant combinations, screened for their anti-methanogenic activity using *in vitro* gas production test, a mixture of plant products and yet another combination of four plants appeared to have good potential as feed additives in the ruminants. Use of nitrate reducing bacteria as probiotic exhibited 80 to 99 per cent reduction of *in vitro* methane emission. There was no accumulation of hydrogen, nitrate or nitrite in the incubation medium, and no adverse effect on *in vitro* true digestibility of feed. Supplementation of Chicory root powder, *Lactobacillus acidophilus* as symbiotic in the diet of dogs proved beneficial in terms of fibre utilization, calcium and lipid metabolism, hind gut fermentation and cellmediated and humoural immunity. *Helianthus tuberoses* L.*Jerusalem artichoke Hathi choke*; a novel phytogenic-origin probiotic, showed immense potential for use in the diet of dogs for enhancing their digestive health, lipid profile and immunity.

UNCONVENTIONAL FEED

Detoxification procedures for jatropha, *karanj* and neem seed cakes were standardized in an industrial process. In growing lambs, replacement of soybean meal with detoxified jatropha cake up to 37.5 per cent showed no adverse effect on DM intake, nutrient utilization, bloodbiochemical profile, growth rate and feed conversion efficiency. Nutritive value of *Robinia pseudoacacia* in terms of high organic matter, crude protein and low lignin content was comparable to *Grewia optiva*, and showed higher nutritional values than other tanniniferous forages. Combined treatment with curry and bel leaf powder had positive effect in restoration of cyclicity and fertility in acyclic goats and buffaloes.

CHELATED MINERALS

Cell-mediated immune response was significantly higher in kids given organic zinc (40 ppm zinc as zinc methionine) as compared to zinc sulphate.

NUTRITIONAL MANIPULATION

Feeding growing Frieswal bull calves with three different levels of rumen degradable protein and rumen undegradable protein for achieving a targeted body weight gain of 700 g and 1,000 g/day, yielded only 608 g and 772 g of average daily gain suggesting that there is a need to develop our own feeding standards for higher body weight gains.

COMPLETE PELLETS

Loin eye area increased linearly with increased level of concentrate mixture in different supplementation group. Kids fed diet without any concentrate mixture had lower yield of separated lean, fat and bone content. Different levels of supplementation improved meat bone ratio of Barbari kids under intensive system.

YAK

Two complete feed blocks (CFB) were prepared based on tree leaves (50 per cent tree leaves, 43 per cent concentrate and 7 per cent molasses) and paddy straw (50 per cent paddy straw, 43 per cent concentrate and 7 per cent molasses). The result indicated that tree leaves based CFB supplemented with limiting micronutrients can be fed during winter to overcome the weight loss in yaks.

REPRODUCTION

CATTLE

Good quality fresh and post-thaw semen of crossbred bulls had significantly higher percentage of HOST positive spermatozoa with low LDH and GOT enzyme leakage as compared to poor quality semen. Frozen semen doses (166,610) of Frieswal bulls were produced and 73,205 doses were distributed that included 50,725 to military farms, 9,000 to field progeny testing project, and 13,480 to various developmental agencies and farmers for cattle improvement.

BUFFALO

Buffalo sperm membrane is highly prone for lipid peroxidation as compared to cattle, and in a study on nutrition and male fertility, it was observed that PUFA feeding may reduce lipid peroxidation and improve frozen semen fertility. Studies on nutrition-endocrine interactions in attainment of puberty in female buffaloes showed that supplementation of higher dietary energy (through bypass fat) has a positive effect on early maturity of hypothalamo-pituitary—ovarian axis in terms of early and better LH secretion and ovarian follicular activity in female buffalo calves.

Studies on physiological reasons for low reproductive efficiency and early embryonic mortality in buffaloes showed that antioxidant enzymes, Cu and Zn-SOD are induced significantly during early pregnancy and may be important for overcoming oxidative stress that otherwise would lead to early embryonic mortality. LPS and TNFα appear to modulate prostaglandin production in endometrium and corpus luteum and are important for protecting and enhancing function of corpus luteum to prevent early embryonic mortality.

IGF-I also stimulated progesterone secretion from the corpus luteum. Five proteins, specific to uterine secretions of pregnant buffalo were recognized and it may be possible to use one of these for early pregnancy diagnosis to monitor pregnancy and prevent early embryonic mortality.

GOAT

Study on effect of dextran, a new cryoprotectant, revealed that dextran alone did not prove to be an efficient cryoprotectant, but 3 per cent glycerol dextran in combination with 6 per cent glycerol resulted in 39.33±1.45 per cent post-thaw recovery. Hormone delivery system for synchronization of heat in goats was standardized by using sponges of different sizes and shapes. The estrus response, onset of estrus and duration of estrus following use of intra vaginal sponge pessaries for induction of multiple birth during peak breeding activity were 100 per cent, 29.14±4.42 h and 25.71±0.37 h, respectively, with 57.14 per cent conception rate.

The kidding rate following induction of multiple birth was 2.0. During low breeding activity season (summer) the estrus response, onset of estrus and duration of estrus were 71.42 per cent, 60.0±16.54 h and 26.4±11.63 h, respectively, following use of intra vaginal pessaries. Oocytes were recovered by follicle puncture technique for *in vitro* maturation, fertilization and culture from goat ovaries. The average recovery of oocytes using follicle puncture was 1.76/ovary. The oocytes were cultured in tissue culture medium (TCM-199) supplemented with EGF (10 ng/ml) and fertilized *in vitro* with a cleavage rate of 41.55 per cent. Embryos were bisected through microtools. Out of 25 bisected cells 3 single cells divided into double cells whereas one bunch of 3 cells developed into 4 cells. Adaptability study on goats revealed that water intake of goats kept under asbestos sheet roof sheds was significantly higher compared to the thatch roof shed during both the moderate and cool seasons. Milk production of animals kept inside the thatch roof shed was significantly higher than the animals inside the asbestos roof shed.

CAMEL

Physiological parameters (rectal temperature, respiration rate, pulse rate) were higher during evening as compared to morning. The morning and evening variation in the THI value was significant for all weeks in different months from October to March. The Benezara coefficient of adaptability (BCA) of camel revealed that BCA was significantly higher during evening as compared to morning. Non-pregnant female camels were divided in groups A,B,C.

Each group was tested with poll gland secretion collected in breeding season (PGS-BS), poll gland secretion collected during non-breeding season (PGS-NBS) and water. The results indicated that the PGS-BS evoked greater olfactory, contact and flehmen response than PGS-NBS and water. Nine female camels, which did not conceive during breeding season were given injection of highly purified hCG for infertility treatment at the time of mating. Pregnancy was diagnosed in three out of nine camels.

YAK

On application of the vaginal sponge containing 600 mg progesterone for synchronization of oestrus followed by AI in yaks resulted in 100 per cent oestrus response with 60 per cent conception rate. The yaks do not experience cold stress (environmental stress) during winter but they are subjected to heat stress (during summer) when the ambient temperature exceeds 10°C at the altitude of 2,750 m above mean sea level. The body weight gain was significantly higher during summer than winter in adult yaks. The lactating yak cows lost about 5.84 per cent of their body weight during winter along with reduced milk yield mainly due to shortage of feed and fodder. Estrous was displayed year round but the highest incidence (around 65 per cent) was restricted to warmer part of the year. Physiological responses, *viz.* rectal temperature, respiration and pulse rates of yak calves, bulls and lactating cows were recorded during winter and summer. The physiological responses of calves were significantly higher than bulls and lactating cows during summer. Standardization for heat shock protein 70 (Hsp 70) assay was carried out in yaks.

MITHUN

Different protocols for preservation of mithun semen indicated that it is possible to preserve mithun spermatozoa at refrigeration temperature in tris-egg yolk diluents till 108 h, and doubling the concentrations of fructose in the diluent buffer increased the progressive motility, live sperm count and decreased total sperm abnormalities. Hence, addition of extra-energy source (fructose) might increase the storage time of mithun sperm at refrigerated temperature up to 5 days.

PIG

Gloved hand method of semen collection was standardized. Semen preservation was done for Hampshire, Duroc and Ghungroo pigs for first time in Asom. The AI technology was used in the farm and in the nearby villages and highest litter size (15 piglets) at birth was recorded from a Ghungroo sow in the farm.

POULTRY

Exposure of birds to 650 nm and 475 nm wavelengths of light improved egg production by 6.55 and 3.05 per cent, respectively, compared to 450 nm (incandescent) light.

LIVESTOCK SECTOR AND THEIR CAUSES

OVERGRAZING

Overgrazing of rangeland reduces the density of vegetation and the amount of organic matter generated. This, in turn, increases soil erosion from wind and water and decreases soil fertility through loss of nutrients. In arid and semi-arid areas these impacts may also contribute to desertification. Fortunately, ecosystems in these areas demonstrate considerable resilience and often recover when grazing pressure is reduced, either through traditional methods or through modern management practices.

POLICY AND LEGAL PROBLEMS

In areas that traditionally rely on grazing, the health of rangeland is generally best maintained by traditional pastoralist practices, which regulate grazing location and herd size in accordance with drought cycles and the supply of fodder.

In fact, government policies or donor interventions that disrupt or discourage these practices may be a root cause of degradation. A variety of government policies may restrict the movement of livestock within a range area and prevent livestock managers from moving stock from areas that have been depleted of fodder to better supplied areas.

Two particular policy-based problems are:

1. *Land Tenure Insecurity:* Lack of confidence in secure title to rangeland (especially on communal lands) has been shown to reduce the incentive to manage the land sustainably. Many national governments have either implicitly or explicitly claimed ownership of range and wildlands and ignored traditional or customary claims.
2. *Privatization of Communal Resources:* Where national governments have privatized, or are privatizing, formerly state-owned or communal lands, new owners may erect fencing or prevent herds from crossing or grazing on their property.

WELLS AND BOREHOLES

Traditionally, access to water on critical grazing lands has been controlled to limit livestock populations and prevent herds from outgrowing the forage supply in dry areas. Thus, new wells or boreholes, generally sponsored by donors or governments, may undermine traditional livestock management systems practices by allowing herds to grow beyond sustainable levels for surrounding areas.

Overgrazing and degradation are most noticeable in the immediate vicinity of the boreholes or wells, but their effects can extend (in gradually decreasing severity) over a considerable radius. Boreholes also reduce pressure on livestock owners to decrease herd size during drought and may discourage movement of herds to other rangelands, disrupting historic wet season/dry season grazing patterns. Larger herd sizes and reductions in pastoral movement may prove to be a recipe for severe degradation of soil and vegetation.

WET-SEASON GRAZING

Poor timing in the use of rangeland can also damage the soil. Wet-season grazing can compact the moist earth, reducing its ability to absorb moisture This increases erosion from water run-off.

POOR BALANCE OF LIVESTOCK SPECIES

Each species or breed of livestock has foraging preferences and will graze favoured areas and plants while neglecting others. Browsing animals, such as goats and camels, prefer the leafy tops of shrubs. By contrast, grazers tend to consume ground-level grasses and leafy plants. A poor balance between browsers and grazers can change the mix of plants in ways that degrade the area. For example, too many grazers can diminish the number and populations of herbaceous plant species and allow woody plants to become dominant, changing, possibly irreversibly, the character and utility of the ecosystem.

DAMAGED HABITAT AND REDUCED BIODIVERSITY

Livestock production can damage habitats and reduce biodiversity in wildlife and domestic stock, in vegetation, and in aquatic and wetland ecosystems.

HARM TO WILDLIFE AND DOMESTIC STOCK AND LOSS OF WILDLIFE HABITAT

The loss of habitat caused by livestock production in grazing and mixed farming systems may be one of the greatest threats to wildlife. Human population growth and density, and the accompanying increase in livestock, often leads producers to expand livestock grazing ranges into wild lands and the conversion of wild lands to mixed farming use.

SLAUGHTER OF WILDLIFE BY LIVESTOCK MANAGERS

Another danger to wildlife is intentional slaughter by livestock managers. Fear that the wildlife will prey on livestock and damage crops is a common motivation, as is the belief that the wildlife are competing with livestock for fodder, the desire to prevent spread of disease to livestock, and concern for human safety.

EXTINCTION OF LOCAL LIVESTOCK BREEDS

Systematic livestock production may result in loss of genetic diversity in livestock species. This is unfortunate because genetic diversity is a measure of a species' robustness. Local breeds may have traits conferring resistance to emergent or future pathogens, or have other favourable adaptations to local environments.

The consistent replacement of local breeds with more productive imported ones can contribute to the extinction of that breed and of all the genetic diversity harboured within its population.

HARM TO VEGETATION

CLEARING OF FOREST AND WILD LANDS

Vegetation is typically altered or destroyed when forests/wild lands are cleared or are burned to promote new growth. This changes local ecosystems and may contribute to global warming. Fires to burn vegetation are dangerous and degrade air quality.

LOSS OF RANGELAND FERTILITY

Ironically, mixed farming systems may reduce the fertility of rangeland while helping to solve a farmland problem. Traditional farming practices cause a net loss of nutrients in farm soils; that is, when crops are harvested and sold (or even when human waste is deposited in a latrine) nutrients that make the soil fertile may be lost.

Mixed farming reduces the extent of this loss, by transferring nutrients from the range to the farm in the form of manure. The gain in fertility for the farm is, of course, a net loss for rangeland. Over time, the altered nutrient balance can reduce the productive capacity of the range and/or lead to changes in the composition and density of plant species.

LOSS OF FARM FERTILITY

Damage from mixed farming often occurs when poverty and population growth pressure change the crop/grazing land ratio, where other nutrient sources are not available. As the land area available for each grazing animal shrinks, overgrazing becomes prevalent leading to soil erosion and nutrient loss. Harvesting livestock further reduces nutrients available for crop production.

DAMAGE TO RIPARIAN SOIL AND VEGETATION

Livestock in grazing and mixed farming systems often graze very heavily in riparian areas along streams and lakes. Results include trampling, loss of vegetation, soil disturbance, and soil compaction, erosion and/or sedimentation which can severely damage riparian habitats, increase siltation and adversely affect watersheds. Pesticide contamination from treatments to protect livestock from insectborne infections (*e.g.*, livestock "dipping") may ultimately reach the aquatic environment. Here it can be toxic to aquatic organisms, as well as people or animals who depend on these sources for drinking water.

INTRODUCTION OF INVASIVE PLANTS

New breeds or fodder crops can introduce invasive non-native plants into a region. The manure, coats and hooves of newly introduced breeds can carry plant seeds. Most non-native plants are not invasive, but when they are, the results can be devastating.

CONTAMINATION FROM MANURE

Livestock manure contains relatively high concentrations of nutrients, solids, enteric bacteria and other microorganisms, and organic material. The manure from industrial livestock operations is often discharged or "leaked" into lakes or streams, because it cannot be economically transported to replenish crop fields. When this occurs, the nutrients can cause eutrophication (rapid plant growth in water bodies), solids can create sedimentation, and organic material leads to oxygen depletion (BOD) of the water. Manure from mixed farming, if applied in a concentrated fashion, can lead to similar problems.

DEGRADE WATER QUALITY AND REDUCE WATER SUPPLIES

Where water is scarce, either chronically or seasonally, the diversion of water to sustain livestock potentially limits its availability for other purposes. This is of particular concern in arid and semi-arid regions, where the construction of boreholes to supply livestock can lead to unsustainable withdrawal rates and the dangerous depletion of aquifer reserves. Stockpiled manure can contaminate bodies of water, causing myriad adverse effects. These include eutrophication, oxygen depletion, sedimentation, contamination with enteric bacteria and possibly other pathogenic organisms, toxic pollution from pesticides, and contamination of groundwater and aquifers with both nitrates and pesticides. Moreover, high concentrations of nitrate in potable water supplies represent a potential health hazard, especially for children.

HARM TO HUMAN HEALTH

Using water for farm animals may make it less available for the many uses that influence health, such as bathing, washing, cooking, and drinking. Moreover, excessive contamination by enteric microorganisms, toxic pesticides or nitrates in may render water unfit for human consumption and may be especially dangerous to children.

Pesticides or other vector control treatments used on livestock represent threats to the health of livestock managers, their families, and others exposed directly or through water use.

These substances may be toxic, cause birth defects, alter children's proper development, promote cancer, or slowly poison one or more organ systems.

CLIMATE CHANGE/GLOBAL WARMING

Approximately 17 per cent of global methane is produced by livestock digestion and manure. Methane is a greenhouse gas that has 56 times more global warming potential than carbon dioxide over a 20-year time horizon. Furthermore, clearing wild or forest land for new fields reduces the land's ability to act as a carbon sink. Setting fire to vegetation to clear land or promote new growth for livestock to graze on may also contribute to global warming.

ODOUR

Concentrated manure stored at industrial livestock facilities can generate strong and unpleasant odours, damaging the quality of life of nearby residents. This problem is most evident when facilities are located in densely populated areas.

SECTOR PROGRAMME DESIGN—SOME SPECIFIC GUIDANCE

The following questions and suggestions are intended to help project designers and managers identify factors and practices that may cause—or prevent—adverse environmental impacts. Bear in mind that the first priority of most livestock managers and farmers is household food security and family welfare. Sustainable practices must always be balanced against these immediate demands.

CONSIDER CLIMATE, TERRAIN, AND ECOSYSTEM

Since environmental impacts from livestock production vary, depending on the specific climates, terrains and ecosystems involved, project designers need to address these characteristics during the initial design phase:

- What is the climate in the project area (arid, semi-arid, temperate, subhumid, humid)? What do historical records of rainfall patterns and flooding indicate? Is the proposed livestock management practice compatible with this climate?
- What terrains are found in the project area(s) (alluvial plain, highland, rocky desert, wetland, etc.)? Do they have any known vulnerabilities to livestock grazing? For example, are there many unprotected streams or rivers? Are there slopes with limited topsoil sensitive to erosion?
- Will the project encompass or border on protected or ecologically sensitive areas? Are there any threatened or endangered species in the area? Would the proposed project directly or indirectly threaten wildlife or native vegetation? For example, does the project require expansion of grazing into protected areas or make livestock more vulnerable to wildlife predators, triggering reprisals by farmers?

EVALUATE POLICY, LEGAL, CUSTOMARY AND CULTURAL CONTEXT

The policy, legal, and cultural contexts of a project merit attention, since, these factors may limit programme options or erect substantial barriers to success.

POLICY/LEGAL

- Do livestock owners and managers have legal and recognized ownership and responsibility for land and grazing resources? What are the current or proposed national land tenure policies, and how are they implemented at the local level? Are they effective in encouraging sustainable management of grazing land and resources?
- What is the tenure status of current or proposed rangeland—Is it owned by individuals or the community? Does the government have claims?
- What wildlife protection laws exist and how are they implemented in the region?

CULTURE/CUSTOM

- What role does livestock play in local culture and customs, and how might the proposed project affect these practices? Would the proposed project disrupt traditional grazing patterns?
- If there are customary land tenure arrangements, what are they and how would the proposed livestock management system work within these arrangements? For example, will livestock herders—who often come into conflict with farmers, particularly during droughts—have a means of working out disputes with farmers?
- If livestock management arrangements are communal, how would these be affected by and/or affect the proposed development activities?

ASSESS CURRENT AND PROPOSED SPECIES AND BREEDS

Introduction of a new breed into an area should be approached with caution. The new breed may bring with it diseases that can decimate local livestock herds and wildlife. In addition, the foraging habits of a new breed may lead to a sharp decline in available forage and biodiversity.

A new breed's reproductive habits can lead to a herd's uncontrolled growth.

Weeds can be accidentally introduced along with a new animal species, and they may displace desirable vegetation. The long term full costs and benefits of introducing a given new livestock species into a particular environment should be assessed.

For example, large animals roam over extensive areas in search of food, often require a greater financial investment, can be more difficult to control, and have lower reproductive potential than small animals. It is important not to underestimate the value of breeds that are well adapted to the environment. Livestock tend to overgraze favoured areas and plants while neglecting others.

Native plants may not be able to survive heavy grazing. While unforaged plants tend to lose vigour and nutritional value as they mature. The introduction of new plant species (whether accidentally or intentionally) may quickly result in replacement of native plants. Even when grazing pressure is reduced, exotic plant species sometimes retain their dominance. Ask the following questions when introduction of a new breed or species is proposed:

ABOUT CURRENT SPECIES AND BREEDS

- Which wild and domestic species are already present in the area, and in what numbers?
- How have they been used in local farming systems and traditions?
- What are the feeding preferences of local livestock and wildlife? What is the balance between browsers and grazers? Do domestic species compete for resources with one another and with wildlife?
- Have population sizes of wild or domestic species changed recently?

- Could local breeds satisfy the project's needs?

ABOUT PROPOSED SPECIES AND BREEDS

- If new species or breeds are being considered, how will their production complement or conflict with local species or breeds, wildlife, and other local resource users?
- How would they fit in local herding systems?
- Are they well suited to the local climate and environment?
- Are they resistant to local livestock diseases?
- Have alternative species or breeds been considered for possible introduction?

EVALUATE CURRENT AND PROPOSED LIVESTOCK MANAGEMENT PRACTICES

To maximize forage productivity, it is best to combine or alternate various livestock breeds on a range. Their differing food preferences can help to keep plants productive by minimizing overgrazing of a particular favoured area and allowing less preferred plant species time to mature. It is prudent to make superior forage available to those animals with the highest needs. When forage is limited, livestock managers may decide that young and milk-producing animals must have first access to new pastures and ranges with a wide variety of abundant forage.

Managers should investigate the value of various systems of rotating livestock. Rotation allows land to be grazed continuously throughout the year. Livestock can be rotated between fields or ranges to prevent the buildup of disease and to vary grazing pressures. Through either fencing or herding, they can be relocated into croplands to consume crop residues. Assessment of seasonal grazing patterns should include potential impact on soils.

Dry-season grazing can benefit the land by breaking up crusted soil and working seeds into the ground. By contrast, grazing on moist soil can cause considerable soil compaction, which reduces the soils ability to absorb moisture and can result in increased erosion from run-off during the rainy season. Many of the environmental impacts from livestock production are associated with particular practices of livestock managers.

Thus it is critical to understand current practices and how the proposed project might alter these practices or promote new ones:

- Who are the local community's livestock managers?
- What practices do a family or community use to control the size and composition of livestock herds?
- How do livestock managers currently control livestock movement? Will the proposed project change these movements in a way that might harm the environment?

- Does the proposed project require the construction of fences? If so, will they interfere with wildlife migration or transit of livestock belonging to other communities? Could the fences lead to overgrazing and land degradation? Will the fences be built with local materials? Would living fencing be practical? Would solar-powered electric fencing be technically and economically feasible?
- Are streams and riverbanks currently protected from livestock damage? If the proposed project will open new areas to grazing, will water supplies need to be protected?
- Must steps be taken to prevent new livestock and associated animals (*e.g.*, dogs) from transmitting disease to wildlife? Is there a vaccination/ animal disease control programme available for this purpose?
- Will the project involve construction of improvements (*e.g.*, boreholes or other infrastructure)? Could these lead to unplanned changes in herding patterns and overgrazing?

ASSESS DEMAND AND USE OF LIVESTOCK PRODUCTS

- Who is marketing livestock and livestock products?
- Is the demand for livestock products coming from local or outside populations? How rapidly is it increasing or decreasing? How stable is the demand?
- In preparing livestock products, are people using technologies which reduce impacts on the environment, open additional markets, or improve health and nutrition (of people and animals.)

ASSESS LIVESTOCK ECTOPARASITE MANAGEMENT

Epidemic and endemic diseases continue to be a major constraint to livestock productivity in large parts of the developing world. Although vaccines have controlled many of the epidemic diseases, they continue to cause severe economic losses through morbidity and mortality.

These diseases include the infections caused by vector-borne haemoparasites and helminths. Existing technologies, such as chemotherapeutic agents and live vaccines that were previously successful in controlling these diseases, are no longer effective—because of acquired resistance or weakened delivery services. Appropriately designed alternatives are often lacking.

CONSIDER POPULATION PRESSURE AND DISEASE BURDEN

Two factors that may affect the outcome and impact of projects population growth and fatal or debilitating epidemic diseases. Population growth may lead to conflicting uses of grazing lands or reduce individual farms or rangeland to sizes that cannot sustain livestock. Fatal or debilitating epidemic diseases may weaken effective dissemination or replication of proper livestock management techniques.

- What is the current and projected population growth rate in the project area? How might this affect project sustainability in the future?
- What is the current extent of the HIV/AIDS epidemic in the region? How might this affect the composition of the population (size, ethnic makeup, age/gender distribution) and family structures necessary for project sustainability? How will development and livestock technical support services be affected?
- Are there other epidemic diseases in the region, such as sleeping sickness, that might adversely affect project implementation?

LIVESTOCK PROTECTION

FOOT-AND-MOUTH DISEASE

During the period under the report, 799 outbreaks were recorded throughout the country. Maximum outbreaks were recorded in the eastern region where there was increase in the number of outbreaks compared to last year.

Disease incidence in West Bengal directly affects the situation in NE region due to animal migration. Drastic reduction in outbreaks was noticed in southern region. FMD cases were not reported in Tamil Nadu, whereas Himachal and Punjab recorded a single case of FMD each. Maximum incidence of disease was reported in March and from August to November.

Out of 1,624 clinical specimens collected from 799 outbreaks, the virus could be detected in 1,067 samples (serotype O-991, A-38 and Asia 1-38) by using sandwich ELISA and multiplex PCR. Serotype O was responsible for maximum number of outbreaks (568) followed by A (33) and Asia 1 (9). Increase in serotype O incidence was noticed in all regions except northern and western region. Serotype A, which was absent in northern and central region last year appeared this year, making it the second most prevalent serotype affecting geographically all the regions of India.

Serotype Asia 1 was completely absent in southern, central and north eastern regions. Incidence of Asia1 serotype increased in western region. Though majority of the outbreaks involved cattle, disease was also reported in buffaloes, pigs, sheep and goats. A major shift was observed in genetic lineage of serotype O viruses circulating in India. The 'Ind2001' lineage viruses gained upper hand after a gap of 8 years and outcompeted PanAsia II lineage. The cases due to PanAsia II, PanAsia 1 and Branch C-II viruses were not completely absent during the year. Branch-C II lineage viruses responsible for outbreaks during 2008 in West Bengal were recorded in Tripura, Asom, West Bengal and Odisha during 2009–10. In serotype A, it was observed that both VP359 deletion and non deletion mutants belonging to genotype VII are cocirculating in the field in recent times.

Precisely, the viruses from Punjab and Uttarakhand shared ancestry and clustered in the same VP359 deletion lineage, VIIf, whereas viruses from two different outbreaks from Odisha and those from Andhra Pradesh clustered together and revealed no deletion in the VP3 coding region. During the period, outbreak due to Asia1 serotype was recorded in Gujarat, Madhya Pradesh, Maharashtra, Uttar Pradesh and West Bengal.

The Asia 1 field isolates were grouped with lineage C reiterating the supremacy of this lineage in the field since 2005. In vaccine matching exercise 33 virus isolates were subjected to one-way antigenic relationship study. All the isolates showed close antigenic match with respective in-use vaccine strains indicating appropriate antigenic coverage.

Virus isolates (133) including 97 type O, 12 type Asia 1 and 24 type A field isolates were added.

In National FMD Virus Repository 1,687 (1,096-O, 317-Asia 1,259-A, 15-C) well characterized field isolates are available. Under National FMD sero-surveillance testing of 11,560 serum samples from 13 states revealed the presence of protective antibody titre in 42.3 per cent, 40 per cent and 30.5 per cent animal population against serotypes O, A and Asia1 respectively. This baseline response has improved due to application of vaccine. The current in-use vaccine strain IND R2/75 is found still the best and covers all the type O circulating outbreak strains in the country.

CLASSICAL SWINE FEVER

Classical swine fever (CSF), the topmost diseases in pigs, is one of the OIE listed diseases. The CSF or hog cholera is associated with high mortality, abortion, mummification of fetuses. It is of great concern in areas, where the pig population is in high density. CSF virus isolates maintained at the repository of PD-ADMAS, Bengaluru, were revived and processed for RT-PCR of NS5B genomic region. Phylogenetic tree based on the alignment of 408 nts of NS5B region was constructed.

It showed three major branches, each representing three previously described genetic groups with further sub-branches.

The majority of the Indian CSFV isolates belong to subgroup 1.1. Clustering of the Indian field isolates away from the vaccine strain and other related strains revealed that the field isolates did not have any vaccine origin.

Phylogenetic analysis based on 52 UTR classified subgroup 1.1 Indian isolates into two minor subbranches, however, no such classification could be seen from the analysis based on NS5b region; nevertheless in the latter analysis grouping of isolates was much more resolved. Lone isolate, IND294/08, is clustered within subgroup 2.2, ak into its grouping based on 52 UTR sequence.

BRUCELLOSIS

Serum samples from goat and sheep received from Gujarat, Rajasthan, Uttar Pradesh and Maharashtra revealed that among them 6.35 per cent were positive from brucellosis. Similarly, 1.8 per cent sheep and goat sera samples for Karnataka were positive for brucellosis. Porcine brucellosis is one of the most important emerging zoonoses and currently, no control programme for porcine brucellosis exists and the epidemiological situation is still unknown.

Indirect-ELISA for screening brucellosis in swine was standardized using smooth lipopolysaccahride antigen from a standard strain *Brucella abortus* S99; 22.6 per cent were found positive, which is much higher than that found in sheep and goat. An indirect ELISA was developed to detect *Brucella* in any of the susceptible species. Expression of four recombinant *Brucella* proteins, *viz.* BP26, BP39, superoxide dismutase (SOD) and *Brucella lumazine* synthase (BLS) is being envisaged to overcome the problem of cross reactivity in diagnosis of this disease.

LEPTOSPIROSIS

PCR based techniques were developed for differentiation of pathogenic and non-pathogenic *Leptospira* as well as for typing. Initially, *Leptospira* genus specific PCR was carried out to differentiate the *Leptospira* from other spirochete organisms, *viz. Borrelia*, *Treponema* and *Leptonema* etc., which amplified 331 bp products from the *Leptospira* organism only.

SERO PREVALENCE OF IBR

Serum samples collected from Andhra Pradesh, Gujarat, Jammu and Kashmir, Karnataka, Madhya Pradesh Maharashtra, Meghalaya, Manipur, Orissa, Punjab, Rajasthan, and Tamil Nadu, were subjected to Avidin Biotin ELISA. An overall apparent percentage positive was 34. The highest prevalence of 67 per cent was in Tamil Nadu, and the lowest in Meghalaya. Region-wise positive percentage were as follows: East zone 17 per cent, south zone 24 per cent, west zone 48 per cent, north zone 37 per cent, north-east 39 per cent, and central zone 25 per cent.

HERBAL ANTICOCCIDIAL AND ANTIDIARRHOEL DRUGS

The extracts CIRG-5 and CIRG-6 from two different plants were evaluated *in vitro* against sporulation of coccidian oocysts and found 96 to 100 per cent effective to check the sporulation up to 96 h. Eight potential plant formulations of herbal extracts were subjected for clinical trials to control diarrhoea caused by *Escherichia coli* in kids of less than 1 month of age. Final product was developed from plant extract prototype CIRG-I and CIRG-2A and final products (formulation) @ of 10 mg/kg bwt for 1–2 days orally was highly efficacious to control EPEC diarrhoea in kid.

VACCINES AND THERAPEUTICS

- Live attenuated orf, buffalopox and camelpox vaccines met the thermo-stability and potency requirements of the biologicals.
- Bi-valent DNA vaccine containing HN and F genes of Newcastle disease virus was evaluated along with genetic adjuvant comprising IL-2, IL-4, IFNγ, and GMCSF gene constructs.
- An aroA gene knockout mutant of *Pasteurella multocida* serotype B:2 was constructed for the development of HS mutant vaccine.
- Chitosan coated cationic PLG nanoparticles were prepared and when tested for the transfection efficiency with commercial transfecting agent, indicated that these nanoparticles can be safely used for delivery of DNA vaccines.
- *Trypanosoma evansi* beta tubulin protein was expressed in both prokaryotic and eukaryotic expression systems.
- Two new target proteins of *Fasciola gigantica, viz.* leucine aminopeptidase and peroxiredoxin coding genes were cloned and successfully expressed in their native conformation in a prokaryotic expression system.
- Immunomodulators of microbial and herbal origin, *i.e. Mycobacterium phlei*, *Lactobacillus acidophilus* and *Tinospora cordifolia* were effective under field conditions to improve productive performance and mitigate the adverse effect of sub-clinical caecal coccidiosis and thereby improvement of body weight gain in broiler birds.
- Pharmacokinetics of gatifloxacin revealed that it has therapeutic potential for use in sheep.
- Medetomidine-butorphanol, thiopental and halothane combination provided better clinicophysiological and haemodynamic stability in general anaesthesia in buffaloes.
- Three TLR genes of yak (TLR 5, 7 and 9) and 2 TLR genes of mithun (TLR 4 and 7) were sequenced completely.

DIAGNOSTICS

- A recombinant lipoprotein 'P48' of *Mycoplasma agalactiae* was evaluated in an ELISA as a rapid immunodiagnostic tool in sheep and goats.
- A genus-specific PCR assay was standardized for rapid detection of *Arcobacter* sp. targeting 16S rRNA for confirmation of the isolates of this emerging pathogen.
- A 36 kDa protein was identified in different serovars of *Salmonella* during biofilm formation.
- Magnetic nanoparticles (MNPs) were synthesized by two different processes and were characterized for structural, morphological, size and magnetic properties. The MNPs were functionalized for

immunoassay applications and evaluated for cellular toxicity in *in-vitro* experiments.

- A specific and sensitive diagnostic marker was identified for early detection of pancreatitis in dog.

MOLECULAR CHARACTERIZATION OF PATHOGENS/RECEPTORS

- Indian H5N1 isolates were characterized genetically and grouped with EMA *3/2.2.3 subclade,* which branched from two distinct nodes in the phylogenetic tree.
- The alignment of deduced amino acid sequences showed that all the Indian H5N1 viruses characterized contained the multiple basic amino acid motif at their HA cleavage site, which is characteristic of highly pathogenic avian influenza viruses.
- Two viruses from 2010 outbreaks in West Bengal contained amino acid asparagine at position 31 of the M2 protein, suggesting that both viruses are resistant to drug amantadine.
- An isolate of buffalo pox from man and animal recovered from an outbreak from Pune was characterized for B5R, H3L, D8L and A27L genes.
- An orf isolate from camel in a sporadic infection was characterized for B2L, GIF and VLTF-4 genes.
- Genotyping and partial sequencing of rotaviruses from man and pigs revealed human-pig reassortants as well as presence of novel combinations in India.
- Complete HN gene of velogenic Newcastle disease virus containing kozak sequence was amplified and cloned into pcDNA3. 1 eukaryotic expression for enhanced expression.
- Phylogenetic analysis by AP-PCR, PFGE and ribotyping showed considerable similarity between human and fish isolates of pathogenic and pandemic *Vibrio parahaemolyticus,* which reflected the public health risk potentials of fish.
- Molecular characterization by PCR-SSCP analysis of *Oesophagostomum* spp. revealed two distinct conformers with six nucleotide polymorphisms (SNPs) among male and female parasites.
- Probability of evolutionary changes among prevailing genotypes of *Oesophagostomum* spp. was predicted by analysing secondary structure of RNA.

QUALITY CONTROL AND PRODUCTION OF VETERINARY BIOLOGICALS

- 223,500 doses of RD 'F' strain vaccine; 166,000 RD 'Mukteswar' strain vaccine; 132,500 doses of fowl pox vaccine; 184,195 doses of lapinized swine fever vaccine; 423,800 doses of tissue culture sheep pox vaccine; 160,600 doses of PPR vaccine; 57,925 doses of *Brucella abortus* strain-19 (live) vaccine; 3,640 ml HS adjuvant vaccine; 73,500 doses of tuberculin PPD; 50,000 doses of Johnin PPD; 25,250 doses of mallein PPD; 55,250 ml of *Brucella* agglutination test antigen; 6,900 ml of *Brucella abortus* Bang Ring antigen; 19,860 ml of Rose Bengal Plate Test antigen; 219 ml of *Brucella* positive serum; 9,000 ml of *S.* Pullorum coloured antigen; 45 ml *S.* Pullorum positive serum; 87 ml *Salmonella* 'O' serum and 3,750 ml of *S.* Abortus equi 'H' antigen were produced and quality tested.

- 7.32 million monovalent doses of FMD vaccine were produced at Bengaluru campus.
- 435,100 doses of PPR vaccine were produced; similarly, 33 PPR c-ELISA, 22 PPR s-ELISA and 100 doses of goat pox vaccine were supplied by Mukteswar campus.

SERO-MONITORING OF IMPORTANT EQUINE DISEASES

During 2009–10, equine disease sero-survey was conducted in Rajasthan, Haryana, Punjab, Uttar Pradesh, Uttarakhand, Madhya Pradesh, Chhattisgarh and Sikkim. A total of 176 out of 1,045 equines (16.84 per cent) had antibodies to equine influenza virus while 28 (2.67 per cent) were positive for EHV-1 and 107 (10.43 per cent) were positive for antibodies to Japanese encephalitis virus. Antibodies to *Babesia equi* were detected in 320 (38.09 per cent) of 840 equines, while *T. evansi* antibodies were detected in 49 (5.4 per cent) out of 906 equines tested. One case of equine infectious anemia was detected in Uttarakhand. EIA-positive equine was reported after a gap of 11 years from India. None of the 1,045 equines tested showed antibodies to *Brucella* and *Salmonella* Abortus equi 'H' antigen.

JAPANESE ENCEPHALITIS (JE) VIRUS ISOLATION

JE, a mosquito-transmitted disease caused by JE virus, is causing disease in equines, pigs and humans. It is prevalent in eastern and southern Asia. Serum samples from clinically affected horses and pooled mosquito vector samples from the surrounding field were found positive for JEV RNA by RT-PCR. Virus was isolated from one of the horses. This is the first report of isolation of JEV from a horse in India. The phylogenetic analysis of the envelope gene (1500 bp) sequence of the isolated virus, JE/eq/India/H225/2009 indicated that the isolate belonged to genogroup III of JEV, and that the equine isolate clustered together with Vellore group of JE isolates from India.

CAMELPOX VIRUS ZOONOSIS

Camelpox virus (CMLV) causes skin eruptions in addition to severe generalized exanthema mostly in camels and outbreaks occur frequently in camel-rearing north-western region of the country. However, their zoonotic potential has not been proven conclusively. The disease was confirmed on the basis of isolation of the virus from diseased camels, detection of the CMLV antibodies in the serum of diseased camels and human patients by serum neutralization test and amplification of CMLV-specific three full-length envelope protein genes (A27L, H3L and D8L) and one partial gene (C18L) of the isolated viruses from camels and human specimens.

The phylogenetic analysis indicated that CMLV isolates cluster closely with variola virus (VARV) and vaccinia virus (VACV). The CMLV zoonosis indicates the changing pattern of host specificity of poxviruses and certainly points towards a declining cohort immunity against Orthopoxviruses in humans, which could be a public health concern. This is the first confirmed report of camelpox virus zoonosis.

CYSTEINE PROTEINASES FROM TRYPANOSOMA EVANSI

Parasite derived cysteine proteinases were proposed as potential targets for immunotherapeutic, chemotherapeutic and diagnostic reagents for parasitic diseases. The proteinases of camel isolate of *T. evansi* were analysed by gelatin substrate SDS-PAGE containing collagen antigen confirming the presence of cysteine proteinases in *T. evansi*. Some of these cysteine proteinases were reactive with experimental/field serum sample from equines with *T. evansi* infection. This indicated the potential of these enzymes in serodiagnosis of this economically important disease of livestock.

THE ENVIRONMENTAL IMPACT OF INTENSIVE LIVESTOCK PRODUCTION

The intensification of livestock production has had not only a significant effect on animal welfare, but also a considerable impact on the environment. In this regard, the disposal of the enormous quantities of animal waste produced by animals confined in intensive livestock production units has created the greatest environmental problems. The reduction of manpower levels and the need for efficient animal management have meant that 'factory' farmers have often abandoned the use of straw as bedding in livestock houses; the animals instead stand on slatted or solid concrete floors. The faeces and urine which these animals produce must therefore be disposed of differently.

When straw bedding is used, manure is collected in dung heaps, exposed to air it gradually ferments, bacteria are thus to some extent destroyed and the excreta eventually becomes fertiliser to be spread on the land. Slurry, on the other hand, must be stored in tanks, where it may decompose and sometimes produce a variety of noxious smelling gases. This method does not kill pathogens quite as effectively as the traditional method.

While straw bedding most closely resembles the natural surface on which most animals lie—and is thus promoted as the ideal floor surface on which animals should be kept—it can provide the conditions under which disease organisms can flourish. If poorly maintained, such bedding when excessively dirty and wet not only causes distress to the animals, but can also lead to environmental mastitis in dairy cattle, 'farrowing fever' in pregnant sows and an increased exposure of poultry to the dangerous *E.coli* bacteria.

Slurry can be returned to the land as fertiliser, but its storage and application can be particularly problematic. It cannot be spread when the ground is too wet and must, therefore, be stored in huge tanks, sometimes for several months. The failure of these storage tanks would have a significant impact on the local environment, most particularly if pathogens, such as salmonella, *E.coli* and leptospira, which are dangerous to both human and animal health, were still present in the slurry. To reduce the risk of infection, farmers have been warned against spreading slurry on grazing land.

If they must do so, they are advised to allow the slurry to adequately ferment and to leave the land on which it has been spread unpastured for at least six weeks afterwards. The main problem with modern farms is that they do not always have enough land to absorb the waste that the animals produce. The waste must, therefore, often be transported elsewhere at considerable expense to the livestock holder. As Johnson suggests, waste disposal can be the main factor which determines how many animals can viably be kept in a livestock production unit, rather than animal welfare. To illustrate this problem, he describes the crisis faced in the Netherlands during the late 1980s.

The great concentration of factory farms had put so much pressure on land that by January 1987 the Dutch authorities placed a ban on the further expansion of livestock numbers. With farms producing around 100 million tonnes of effluent per annum, twice the amount of liquid waste accumulated than could safely be spread on the country's 2 million hectares of farmland. As a consequence, the transport of slurry to areas away from factory farms was encouraged and subsidised by the Dutch government and large-scale dumping of slurry was banned due to the potential pollution and health risks it presented. Dutch farmers are now required to pay the full costs of disposal through sewage plants if they are unable to find appropriate land to spread their farm waste on.

Like slurry, the production of silage—a grass crop harvested when green and then partially fermented to make animal fodder—can present serious environmental problems. The anaerobic fermentation which accompanies silage production can potentially produce a waste liquid which has been estimated to be 200 times more polluting than domestic sewage, particularly when the grass is wet. Such waste presents great pollution problems if it escapes the silos in which it is held and enters natural waterways. The aquatic life of rivers and streams can be either poisoned or starved of oxygen as a result of the introduction of this effluent. The contamination of natural waterways is one of the most serious environmental consequences of modern agricultural practice. Aside from the problems posed by slurry and silage effluent, intensive livestock production contributes towards nitrate pollution. In addition to this, the additives used in animal feed and the antibiotics which the animals are routinely given are potential sources of environmental contamination. Although the blame for environmental pollution is often apportioned to industry, rather than agriculture, the intensification of farming methods—most particularly with respect to the waste it produces—presents a significant threat to the environment.

In addition to the phenomenal impact on animal welfare and the environment, the intensification of livestock production has created new opportunities for disease organisms to flourish and threaten the health of both animal and human populations. As this book has thus far clearly illustrated, our interactions with other animals have always brought with them the potential risk of infectious disease. However, the increasing exploitation of and dependency on animal resources, which has occurred through the intensification of livestock production, has meant that disease *risks* which did not exist, or were unthinkable, forty to fifty years ago are today commonplace. In short, the increase in stocking densities has led to the increase of opportunities for disease transmission.

The housing and living conditions of intensively farmed animals—most notably pigs and poultry—potentially provide significant exposure to bacteria and viruses, thereby enhancing the risk of disease. Disease risks can be greatly reduced with good stockmanship. On 'factory farms', however, animal management is primarily linked to maintaining or increasing productivity. In the past, stockmen could attend to animals more individually and were familiar with the small herds they tended.

Today, stockmen are not only fewer in number, but are also required to manage huge numbers of livestock as efficiently and cost effectively as possible. The reduction of manpower levels in livestock farming combined with the increasing numbers of animals kept on site can lead to inefficiency in disease control and hygiene which, in turn, may result in the outbreak of infectious disease amongst the animal stock and—depending on the nature of the disease in question—possibly even transmission to the humans who come into direct contact with them. This is, of course, not to mention the risk of contaminated animal protein entering into the food chain and putting human consumers at risk of infection.

One of the disease management techniques adopted in intensive livestock production is to accept the pervasiveness of infection. Antibiotic drugs and vaccinations are thus routinely administered in order to keep stock healthy and maintain productivity. When disease outbreaks occur, it may not be viable to cull stock given the size of the animal population housed. Similarly, farm buildings may have been too poorly constructed and cannot be properly disinfected or cleaned, the animal's excreta, as suggested above, creating potential hazards for both animal and public health.

Management of disease purely through drug intervention is far from ideal. Moreover, it arouses considerable consumer concern about the quality of the animal produce which ends up on our dinner plates. The use of growth hormones and the possibly harmful effects of veterinary drug residues in food have become an increasing source of consumer anxiety in recent years; as has the suggestion that the persistent administration of antibiotic drugs may ultimately lead to the evolution of pathogens resistant to such substances.

For fifty years, antibiotics have been added to animal feeds in order to improve feeding efficiency and growth rates. Since the 1960s, scientists have looked into the possibilities of resistance risks and by 1970 the European community had made moves to prevent the antibiotics that are used in human medicine being used as animal growth promoters. Today, the issue of resistant risks has again become the topic of political and scientific debate, stimulated in part by the development of a veterinary antibiotic called *avoparcin* which is closely related to an important human medical antibiotic known as *vancomycin*. There is a concern that such veterinary antibiotics may promote the emergence of bacterial strains that have a genetic resistance to antibiotics.

Although there is currently little microbiological consensus on this topic, some scientists have gone so far as to suggest that the widespread use of antibiotics in livestock production, particularly those used to treat infections, may well be responsible for resistant strains of salmonella and campylobacter that are being found in increasing numbers of food poisoning cases in humans. In recent years, the problem of animal disease control has also been tackled from another angle, that rather than seeking to alter the living conditions of the animals, instead seeks to alter the animals themselves. In this context, attempts have been made to rear strains of 'minimal-disease' animals which are kept in highly controlled environments. Poultry flocks today can be bred to be 'specific-pathogen free', which essentially means that particular disease-producing agents have been eliminated by breeding and rearing birds in sterile conditions. Such isolation, however, can render such birds vulnerable to other microorganisms that might otherwise have been innocuous.

Modern intensive livestock production is a risky business and sometimes it is human health that is at stake. In spite of the industry's endeavours to produce animal products that are safe for human consumption, as indicated above, food-borne infections still pose a threat to public health. The intensification of poultry rearing is the main culprit for bacterial infections in humans. Salmonella infection is one of the ever-present hazards of broiler rearing. It is in fact widely accepted that salmonella will persist in poultry stocks, although it can be kept at bay by good feeding and housing practices.

The pervasiveness of this bacteria in poultry is largely due to contaminated feed, although people and wild animals or birds can also be the vectors of transmission. Intensively reared poultry, like other farm animals which can carry the salmonella bacteria, are generally fed on high-protein feeds which can contain slaughterhouse waste and other animal by-products such as dried poultry manure. If such feed is inadequately sterilised, disease organisms can persist and can be passed on to other animals. Sterilisation is a costly process which occurs through heat treatment. However, irradiation of animal feed is not always efficient because strains of salmonella have emerged which are heat-resistant and continue to contaminate feed.

The transmission of salmonella requires 'the transfer of faeces or intestinal contents from an infected individual to the digestive tract of a susceptible one'. Like animals, humans can become infected by salmonellosis through the ingestion of contaminated animal produce. The disease can cause gastroenteritis, septic shock syndrome and focal infection in humans, and can even be fatal to vulnerable individuals, such as infants or elderly people. It should be emphasised that good hygiene and effective cooking methods from the abattoir to the kitchen can reduce the risk of infection for the human consumer.

The modern systems of livestock rearing, mass animal transportation and conveyor belt style slaughterhouses, however, increase the possibilities of cross-infection and the contamination of human food. Although the most common bacterial infection, salmonella is not the only disease which may be contracted from the ingestion of animal produce. Bovine tuberculosis, brucellosis, trichinosis, campylobacteriosis, listeriosis, staphylococcal enterotoxicosis, clostridial toxicosis and toxoplasmosis may be transmitted through tainted animal food and cause human illness.

Aside from food-borne infections and intoxications, consumer health is generally not affected by the diseases which farm animals carry. The people at greatest risk from agricultural animals and their products are in fact those who work with them; in other words, farm workers, veterinarians and individuals who handle animal carcasses occupationally. In Western Europe today, the major zoonoses, such as brucellosis (contagious abortion), tuberculosis and anthrax, are uncommon. Other zoonotic diseases, however, present serious occupational hazards to those in the meat, dairy and animal product industry.

For example, leptospirosis—commonly known as Weil's disease—is a disease which can seriously affect dairy farmers, both economically and physically. It is a contagious bacterial disease which can produce a high abortion rate in cattle late in pregnancy; it can be responsible for high mortality in calves, reduced milk production and contaminated milk. The organisms that cause leptospirosis are transmitted through urine and live in water or moist soil. Cattle—and other livestock—become infected through being splashed with infected urine, grazing on contaminated pasture or by drinking infected water.

Other wild species, particularly rodents, are responsible for transmitting the disease to cattle, and humans can be infected by bathing in contaminated water or direct contact with infected urine. The effect of leptospirosis on humans varies from inapparent to acute febrile illness which can be accompanied by fever, hepatitis, jaundice and meningitis—amongst other maladies. Good animal management, vaccination and reduction of contacts between livestock and rodents reduce the risks of infection by leptospirosis. There are many more health hazards for those who come into routine contact with food-producing animals. Other occupational hazards for humans are presented by rickettsial diseases, fungal infections (*e.g.* ringworm), chlamydia infections and respiratory viral diseases, in addition to bacterial infections such as those described above.

The fact that intensive livestock production today happens within closed systems, away from public view, has much to do with the reduction of infection risks. Modern livestock housing is essentially designed to keep out disease-causing organisms and their vectors of transmission and to keep disease-free animals in. Such systems are far from foolproof; infection may, for example, occur through the drinking water or animal feed. Furthermore, farmers, stockmen and veterinarians may introduce disease to livestock. It is commonplace when one enters a livestock production unit for one to have to don special overalls and boots in order to minimise the possibility of introducing new pathogens to the animals from outside. However, keeping all visitors out, apart from stock and veterinary personnel, not only aids disease prevention and control, but to some extent prevents consumers from gaining a first-hand glimpse of the conditions under which their food is produced.

Modern consumer society demands a continual, cheap and dependable supply of animal products which are pre-packaged and ready to enjoy: yet never before have the majority been so divorced from the production of the animal proteins which sustain them. The urban consumer is estranged from the origins of the food he eats and the leather he wears and has limited involvement in or knowledge of the circumstances of their production. The consumer is very much reliant on the producers and retailers to offer fresh and uncontaminated animal products—at a low price—on demand. The 'farm fresh' animal products which are nowadays found in urban supermarkets are however unlikely to have been produced locally, as was formerly the case. Animal produce may originate from farms and factories situated hundreds of kilometres from the consumer's domicile. The transport networks and storage capacities which today so efficiently ensure the swift despatch and delivery of farm-fresh animal produce to the consumer were not possible until recently.

Previously the movement of livestock and carcasses was fairly restricted, both within and across provincial or national borders, not only through lack of adequate transportation and refrigeration, but also due to the fear and possibility of the spread of contagious disease. Once a sufficient level of understanding and knowledge of transmission and control through governmental intervention had been reached, disease could be better contained, cross-infection prevented and animals could be more safely moved from one location to another.

Yet although governmental regulations pertaining to animal disease control have existed both nationally and internationally since last century, it is still possible for contaminated animals to slip the net and infect others (be they animal or human). Given that infected livestock or tissue can easily and quickly be moved from one (international) location to another, new opportunities for disease-producing organisms can develop in areas and amongst animal-human groups previously unaffected. The risk of disease can be further exacerbated by the failure to properly disinfect vehicles used for animal transportation, and delays in the transfer of live animals or meat from one place to another facilitate contagion further. Since, on an international level, animal movements are more likely to increase than decrease as a result of political and economic accord, the potential problem of animal disease transmission could, in the future, be one of increasing significance.

VETERINARY REGIME DURING THE TWENTIETH CENTURY

As the above discussion suggests, the possibilities for the transmission of infectious diseases are very real, most particularly given the risks posed by intensive systems of agricultural animal management and animal transportation. It is largely for this reason that veterinary medicine has become so important during the twentieth century in maintaining and monitoring animal health, to keep the spread of infectious disease in check and to ensure that farmers adhere to and understand the guidelines and regulations concerning disease prevention and control. Yet while disease *risks* may today be far greater than ever before, the danger of humans *actually* contracting diseases from contact with animals or the consumption of animal produce is in fact relatively small.

Throughout the course of the twentieth century, the veterinary regime has continued to move from strength to strength. The state veterinary services that were instituted in the industrialised nations of Europe during the late nineteenth century have played an increasingly important role in the control and prevention of animal disease. Legislative measures to combat the effects, spread and introduction of contagious disease have been enforced with increasing vigour.

The successes of earlier campaigns and stringent legal controls to stamp out the animal plagues that devastated European livestock populations provided the basis for future legislation. For example, veterinary inspection and the rigourous enforcement of all regulations pertaining to the importation, transport, sale, slaughter and destruction of livestock successfully led to the eradication of contagious bovine pleuropneumonia in the Netherlands by 1887 and Britain in 1898.

The introduction of measures to financially compensate farmers for their losses through compulsory slaughter policies undoubtedly provided an incentive to abide by governmental regulations and allowed veterinary officials to perform their duties without hindrance. Eventual success in eradicating disease helped to persuade livestock producers to accept the ruthless and restrictive regulations that the state sought to enforce. As a consequence, state veterinary intervention has become institutionalised within modern livestock production. Moreover, during the twentieth century, the appreciation of the need for animal disease control and prevention has more or less been incorporated into the habitus of livestock producers.

Scientific developments have, of course, played a crucial role in the prevention and control of animal disease. The development of the science of bacteriology, advances in pathology, parasitology, immunology and microbiology and so forth have provided important contributions to advances in animal and public health. By the turn of the century, European governments had become increasingly aware of the need for better scientific understanding of animal disease if they were to reduce the risks that it posed. In the Netherlands, for example, the State Serum Institute was founded in 1904-by the pioneering veterinarian and bacteriologist Jan Poels—with a view to steering the work of the established Dutch Veterinary Service towards more scientific means of preventing and curing livestock disease.

The Serum Institute experimented with the production of a serum against foot-and-mouth disease and an anthrax vaccine. Moreover, it conducted bacteriological and chemical investigations into milk and dairy produce. Interestingly, this new institute operated specifically outside the dominion of the veterinary school, being instead an organ of the state. In 1933, nearly three decades later, the State Veterinary Research Institute was established, primarily to study and develop a vaccine against foot-and-mouth disease—a highly infectious disease that has perhaps wreaked the greatest havoc on European livestock and the livestock industry during the twentieth century. By 1946, the Institute's director, H.S. Frenkel, had succeeded in developing a method by which a vaccine against foot-and-mouth disease could be produced in large quantities. This method, however, only proved effective until 1951, when a more virulent strain of the disease struck Dutch livestock.

From 1945 onwards, the structure of the state veterinary machine became increasingly more complex, as government-subsidised Animal Health Services were established in each province within the Netherlands. The earliest regional animal health service had in fact already been established in Friesland in 1919. Its early work led to significant progress in the battle against bovine tuberculosis. The new regional services were modelled on their Friesian predecessor and strove for the coordinated control of animal disease, involving not only scientific and governmental veterinary advisors, but also representatives of local farmers and dairy organisations and practising veterinarians.

The eleven regional Animal Health Services were in turn coordinated by the Industrial Board for Agriculture, who established an Animal Health Committee to deal with the allocation of funds and organisation of schemes for disease control. After the Second World War, the regeneration of the meat and dairy export industry became an important part of Dutch economic recovery. A key aspect of this was a national plan to rid Dutch cattle herds of tuberculosis, thereby ensuring the safety and quality of the national product. A considerable portion of the Marshall Plan funds that were made available to the Netherlands to assist post-war economic reconstruction was devoted to achieve this end. By 1956, the disease had been conquered due to the efforts and cooperation of the Animal Health Services, practising veterinarians and the meat inspection service.

On the basis of this national plan's success, similarly successful attempts were made to control and eradicate brucellosis from the Dutch livestock population. By 1957, the Serum Institute and the State Veterinary Institute amalgamated to form the Central Veterinary Institute, which functioned under the auspices of the Dutch Ministry of Agriculture and Fisheries. The Veterinary Service was effectively given the governmental mandate, through the establishment of legislation such as the Livestock Act and the Veterinary Medicines Act, to prevent animal disease and preserve public health.

The work of the Central Veterinary Institute continued to grow and diversify throughout the 1960s and 1970s. In 1986, it consolidated and moved to new premises in Lelystad. By 1994, the CVI amalgamated with a number of other agricultural animal research institutes to create the Institute for Animal Science and Health.

Today, this institute employs over 600 people and conducts research into animal health, breeding and reproduction, animal welfare, stress and management, animal nutrition, and so forth. Moreover, it is responsible for diagnostic work, the production of vaccines, specific-pathogen free animals, diagnostic preparations, etc., and the control and standardisation of veterinary medicines.

The developments in state-directed animal disease control in the Netherlands can be mirrored in the rise of the State Veterinary Service in Great Britain. There too was an understanding for the necessity of improved scientific knowledge and diagnosis of animal disease.

The first governmental diagnostic laboratory was set up in 1894 in response to an outbreak of swine fever. This was followed by the establishment of the Central Veterinary Laboratory at Weybridge, Surrey, in 1913. Initial research was conducted into spontaneous abortion, swine fever, scrapie, foot-and-mouth disease and diseases of poultry.

As farmers began to increase the size of their flocks, as the process of intensification of production began to take off, the laboratory found itself increasingly dealing with the effects and greater incidence of poultry disease. During the 1930s, the laboratory began to research and produce tuberculin, in addition to an anti-abortion vaccine. The Agricultural Act of 1937 facilitated further expansion of the laboratory. New research departments of bacteriology, biochemistry, parasitology, pathology and poultry disease were established and a further laboratory was set up in Scotland.

By 1960, the Animal Health Division successfully completed its programme to eradicate tuberculosis from the British cattle population. During the 1960s, the new challenges to animal health presented by the changing methods in livestock production gave extra impetus to the fight against animal disease.

The intensive rearing of calves, for example, led to an increasing incidence of respiratory disease. This unforeseen consequence of new methods of animal husbandry required urgent bacteriological and virological investigation. Similarly, the development of artificial insemination and its widespread use in cattle presented new problems for animal disease control. The movement of inseminators between farms and the importation of and trade in semen presented new hazards for the transmission of infectious animal disease. In 1955, legislative measures were taken to prohibit importation of foreign animal semen without a licence and the Animal Health Division took steps to screen the health of bulls.

In addition to the Central Veterinary Laboratory, the Veterinary Investigation Service was established in 1922, with regional centres throughout Britain. This service essentially provided local laboratory facilities on which practising veterinarians could rely to help diagnose and investigate animal health problems in the herds that they attended.

Initially, the Veterinary Investigation Centres were attached to agricultural colleges, but in 1946 they were incorporated into the Ministry of Agriculture, Fisheries and Food. The interaction between local veterinarians, veterinary investigation officers, the researchers of the Central Veterinary Laboratory and legislators in fact provides a good illustration of the increasing complexity of animal disease control during the twentieth century. At the grassroots level, practising veterinary surgeons observe and report the local incidence of animal disease.

This in turn is investigated by veterinary investigation officers who conduct field investigations and epidemiological studies to gauge the extent of animal health problems. In the meantime, the Central Veterinary Laboratory attempts to understand the aetiology of the disease and explores how it may effectively be controlled.

These combined investigations may in turn lead to the inception of statutory control measures, vaccination, test and slaughter programmes to contain and eradicate animal disease. In Britain, this kind of cooperation eventually led to the eradication of brucellosis and is today responsible for the investigation and extensive monitoring of new pathogens and diseases, *e.g.* bovine spongiform encephalopathy. In October 1995, the Central Veterinary Laboratory and the Veterinary Investigation Service merged to form the Veterinary Laboratories Agency in a further bid to improve the collaboration necessary to ensure animal disease control and the provision of advice to the Ministry of Agriculture, Fisheries and Food and other veterinary health workers.

Throughout the twentieth century, the necessity for collaboration and cooperation has extended far beyond the activities of local and nationally organised veterinary organisations. International cooperation on matters of veterinary public health first began during the nineteenth century as a result of the devastation caused by the cattle plague. In 1924, the *Office International des Epizooties* was established in Paris. This intergovernmental organisation initially sought to promote and coordinate research into the control and surveillance of animal disease, and has remained active on this front to this day; its original membership of 28 having gradually been expanded to some 144 member states.

During the post-war period, additional international fora for dealing with the economic and health problems associated with animal disease were institutionalised. In Europe, the European Union—and its former incarnations, *i.e.* the EEC and EC—have provided the main political arena within which European governments can jointly discuss and legislate on matters pertaining to veterinary public health and animal welfare. The European Economic Community (EEC) was originally established in 1958- following the signing of the Treaties of Rome -with a view to achieving economic and political union amongst its member states, namely Belgium, Germany, France, Italy, Luxembourg and the Netherlands. In 1973, Denmark, Great Britain and Ireland became members of the EEC. In later years, Austria, Greece, Finland, Spain, Sweden and Portugal joined the ranks of states seeking European economic union; the organisation is set to expand yet further in future as former Eastern Bloc countries vie for membership.

Since its inception, the European Union has had a critical impact on agricultural practice and the trade in live animals and animal produce within Europe. The Common Agricultural Policy was instituted during the early 1960s with a view to increasing agricultural productivity, ensuring a fair standard of living for the agricultural community, stabilising markets, ensuring food supplies and providing consumers with food at reasonable prices.

It has been perhaps the most important Common Policy that operates within the Union and has served as an important precursor to the 'single market' that is today intended to guarantee the free movement of goods, services, capital and people between the fifteen member states. The Common Agricultural Policy greatly influenced the economics of livestock farming within Europe and initially encouraged farmers to intensify production.

Ultimately this led to massive overproduction and the creation of huge surpluses, which in turn depressed market prices. 'Butter and beef mountains' came to symbolise the crisis in European agriculture; livestock numbers needed to be reduced and milk quotas were introduced. The Common Agricultural Policy has consequently undergone significant reforms and today it even encourages the *extensification* of production methods, particularly in the beef sector.

Free trade between member states remains at the top of the European Union's agenda, and trade naturally includes the trade in livestock and animal produce. As the epizootics of the past clearly evidenced, the movement of animals across national borders presents potentially great hazards to both public and animal health. Each member state has its own internal legislation and procedures regarding the control and prevention of livestock disease; the efficacy of which is entirely dependent on the efficiency of and cooperation between livestock producers, veterinarians and governmental agencies. However, membership of the European Union has required that each country's system of animal disease control be brought in line with those of other member states, in order to ensure a harmonisation and uniformity of disease prevention. Moreover, a system of veterinary inspection and veterinary and live-stock regulations have necessarily been introduced in an attempt to ensure the health of animals and the safety of meat that are crossing national borders.

With regard to livestock, each animal that is traded between member states, or enters the Union from a non-member state, must be accompanied by a veterinary health certificate that has been signed by an official veterinarian. Two types of certificates exist: one for animals intended for breeding and production; and another for animals intended for slaughter. The health requirements for breeding animals are much more stringent than for those destined for slaughter.

These certificates detail the animal's origins, destination, means of transport, age, breed, type, means of identification and whether they passed through a market or assembly point. Such certificates are valid for ten days after the animal has been loaded for transport. In this way, diseased animals can be traced along with those that they came into contact with during transport.

The health certificate states that the animal has been examined and has been passed as healthy by an official veterinarian and that it originates from a herd or region free from specified diseases. Animals may have to undergo further tests depending on their origin, type and destination. Traded livestock must meet a number of welfare considerations determined on a Union-wide basis. Likewise, the rules for the trade in meat have been harmonised throughout the European Union, in order that animal flesh is produced, preserved and transported according to rules of hygiene identical to every member state. Certificates issued by official veterinarians at the place of meat production provide a guarantee that the meat is fit for consumption and serve as a valid transport document which testifies to the safety of the product. Produce from outside the European Union is subject to similar checks at border inspection posts and is required to meet the standards deemed acceptable within the Union.

The veterinary regime has thus been further consolidated on an international level and has led—at least in principle—to the standardisation of the control of animal disease, welfare and health care and the quality of meat and other foods of animal origin within Europe. The legislators in Brussels have ensured that member states adopt uniform policies with regard to livestock production. European legislation, therefore, has increasingly come to determine the nature of national legislation.

For example, since 1988, EU countries have maintained a ban on all growth-stimulating hormones, although such products are still widely employed in the USA. Attempts to eliminate the illegal trade in such hormone products even led to the murder of a Belgian veterinary inspector in 1995. An illustration of the impact of the EU's veterinary regulations on member states can also be provided by events surrounding the bovine spongiform encephalopathy (BSE) crisis. The severity of the legislation has greatly disabled the British livestock and meat trade and has also had significant repercussions for other countries.

It seems likely that the public health issues and trade problems surrounding BSE will continue to feature prominently on the European economic, agricultural and veterinary agenda in Brussels in the future. Similarly, the way the 1997 outbreak of classical swine fever in the Netherlands has been dealt with has to a great extent been affected by EU regulations. While vaccination may provide the key to ending the crisis, the EU has committed itself to a policy of eradication of classical swine fever and, at present, forbids the use of a vaccine which may be effective against *the* disease. The Dutch pig industry has been crippled by this outbreak, export embargoes have been instituted and over three million pigs have been destroyed, not necessarily because they had contracted the disease, but to prevent overcrowding of pig stalls.

While animal disease control and prevention measures have become increasingly stringent throughout Europe during the twentieth century, leading to the virtual eradication of the majority of animal plagues that decimated livestock and played havoc with the European economy in centuries past, we cannot forget that the modern global economy and the routine movement of individuals—both human and animal—across and between continents also have a bearing on veterinary public health within the bounds of Europe. Diseases such as tuberculosis and classical swine fever, for example, are known to have been transmitted to western European livestock from live animal imports or animal produce from Eastern Europe. Similarly, imports from countries where animal disease control is not so advanced or strict may still pose a threat to public health since spores of diseases such as anthrax, which have more or less been eradicated in advanced Western societies, can still be introduced through the import of hides, animal foods etc. from regions where the disease still persists. It must also be noted that while a highly effective veterinary regime exists within Europe, veterinary policing is not always as efficient as it should be and regulations are not always well enforced.

Essentially, it only takes one weak link in the chain of interdependencies involved in the process of transforming animals into safe food to lead to inefficiencies and failures in animal disease control. Again, the recent BSE crisis provides a clear example of this. Inadequate compliance with and policing of the meat and bone meal bans, first introduced by the British government in 1988 to prevent the feeding of mammalian protein to ruminants, appear to have allowed the epidemic to continue, when it might otherwise have been curtailed. One may thus conclude that any animal disease control measures that the legislators introduce can only be as good as their policing.

The control of animal disease today is far from simply a local or regional problem, it has instead become a global one. The involvement of veterinary experts in the policy making and execution of the World Health Organisation (WHO) and the Food and Agriculture Organisation (FAO) certainly attests to this. However, although advanced industrialised societies have the animal disease problem more or less under control—producing significant animal protein surpluses in spite of continuing population growth—the (rural) populations of developing nations across the globe continue to suffer both the economic and health effects of animal disease. Furthermore, animal diseases can affect livestock populations and food production to such an extent that they do not even return a profit. Livestock disease therefore not only generates economic problems, but also wastes resources which could help to solve the problem of world food shortages.

While the focus of this chapter is far from global, it is still important to be aware of the fact that the relationship between humans, animals and the environment in which they all live is a continually changing one which is not confined within national borders. The complexity and scale of interdependencies between humans and other animals is such today that failure to keep disease-causing microorganisms adequately in check could be potentially devastating for both the physical and economic health of world populations. The necessity of an effective and globally-oriented veterinary regime in present-day society is, therefore, manifest.

2

Physical Structure of Veterinary

Birds and mammals both possess a constant temperature which is ordinarily well above that of their surroundings. They should, therefore, be equipped with an arrangement for preventing too rapid loss of heat; and this is provided by the feathers in one case, the hairs in the other. Swimming birds, almost without exception, are webfooted or lobe-footed.

The Jacana, which seeks its food on the floating leaves of water-plants, has enormous toes which distribute its weight as do skis or snow-shoes on the thin crust of snow. Indeed, in general the feet of birds are remarkably well adapted to the life which their owners lead. The egg-eating snake, Dasypeltis, possesses a special mechanism by which it can temporarily dislocate its jaws to swallow an egg whole. In addition, one of its neck vertebrae bears a downwardly directed spine which protrudes into the gullet, and is used to crack the egg where none of its contents will be wasted. The larvae of crabs live near the surface of the sea, and are provided with long spines which increase friction and make them less ready to sink. But the most striking examples are those known as convergence, where a similar mode of life produces similar effects on quite unrelated animals.

During the interval, progress has been steady and gradual in both departments of naval warfare, each advance in efficiency of guns being the stimulus for new invention in the methods of protection, and vice versa. It is when there is a general increase of the animal's powers of control that we speak of progress; when the increase is in one special particular only, we speak of specialisation. For instance, the horse is specialised for running, the mole for burrowing, the bat for flying, the whale for a marine life, the lion for catching and devouring large animals, the sloth for living in trees.

Each of these animals is well adapted for its particular mode of life, but each is by that very adaptation quite cut off from leading the life of any other. During the late Secondary period there existed similarly specialised types of animals. For example, the Ostrich Dinosaur was adapted for running, the Ichthyosaur for life in the sea, the Tyrannosaur for preying on large animals, the Pterodactyl for flight, and so on. But all the former list of animals are mammals, all the latter were reptiles. The mammals are higher in that they possess proper temperature-regulation, for instance, and better prenatal care of their young, as well as in many other points. Thus, while the types of specialisation, or of adaptation to particular modes of life, are somewhat similar in the two cases, yet each member of the first group is higher than any member of the second, because its general organisation is on a more efficient level.

Whenever a new group of animals is evolved, it is found that its members soon become specialised in different directions, thus filling up different vacant places in the economy of Nature. This adaptation to different modes of life, while, as we have seen, we call it *specialisation* when we are thinking only of one species of animal, is called *adaptive radiation* when we are thinking of the group as a whole. All fish, for instance, breathe by gills, possess fins as limbs, and have other characteristics in common, so that any member of the group can be at once recognised by a brief examination of its structure.

Yet the detailed form of different species of fish is extremely varied. Besides the ordinary type of active, free-swimming fish, like the herring or trout or mackerel, there are fish flattened in adaptation to life on the bottom, some flattened sideways, like the sole and plaice, others flattened from above downwards, like the skates and rays; there are elongated fish like eels and pipe-fish; there are fish with prehensile tails, like the sea-horse; there are the flying-fish adapted for leaping long distances out of the water; mottled fish of irregular outline adapted for living on rocks; deep-sea fish with wonderful phosphorescent organs for searchlights and huge eyes for perceiving the faintest trace of light; cave-fish without eyes at all; Sucker-fish adapted for sticking tight to the underside of stones, or for being carried about by larger fish without expending any energy themselves; and so on and so forth through almost every conceivable sort of form possible in an under-water existence.

The examples previously mentioned give an idea of some of the adaptive radiation which has taken place in reptiles and in mammals; very similar instances could be taken from any other large group, such, for instance, as the insects. It is interesting to take any such group and to see what part adaptive radiation has played in giving rise to the main sub-groups into which it is divided. There is one particular form of specialisation which we have not so far mentioned; and that is the retrograde form of specialisation known as degeneration.

There are many cases known where animals can be definitely shown to be less highly organised to-day than were their ancestors in the past. The animal known as Sacculina, for instance, is a parasite upon various sorts of crabs. It consists of a mere bag filled with little else but reproductive cells, and sending out a whole series of branched roots which penetrate the crab's body in all directions and suck nourishment out of it. At first sight, the relationships of this unpleasant creature are very difficult to determine.

But when its development is investigated, it is found that it hatches out of the egg as a a free-swimming larva exactly like those found in many Crustacea. It has jointed limbs, an external skeleton made of chitin, and in fact bears all the distinguishing marks that other Crustacea do, be they crabs or lobsters or shrimps or water-fleas. It is, in fact, a crustacean which has become adapted to a parasitic life; and in so doing it has acquired special adaptations, such as the root-like organs, specially suited to that life, while it has lost sense-organs, limbs, digestive system, and everything else necessary for leading a free and independent existence.

It has lost more than it has gained, its organisation has become simpler, its independence less; in fact it has gone down the evolutionary hill, and the direction of its history has been in most ways the opposite of the direction which characterises biological progress. The form in which Sacculina hatches out resembles in general many crustacean larvae; but it is particularly like the early stages of the animals known as barnacles.

Every one who has been to the seaside knows what an acorn barnacle looks like—a little creature attached to rocks or piles, enclosed in a white shell, and capable of sending out of a slit in the top of the shell a regular sweep-net composed of a number of 'arms'—really appendages—with which it drags minute floating particles of food in towards its mouth. Even in the adult state a barnacle shows some resemblance to ordinary Crustacea, especially in its jointed limbs and chitinous external skeleton; the early free-swimming larva clinches the matter and gives complete proof, as in the case of Sacculina, of their crustacean nature and affinities.

The very close resemblance of the larva of barnacles to that of Sacculina points to an especially close connexion between the two sorts of animals; and as a matter of fact, the two are undoubtedly descended from a common stock. The barnacle is degenerate as well as the Sacculina, but it is not so degenerate. It still, for instance, possesses organs for capturing and digesting food. On the other hand, it has lost its organs of special sense and of locomotion. Further, it is not adapted to the same general mode of life as is Sacculina; it is not parasitic, but sedentary or sessile.

This settling down and becoming fixed is the other great cause, besides parasitism, of degeneration in animals; as would be expected, the degeneration due to a sedentary life is rarely so great as that due to parasitism, since the sedentary animal does not obtain its food ready-digested as do most parasites. It must not be supposed that because the general rule among animals is that time brings change, that therefore time *invariably* brings change. The common lamp-shell Lingula, for instance, has persisted without any appreciable change whatever in the structure of its shell for the prodigious period of time, certainly over five hundred million years, which has elapsed since the Cambrian epoch in the Primary period. Even when individual species have changed, the general characters of groups have often persisted with very little modification, as, for instance, those of dragonflies since the coal-measures, of shark-like fish since the Silurian. In some cases this may mean that for some unknown reason the species or the group has lost the power of varying to any considerable extent.

More often, probably, it so admirably fills one particular niche in the order of things, and a niche which stays the same throughout the geological periods, that it pays for the animal or the group of animals to stay as they are, leaving it to other groups or to other branches of the same group to colonise new niches and to progress towards fuller existence. To sum up, we may say that two main types of evolutionary change result in animals (and also, as a matter of fact, in plants) from the struggle for existence and the constant appearance of inheritable variations. In the first place, once a new type or plan of structure has been evolved, it undergoes adaptive radiation; in other words, there are developed a number of separate species all built on the same general ground-plan, but adapted to different and usually incompatible modes of life.

In the second place, new types and new plans are continually appearing as time goes on, and progress is marked by the fact that among the later-evolved types there is to be found greater complexity of organisation, greater control and independence of environment, than among the earlier. It might perhaps be thought that specialisation was often the same thing as progress. Specialisation, however, implies close adaptation to one particular mode of life, while progress means greater general efficiency.

If we look into the actual history of animals in the past, we find that specialisation in one direction involves the sacrifice of possible advance in other directions, and is a barrier in the long run to any but a quite limited degree of progress. As a result of its long course of specialisation, extending for tens of millions of years through the better part of the Tertiary epoch, a specialisation all tending towards greater efficiency in running and browsing, the horse stock has, it seems, cut itself off from the possibility of adapting itself to other modes of life—to a life in the water or in the trees, or to a carnivorous habit. There is a limit to the perfection which any line of specialisation can attain. While the horse was growing larger, developing hoofs, reducing the number of its digits, another and wholly different type was being evolved in the person of man.

If it were not that horses are useful to man, and are accordingly domesticated, they would now be wholly or almost wholly extinct. The 'natural' enemies of the horse are the large carnivores. These are built on the same general plan as the horse—the mammalian—and indeed are the results of the specialisation of the same plan in another direction. The same limits are thus set to them as to the horse stock. Before the advent of man, a state of equilibrium existed between the horse and its enemies, the latter not able to destroy the former entirely, the first not able to escape the payment of some toll to the second. But the horse came up against the wholly new biological conditions introduced by the new type, man; it was in direct conflict with man's cunning and tools and his habit of hunting in bands; still more important, it had to compete with the indirect effects of the new development, such as the settlement and cultivation and enclosure of the land. Against all this the horse was powerless. It could not develop far enough or fast enough to adapt itself to such sudden changes, and as a result it is becoming extinct as a wild species.

Over and over again the same thing happens, and the specialised representative of the old type, plant or animal, is extinguished in competition with the specialised representative of the new. For tree-like representatives of the horsetail type, which existed in the Carboniferous period, we have seed-bearing trees to-day; for pterodactyls, birds and bats; for dinosaurs, the large mammals; for the large early amphibians, the Stegocephalia, we have crocodiles; for the abundance, both in numbers and in species, of the larger mammals in the Pliocene, we have the multifarious activities of the swarms of man. The new type seems always to have arisen from some comparatively unspecialised branch of the old, and to have attained its pre-eminence by means of adaptations towards general instead of towards special efficiency.

Man has ousted the other mammals from their dominant position owing to the development of his mind. Through his particular type of mind he is able to deal rationally with the problems that confront him; tools, machines, tradition, civilization, and unexampled control and independence have been the result. The human mind is not merely adapted for solving one or two particular problems; it represents a *method* more efficient than any previously adopted for dealing with any and every problem that may confront an organism. Man's body is not highly specialised, and he seems to have arisen, through a monkey-like ancestor, from some unspecialised early mammalian stock like the Insectivores.

Nor did the early mammals show any signs of specialisation. All the fossil mammals that we know of during the time when the great period of reptilian dominance and specialisation lasted were small, primitive creatures, at first sight not likely to wrest the palm from their powerful rivals. Very similar chains of events may be seen taking place in the evolution of human machines and inventions. Take, for example, the history of transport. The most primitive method was the carrying of single loads by human beings or pack-animals. After that came the invention of vehicles in general and of wheeled vehicles in particular. The wheeled vehicle became specialised ('adaptive radiation') in innumerable ways.

We have the war-chariot; the rapid vehicle for passenger transport and for pleasure; the heavy wagon and cart; the van and pantechnicon; the four-inhand mail-coaches bowling daily at fixed hours along the main roads of the land. A limit was set to the capacity and the speed of such vehicles by the speed and strength of beasts of burden on the one hand, and by the imperfections of road-surface on the other. At the beginning of last century a new type of vehicle was evolved for which these limitations no longer existed. It was discovered how to replace the energy of animals by that of steam, and in large part to overcome the difficulties of surface friction by making the wheels of the vehicles run on metal rails.

As a result, steam locomotives became for certain purposes the 'dominant type' of vehicle within an extremely short period of time. Here again the type itself has been improved, so that we have now for some time been close to the limit of its possibilities. It does not seem possible to run a profitable service at speeds of much over sixty miles an hour. About half a century later a new plan was evolved. The internal combustion engine was produced, and gave certain great advantages, notably in being ready to start at once without the long preparation of 'getting steam up'. It appears to be the fate of new types to lead a precarious existence for a considerable time before they can compete successfully with the dominant types of the period. This was so, as we saw, with the earliest mammals; it was so for the steam-engine; and it was so, to a very marked degree, for the internal combustion engines. They were laughed at when their inventors took them out on the roads; the law laid down that they should be preceded by a man with a red flag; the early defects in construction did actually occasion many a breakdown.

But within thirty years they came into their own. Meanwhile, still another competitor is in the field—the flying machine, a totally new type, abandoning not only the particular device of the wheel, but the whole element to which the wheel was adapted. It looks as if in certain respects, where speed is the main object, the aeroplane would become the dominant type of vehicle; but that it would leave the major part of transport to be dealt with by train and by motor. Another interesting parallel with the evolution of organisms is found in the fact that although there has been progress, although the dominant type of vehicle has altered with the passage of time, yet many representatives of the old types have survived. They have managed to survive by becoming restricted to a few special conditions and places. Packanimals, for instance, while once universal, are now only employed in mountainous or roadless countries.

The American buggy is still of the greatest use over the unmade roads which are still to be found in so many parts of the United States. The horse-drawn vehicle will long hold its own in businesses in which not much capital is available, and speed and great power are not of the first importance. This survival of all or almost all the types that arise in evolution, even though new types arise and supremacy changes hands, is of general occurrence in the development of animals and plants, and is at first sight very puzzling. However, the insight which is gained by looking, as we have done, into the evolution of something familiar and human like the means of locomotion, helps us to understand the more complex and slower-moving processes of organic evolution. Another point which is brought out by the study of the development of some human contrivance such as the means of transport, is the great speed of change now possible in human affairs as against the slowness of change prevailing in lower organisms. The whole period from the stage-coach to the aeroplane is comprised in well under two centuries. The resulting change in human habits has been enormous; to produce comparable changes in the habits of an animal stock there would be needed a period certainly to be reckoned in tens of thousands, possibly in millions of years.

On the other hand, the evolution of machines is a perfectly real evolution. Two different types of machines capable of performing the same general function—such, for instance, as the motor-lorry and the goods steam-engine—do come into a very real competition with each other, and the issue of the struggle is decided by a form of true natural selection, depending in the long run upon which of the two pays the better. Here again the study of machines throws light upon the course of events in animals. It is often supposed that evolution must involve some conscious effort on the part of the evolving organism, that the struggle for existence is a conscious struggle, or that a species in some mysterious way 'learns' how to develop some new improvement in its structure. As a matter of fact, almost the whole of such ideas are purely metaphorical, and arise simply because we read the processes of our own minds into the operations of nature; it is not scientifically correct to speak, for instance, of *purpose* except in relation to human minds.

We see at once that the machines have no idea themselves of the direction of their evolution, that the 'struggle' between them is only a metaphorical struggle, that the selection between them is so far as they are concerned a mere sifting process, the issue of which depends upon the advantages or disadvantages which they may happen to possess. It is the same with organisms. If two races of animals come into competition, the issue is decided by the qualities which each happens to possess; 'natural selection' is a name for the effect exerted by all the forces of the environment with which they come into relation, an effect which acts again like an automatic sieve, and lets some through to perpetuate themselves, keeps back and so extinguishes others. The 'struggle' and the 'competition' are again usually metaphorical only. For instance, when the common housesparrow was introduced into America, it entered into competition with many of the small sparrow-like birds which had developed in that country. But the struggle did not take the form of a war between the invader and the original inhabitants.

It was an indirect struggle, due to the fact that both lived upon the same sort of food, both occupied the same sort of sites. The European sparrow happened to be endowed with qualities which gave it an advantage, and as a result it has spread enormously over the American continent, while many of the native birds have correspondingly decreased. If we wish to use a human metaphor, we can say that its success has been the result of 'peaceful penetration', not of fighting. The difference between machines and organisms, of course, is that the machines are directly designed by man, whereas the place of the designer in animal evolution is (roughly) taken by the variation which seems to be universal in organisms.

It is variation which provides the raw differences upon which the sieve of natural selection can work. Mention must also be made of the theories of evolution which are summed up under the term *orthogenesis*, which means evolution in straight lines. It is frequently found, as for instance in the development of the horses or of the elephants, that evolution as revealed by fossils proceeds straight onward through geological time in a perfectly definite direction-in the horses towards single toes and hoofs, in the elephants towards great bulk, tusks, and trunk. Orthogenesis is sometimes used merely as a descriptive term, to denote this observed fact of straight-line evolution. But by others it is used to mean that there exists some inner necessity for the evolutionary line in question to develop in just that one way and no other. However, a series of fossils, even if beautifully complete, can really give us no insight into the method by which its evolution occurred. Whenever the direction in which the series is evolving is adaptive or seems biologically advantageous, the orthogenetic series can be perfectly well explained by natural selection.

On the other hand, there do exist cases where at least no advantage can be perceived by us in the direction pursued. This is so, for instance, as regards the Ammonites, the extinct cephalopod molluscs which died out near the end of the Secondary period. Near the close of their time on earth, they often evolved orthogenetically into the most bizarre forms, their spiral shell becoming unwound, or irregular. Possibly in such cases a real causal orthogenesis, an inwardly determined mode of variation, is at work. But we are justified in saying that such cases, if they occur at all, are certainly rare.

ORGANIZATION OF ANIMAL BODY

Animals are multicellular heterotrophs whose cells lack cell walls. At some point during their lives, all animals are capable of movement, although not all animals have muscles they use for this. In the most commonly encountered animals, the mobile stage is the adult, although some animals (such as corals and sponges) have sessile (or non-mobile) adult phases and mobile juvenile forms. Both animal and plant evolutionary history show the development of multicellularity and the move from water to land (as well as a secondary adaptation back to water, for example dolphins, whales, duckweed, and elodea).

Animals developed external or internal skeletons to provide support, skin to prevent or lessen water loss, muscles that allowed them to move in search of food, brains and nervous systems for integration of stimuli, and internal digestive systems. Organs in animals are composed of a number of different tissue types. For example, the stomach shown in Figure, has epithelial tissue making linings and secreting gastric juices, connective tissues

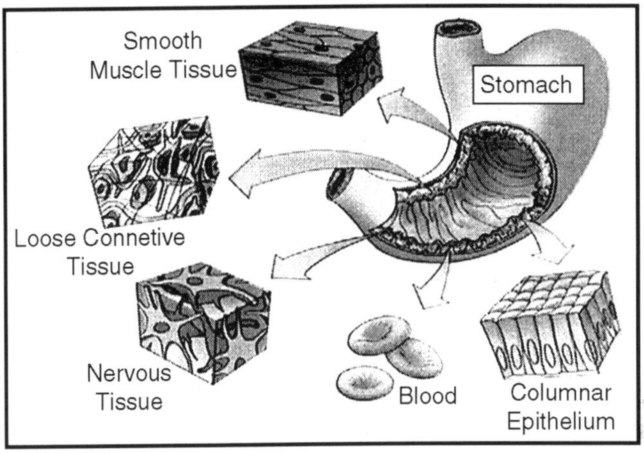

Fig. Cells and Tissues that Comprise the Stomach

Plants are simpler organisms than animals, having three organ systems and fewer organs than do vertebrate animals. Organs are composed of tissues, which are in turn composed of cells. Plants have three tissue types: ground, dermal, and vascular. Animals have four: epithelial, connective, muscle, and bone.

EPITHELIAL TISSUE

Epithelial tissue covers body surfaces and lines body cavities. Functions include lining, protecting, and forming glands.

Three types of epithelium occur:

1. Squamous epithelium is flattened cells.
2. Cuboidal epithelium is cube-shaped cells.
3. Columnar epithelium consists of elongated cells.

Any epithelium can be simple or stratified. Simple epithelium has only a single cell layer. Stratified epithelium has more than one layer of cells. Pseudostratified epithelium is a single layer of cells so shaped that they appear at first glance to form two layers.

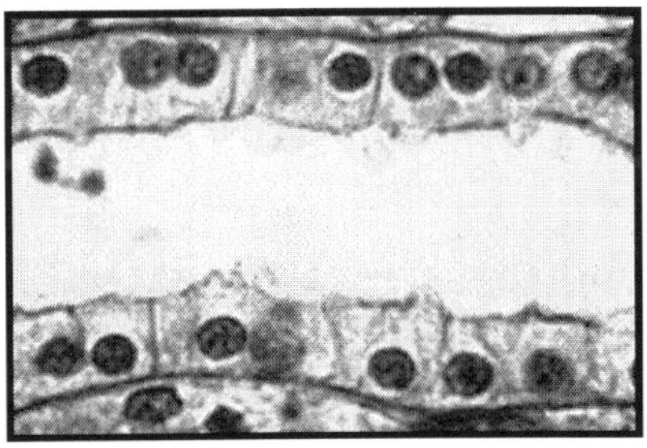

Fig. Cuboidal Epithelium

Fig. Epithelium Lining the Intestine of a Rat, as Seen with SEM

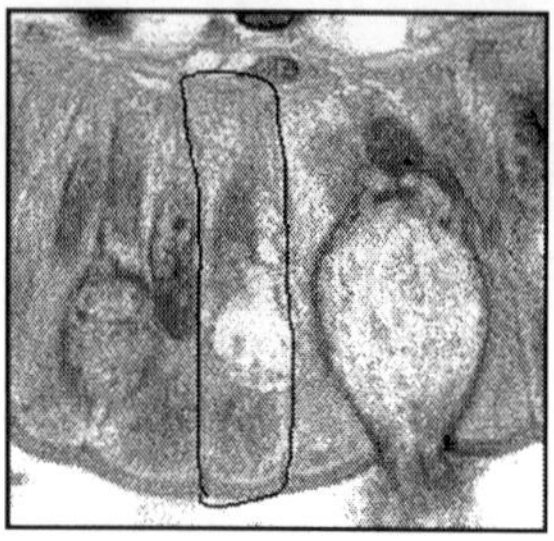

Fig. Columnar Epithelial Cells

Functions of epithelial cells include:

- Movement materials in, out, or around the body.
- Protection of the internal environment against the external environment.
- Secretion of a product.

Glands can be single epithelial cells, such as the goblet cells that line the intestine. Multicellular glands include the endocrine glands. Many animals have their skin composed of epithelium. Vertebrates have keratin in their skin cells to reduce water loss. Many other animals secrete mucus or other materials from their skin, such as earthworms do.

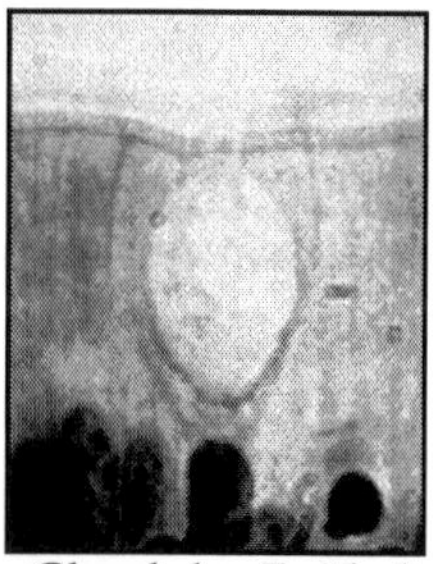

Fig. Glandular Epithelium

CONNECTIVE TISSUE

Connective tissue serves many purposes in the body:

- Binding
- Supporting
- Protecting

- Forming blood
- Storing fats
- Filling space

Connective cells are separated from one another by a non-cellular matrix. The matrix may be solid, soft , or liquid. Two types of connective tissue are Loose Connective Tissue and Fibrous Connective Tissue. Fibroblasts are separated by a collagen fibre-containing matrix. Collagen fibres provide elasticity and flexibility. LCT occurs beneath epithelium in skin and many internal organs, such as lungs, arteries and the urinary bladder. This tissue type also forms a protective layer over muscle, nerves, and blood vessels.

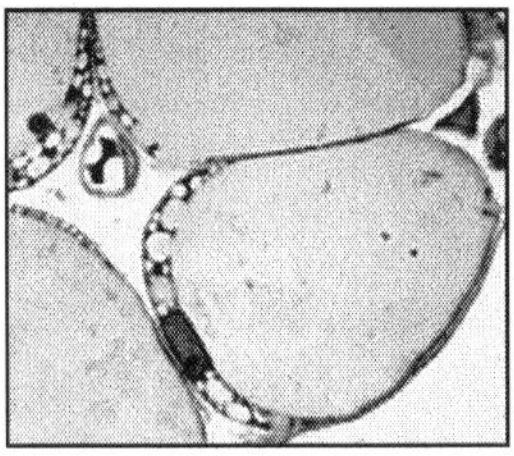

Fig. Adipose Tissue, a type of Connective Tissue

Adipose tissue, has enlarged fibroblasts storing fats and reduced intracellular matrix. Adipose tissue facilitates energy storage and insulation. Fibrous Connective Tissue has many fibres of collagen closely packed together. FCT occurs in tendons, which connect muscle to bone. Ligaments are also composed of FCT and connect bone to bone at a joint. Cartilage and bone are "rigid" connective tissues. Cartilage, has structural proteins deposited in the matrix between cells. Cartilage is the softer of the two "rigid" connective tissues. Cartilage forms the embryonic skeleton of vertebrates and the adult skeleton of sharks and rays. It also occurs in thc human body in the ears, tip of the nose, and at joints such as the knee and between bones of the spinal column.

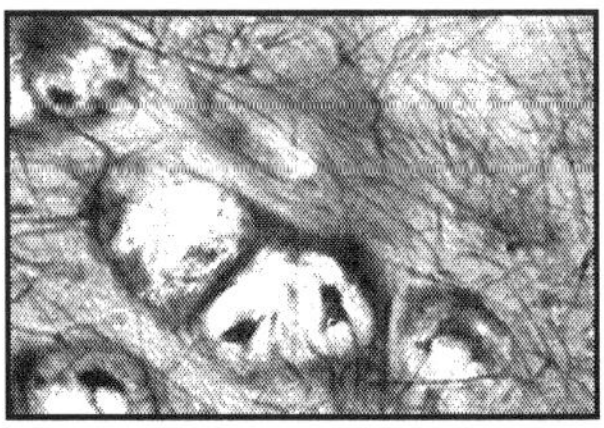

Fig. Cartilage, a Type of "Soft" Connective Tissue

Bone, shown in Figure, has calcium salts in the matrix, giving it greater rigidity and strength. Bone also serves as a reservoir for calcium. Protein fibres provide elasticity while minerals provide elasticity. Two types of bone occur. Dense bone has osteocytes located in lacunae connected by canaliculi. Lacunae are commonly referred to as Haversian canals. Spongy bone occurs at the ends of bones and has bony bars and plates separated by irregular spaces. The solid portions of spongy bone pick up stress.

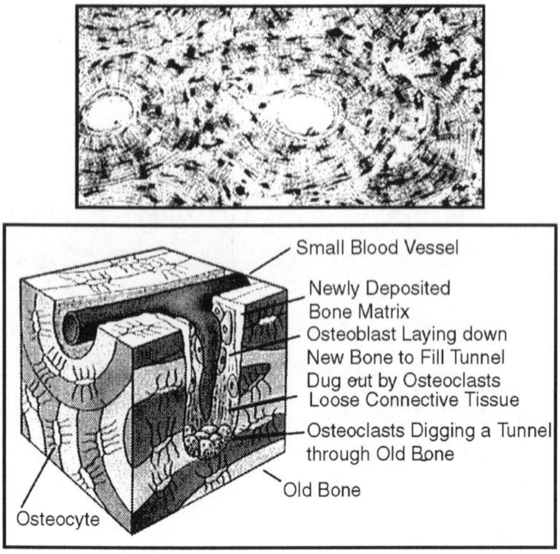

Fig. Bone. The Image of Bone

Blood is a connective tissue of cells separated by a liquid (plasma) matrix. Illustrations of blood cells are shown in Figure. Two types of cells occur. Red blood cells (erythrocytes) carry oxygen. White blood cells (leukocytes) function in the immune system. Plasma transports dissolved glucose, wastes, carbon dioxide and hormones, as well as regulating the water balance for the blood cells. Platelets are cell fragments that function in blood clotting.

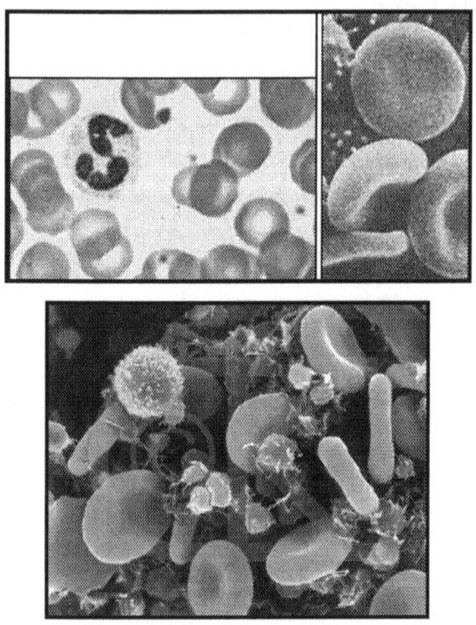

Fig. Elements of the Blood.

MUSCLE TISSUE

Muscle tissue facilitates movement of the animal by contraction of individual muscle cells (referred to as muscle fibres).

Three types of muscle fibres occur in animals:

- Skeletal (striated)
- Smooth

- Cardiac

Muscle tissue and organization is shown in Figure.

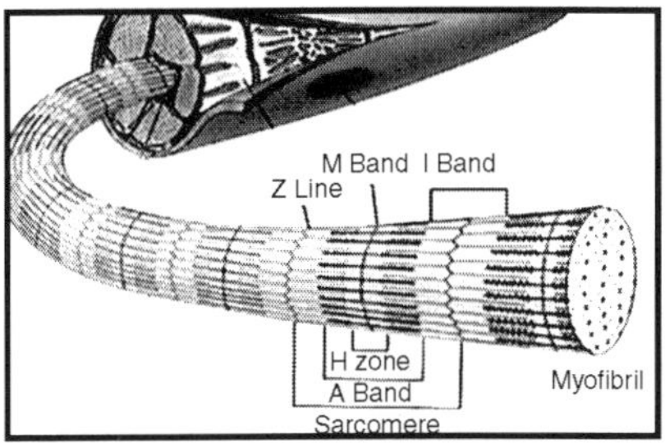

Fig. Organization of Muscle Tissue

Muscle fibres are multinucleated, with the nuclei located just under the plasma membrane. Most of the cell is occupied by striated, thread-like myofibrils. Within each myofibril there are dense Z lines. A sarcomere (or muscle functional unit) extends from Z line to Z line. Each sarcomere has thick and thin filaments. The thick filaments are made of myosin and occupy the centre of each sarcomere. Thin filaments are made of actin and anchor to the Z line.

Skeletal (striated) muscle fibres, shown in Figure, have alternating bands perpendicular to the long axis of the cell. These cells function in conjunction with the skeletal system for voluntary muscle movements. The bands are areas of actin and myosin deposition in the cells.

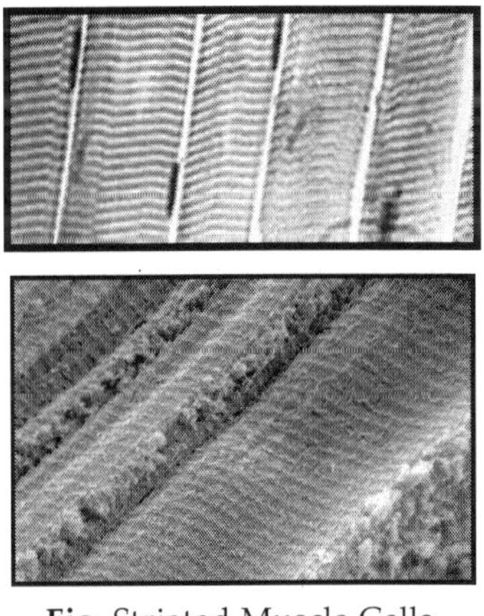

Fig. Striated Muscle Cells.

Smooth muscle fibres, shown in Figure, lack the banding, although actin and myosin still occur. These cells function in involuntary movements and/or autonomic responses (such as breathing, secretion, ejaculation, birth, and certain reflexes).

Smooth muscle fibres are spindle shaped cells that form masses. These fibres are components of structures in the digestive system, reproductive tract, and blood vessels.

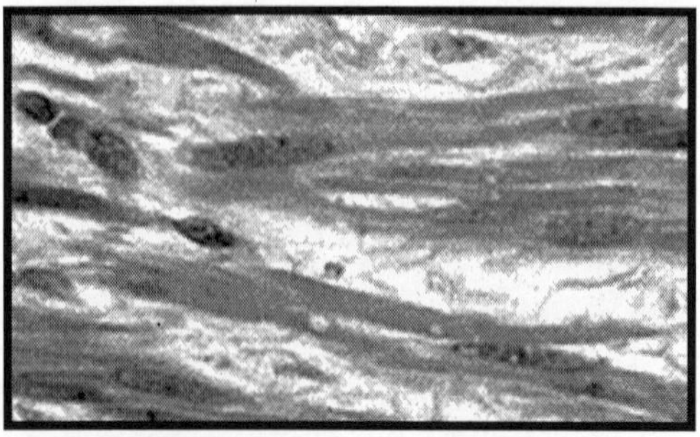

Fig. Smooth Muscle Cells.

Cardiac muscle fibres are a type of striated muscle found only in the heart. The cell has a bifurcated (or forked) shape, usually with the nucleus near the centre of the cell. The cells are usually connected to each other by intercalated disks, as shown in Figure.

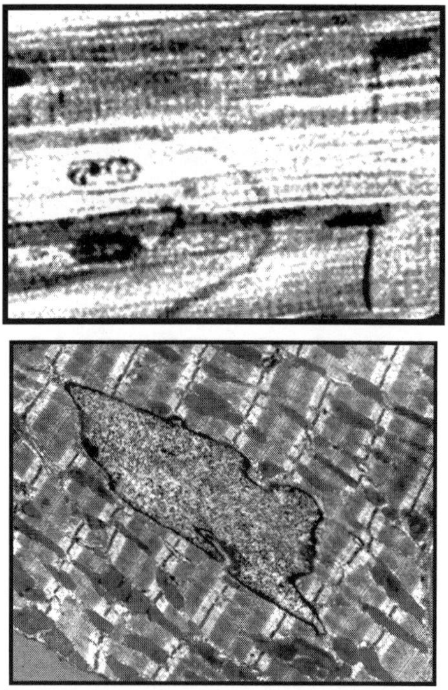

Fig. Cardiac Muscle Cells.

NERVOUS TISSUE

Nervous tissue, shown in Figure, functions in the integration of stimulus and control of response to that stimulus. Nerve cells are called neurons. Each neuron has a cell body, an axon, and many dendrites. Nervous tissue is composed of two main cell types: neurons and glial cells. Neurons transmit nerve messages. Glial cells are in direct contact with neurons and often surround them.

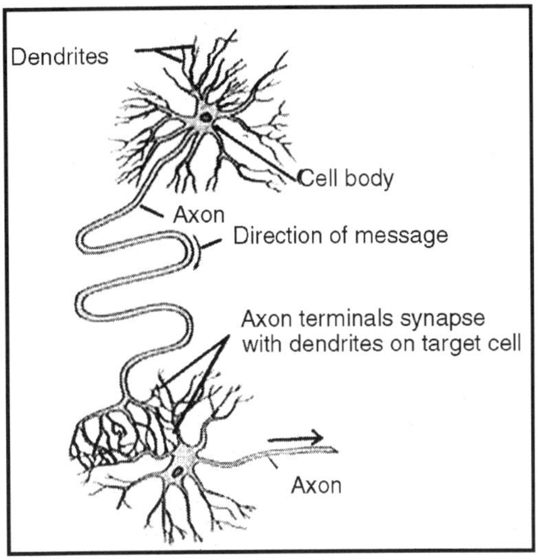

Fig. Organization of a Neuron.

The neuron is the functional unit of the nervous system. Humans have about 100 billion neurons in their brain alone! While variable in size and shape, all neurons have three parts. Dendrites receive information from another cell and transmit the message to the cell body. The cell body contains the nucleus, mitochondria and other organelles typical of eukaryotic cells. The figure below show of large multipolar neuron. The another image shows Pyramidal Neurons from the Central Nervous

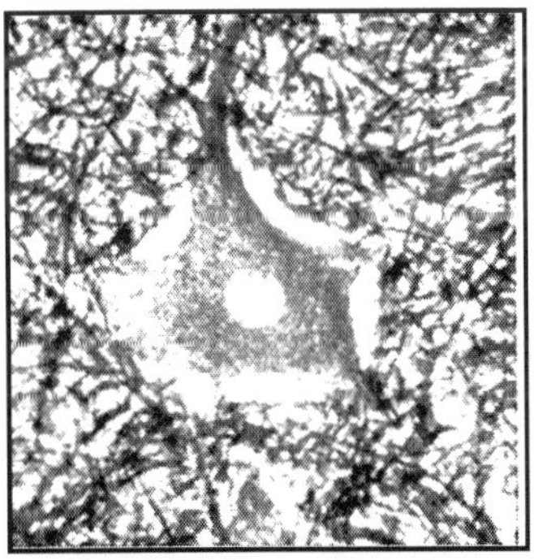

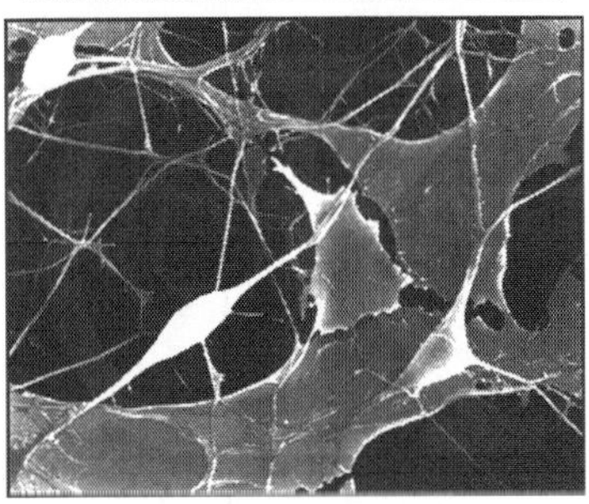

3

Variety of Nutrients in Animals

Animals need a variety of nutrients to meet their basic needs. These nutrients include fats and carbohydrates that provide energy, proteins that furnish amino acids, vitamins that serve as co-factors for enzymes and perform other functions, ions required for water balance and for nerve and muscle function, and selected elements that are incorporated into certain molecules synthesized by cells.

To determine the levels of nutrients that are needed to sustain normal activities, researchers monitor the relationship between nutrient intake, the levels of nutrients maintained in the body, and health. To determine how nutrition can affect athletic performance, researchers alter the intake of specific nutrients and assess how athletes respond. For example, investigators have shown that athletes who ingest a high-carbohydrate diet for three days outperform individuals who ingest a high-fat, high-protein diet.

There are seven major classes of nutrients: carbohydrates, fats, fiber, minerals, protein, vitamin, and water. These nutrient classes can be categorized as either macronutrients (needed in relatively large amounts) or micronutrients (needed in smaller quantities). The macronutrients are carbohydrates, fats, fiber, proteins, and water. The micronutrients are minerals and vitamins.

The macronutrients (excluding fiber and water) provide structural material (amino acids from which proteins are built, and lipids from which cell membranes and some signaling molecules are built) and energy. Some of the structural material can be used to generate energy internally, and in either case it is measured in joules or calories (sometimes called "kilocalories" and on other rare occasions written with a capital C to distinguish them from little 'c' calories). Carbohydrates and proteins provide 17 kJ approximately (4 kcal) of energy per gram, while fats provide 37 kJ (9 kcal) per gram.,[1] though the net energy from either depends on such factors as absorption and digestive effort, which vary substantially from instance to instance.

Vitamins, minerals, fiber, and water do not provide energy, but are required for other reasons. A third class dietary material, fiber (i.e., non-digestible material such as cellulose), seems also to be required, for both mechanical and biochemical reasons, though the exact reasons remain unclear. Molecules of carbohydrates and fats consist of carbon, hydrogen, and oxygen atoms. Carbohydrates range from simple monosaccharides (glucose, fructose, galactose) to complex polysaccharides (starch). Fats are triglycerides, made of assorted fatty acid monomers bound to glycerol backbone. Some fatty acids, but not all, are essential in the diet: they cannot be synthesized in the body. Protein molecules contain nitrogen atoms in addition to carbon, oxygen, and hydrogen.

The fundamental components of protein are nitrogen-containing amino acids, some of which are essential in the sense that humans cannot make them internally. Some of the amino acids are convertible (with the expenditure of energy) to glucose and can be used for energy production just as ordinary glucose.

By breaking down existing protein, some glucose can be produced internally; the remaining amino acids are discarded, primarily as urea in urine. This occurs normally only during prolonged starvation.

Other micronutrients include antioxidants and phytochemicals which are said to influence (or protect) some body systems.

Their necessity is not as well established as in the case of, for instance, vitamins. Most foods contain a mix of some or all of the nutrient classes, together with other substances such as toxins or various sorts. Some nutrients can be stored internally (e.g., the fat soluble vitamins), while others are required more or less continuously. Poor health can be caused by a lack of required nutrients or, in extreme cases, too much of a required nutrient. For example, both salt and water (both absolutely required) will cause illness or even death in too large amounts.

NUTRITION AND DIET OF COW

Now what, necessarily, do cows eat? Calves, when first born, are fed milk, much like human infants. Unlike human babies, however, calves can walk within an hour of being born, and three days later they are introduced to calf grain. Whether or not they start eating it is up to them. Calf grain is mainly comprised of corn followed by other grain products, such as wheat and oats. And, to sweeten the deal, molasses is added, giving the grain a sweet smell.

As humans grow, so do their appetites, which are the same as those of dairy cows. During the four- to five-week range, calves are introduced to hay and cow feed (as opposed to calf feed). Hay is dried alfalfa — so in human terms, it's a salad! Then in the six-to eight-week range, dairy farmers will wean calves off milk — that is stop feeding it to them and start pushing more solids and water.

Now the calves become young adults or teenagers, and they're called heifers. Heifers are put together in a group until they become pregnant. Once a heifer has her first calf, she becomes a cow. In order for a cow to continue to produce milk, she has to have a baby every year. Adult cows are like adult humans — they have a job. Theirs is to produce milk. Cows are fed a number of things, including cow feed, hay, grass (in the spring and summer) and silage. Silage can be either corn or alfalfa. Corn silage is slightly different because it contains the whole stalk of corn. Of course, cows receive tons and tons of water, as it takes five gallons of water to produce one gallon of milk. Happy cows equal more milk!

Corn has been a staple ingredient in lactating dairy cow rations for a hundred years. However, many dairy farms import corn to feed to their dairy cows. The price of this imported corn has been fairly stable for a number of years.

Fig. Feeding Dairy Cow

The government is trying to reduce dependence on foreign oil by supporting efforts to produce fuel (ethanol) from renewable resources like corn or cellulose. There is widespread public support for incorporating ethanol into fuelburning vehicles (gasoline and diesel) because ethanol addition results in greater combustion of these fuels and is nontoxic and biodegradable in the water and the soil. The Congress passed legislation called a renewable fuels standard (RFS) that will at least double the use of ethanol and biodiesel by the year 2012. The number of plants producing ethanol from corn is projected to increase dramatically.

This growth in ethanol production has reduced the amount of corn available for livestock feed and concurrently increased its market price. Increased corn prices have resulted in increased feed costs.

While higher milk prices may compensate for ration cost increases, partially replacing corn with other less expensive feedstuffs may be required over the longer term.

ALTERNATE RATION STRATEGIES

Other Grain Sources of Dietary Starch

Fig. Grain Diet for cow

Grains, like corn, are an excellent source of starch which is highly fermentable by ruminal microorganisms. The propionic acid produced by starch-fermenting bacteria is converted to glucose by the cow's liver; this glucose is used to make milk. In addition, the bacteria provide about 50 to 60 per cent of the protein needs of the cow as they are washed out of the rumen and are digested in the abomasum and small intestine. To optimize milk production, starch should make up 24 to 26 per cent of the dietary dry matter. Barley and wheat are rich sources of dietary starch. Shifting cow diets between corn and barley or wheat needs to be done slowly.

The starch in barley and wheat is more rapidly available than that in corn so quick dietary changes can result in digestive upsets and lowered ruminal pH. Wheat flour can sometimes be bought for less than corn, but it is very dusty and potentially explosive so it is often avoided by mills. Field reports suggest that flour might 'paste up" in the rumen, so it is not a popular feed ingredient. Hominy contains less starch but more protein, fibre, and fat compared to corn, however their energy density is similar.

Replacing all of the corn with hominy will reduce the dietary starch concentration about 3 percentage units. Dietary costs would be expected to be reduced by replacing corn with hominy. If no milk is lost, then profit is improved. Experiments comparing performance of cows fed ground corn versus hominy could not be located. However, many Florida dairies feed hominy successfully. Even replacing half the corn with hominy can be a reasonable strategy. Because hominy is higher in fat and phosphorus than corn, care should be taken to avoid overfeeding these nutrients.

Low Starch Feedstuffs as Possible Substitutes for Corn

Because grain sources rich in starch, such as barley or wheat, may not be economically available year-round, other commodity feeds must be considered.

Ingredients with lower prices than corn should be considered as partial replacements for corn. Most common feedstuffs contain quite a bit less starch than corn. The best starch sources after corn include corn silage at 30 per cent, wheat midds at 26 per cent, wheat bran at 23 per cent, rice bran at 19 per cent, and corn gluten feed at 16 per cent (DM basis). All other feedstuffs are less than 10 per cent starch. Replacing some of the corn with these feeds will reduce dietary starch to less than 25 per cent which is a common, US ration target. How far can dietary starch be reduced without significantly affecting milk production?

Corn Feed

Corn gluten feed (CGF) is a byproduct of the manufacture of corn sweeteners, corn starch, corn syrup, and corn oil using the wet milling process. The corn starch is used to make ethanol. The CGF consists of the corn bran and a steep liquor (fermented nutrients extracted from water used to soak corn grain) mixed in approximately a 2:1 ratio. CGF is sold in both wet and dry forms. It contains 16 per cent starch, 36.1 per cent NDF, and 23.5 per cent protein, the protein being highly degradable in the rumen. Pricing is substantially lower than corn. CGF may serve as a concentrate, replacing only the corn and soybean meal, or as both a concentrate and a fibre source, replacing both concentrate and traditional forages. CGF generally accounted for 10 to 45 per cent of the diet, although some studies went higher. When CGF was fed at about 20 per cent of ration DM, milk production was decreased statistically in 1 study, increased in 1 study, and remained unchanged in 9 other studies. Cows in these studies were milking between 50 and 90 pounds per day.

Increasing the CGF to 30, 40 or even 57 per cent of the diet did not reduce milk in any study with the exception of Staples et al. In two studies, in which CGF was fed between 38 and 40 per cent of the diet replacing both forage and concentrate starting at calving, average milk yield was increased significantly by feeding the corn gluten feed. As more CGF replaced more corn, the dietary starch concentrations dropped, going as low as 15 per cent in some cases. In spite of lowered starch intake, milk yield did not drop significantly.

The bulk of these studies indicate that dietary starch could be reduced by replacing some corn and protein meals with the digestible fibre found in CGF. Dropping dietary starch from 26 to 21 per cent may be an acceptable compromise. Sample diets were formulated to include CGF at 0 per cent, 10 per cent, and 20 per cent of the diet; starch was reduced concurrently from 25.9 to 23.5 per cent to 20.9 per cent in these rations. As CGF increases in diet, the proportion of extruded soy meal increased to keep the ruminally undegradable protein constant. Degradable protein sources had to be reduced to keep the dietary crude protein from exceeding 17 per cent. Whether this approach will maintain milk production is not known. Milk and feed intake will be maintained when CGF is fed at 20 per cent of dietary DM. But feeding CGF at 10 per cent of the diet may be a preferred approach initially to "play it safe." Some signs that may indicate that the dietary starch has gotten too low include; dropped milk production, stiffer manure, increases in milk urea nitrogen, and loss of body condition.

Another concern is that drying CGF can reduce the digestibility of the protein if the drying temperature is too high. This is determined by analyzing CGF for acid detergent insoluble protein (ADIN). Bernard et al.reported that dried CGF had higher concentrations of ADIN than wet CGF. In recent years, high ADIN may not be as big of a problem in CGF based upon analyses reported by Dairy One. The ADIN values were less than 5 per cent of the nitrogen in the majority of samples; ADIN is usually not considered a problem until it reaches 10 per cent of the total nitrogen in the feed. Commodities need to have some consistent level of nutrient density from delivery to delivery in order to keep the daily ration nutrients consistent for the cows. Dairy One lists on their web site the normal range and standard deviation of the nutrients of most of the feeds they have analysed over the past 6 years.

Corn has an average starch value of 70.6 per cent. The standard deviation is 5.1 per cent. This means that two-thirds of the samples that were analysed ranged between 65.5 per cent and 75.7 per cent which is 5.1 per cent units added to or subtracted from the mean of 70.6 per cent. This also means that one-third of the samples analysed were outside this range. This is a fairly narrow range compared to byproduct feeds. Starch in CGF samples has a typical range of 8.6 per cent to 24.0 per cent with a mean of 16.3 per cent. This large variation is likely due to the fact that CGF production is not standardized across mills. Likewise, starch in hominy typically ranges between 42.9 per cent and 63.9 per cent with a mean of 53.4 per cent. By feeding these byproducts in smaller amounts, the potential negative effect of nutrient variation on cow performance is reduced. When contracting for these commodities, an acceptable range in variation between loads should be agreed upon ahead of time. Loads delivered outside this agreed-upon-range would then be refused.

Wheat Midds

Wheat midds contain 26 per cent starch and 18 per cent protein. Price can be substantially lower than corn at the right time of the year. Wheat midds have not been evaluated as a substitute for corn in many experiments. In a recent study, wheat midds were fed at about 7.5 per cent of the diet, replacing a combination of corn and soybean meal.

Corn was reduced in the diet from 34.1 to 28.9 per cent. Cows fed wheat midds tended to eat less feed dry matter but milk production and milk composition were not affected. This reduced feed intake may have been because wheat midds have a high water-holding capacity, thus increasing gut fill with fibre. Cows fed wheat midds appeared to have looser manure than those fed more corn and soybean meal.

In a second study, wheat midds were fed at 0 or 22.4 per cent of the diet, replacing 35 per cent of the ground corn and 30 per cent of the soybean meal. Intake of feed dry matter, production of milk, and milk composition were not different between the two groups of cows which averaged 150 days in milk at the start of the study. However the digestibility of dry matter and neutral detergent fibre were lower for cows fed wheat midds. Therefore wheat midds do not appear to be an effective feed replacement for ground corn except for lower producing cows.

Dried Distillers Grains Plus Solubles (DDGS)

Dry corn put through the dry-milling process will produce ethanol, carbon dioxide, and DDGS. Each 56-pound bushel of corn processed through the dry milling process results in 2.85 gallons of ethanol, 18 pounds of carbon dioxide, and 18 pounds of DDGS. Plants using the dry milling process are less expensive to build than the plants using the wet milling process, thus about 75 per cent of the ethanol produced from corn comes from dry milling.

The production of DDGS in the US increased tenfold between the years 1980 and 2000, increasing from 320,000 to 3.5 million metric tons (1 metric ton = 2205 pounds). Production doubled again between 2000 and 2004 to over 7.3 million metric tons. Every dairy cow in the U.S. would need to consume 7.3 pounds per day of DDGS over a 305-day lactation in order to use up this supply, but of course DDGS is also fed to other livestock and is exported.

During the production of ethanol from corn, a syrup product and a cake product are produced. These products are blended together in different proportions to form DDGS. Depending on the plant, the DDGS may be a mixture of syrup to cake ranging in proportion from 35:65 to 55:45. Because of the variation in the ratio of syrup to cake used by different plants to form DDGS, nutrient concentrations will vary from one plant to another.

However the variation looks to be less than that of CGF. Although DDGS is very low in starch, it is an excellent source of protein (30.3 per cent) and fat (13 per cent) and therefore has an energy density very similar to corn. In addition, the ethanol-making process increases the digestibility of the fibre. Unlike CGF, DDGS is a very good source of ruminally undegradable protein (RUP) but if the temperature of the drying process is too high, the protein may be indigestible.

Purchasers of DDGS should analyse for ADIN regularly. Kalscheur et al. reviewed 24 experiments in which 98 comparisons were made between cows fed diets containing DDGS and those not fed DDGS. Cows fed DDGS at up to

30 per cent of the ration produced as much milk as those not fed any DDGS (about 73 lb/day). In spite of decreasing starch content, milk production was maintained. In a study conducted at the University of Florida, whiskey DDGS were fed at 0 per cent or 20 per cent of a 55 per cent concentrate:45 per cent forage diet. The forage was all corn silage in one set of diets and 50 per cent corn silage:50 per cent rhizome peanut silage in another set of diets. The DDGS replaced around 40 per cent of the corn and soybean meal. Milk production was increased from 59.0 to 60.7 pounds per day without changing milk fat test.

The effect of DDGS was the same regardless of whether the forage was totally corn silage or an equal mixture of corn silage and rhizome peanut silage. However, a maximum feeding level of DDGS might be 15 per cent in order to prevent the overfeeding of protein, RUP, unsaturated fat, and phosphorus. Increased corn oil in the DDGS can also lead to decreasing milk fat, so fat content of DDGS should be monitored regularly.

Feeding DDGS at 20 per cent of the diet will increase dietary fat by 2.6 per cent. This will likely be a concern if the diet contains other supplemental fats. Dietary fat should be kept below 6 per cent. The feeding of one pound of DDGS can replace about 0.6 pounds of corn and 0.4 pounds of soybean meal.

Soybean Hulls (SBH)

During the processing of soybeans for oil and meal, the hulls are separated and ground. The resulting soybean hulls consist largely of the outer covering of the soybean so they are high in fibre and contain moderate amounts of protein but very little starch. Protein concentration will depend upon how well the hulls are cleaned.

Positive characteristics include a high content of lysine, a low content of phosphorus, and a highly digestible fibre. Fat content can vary widely. The availability of SBH is likely to increase in coming years as more acres are planted to soybeans in order to use soybean oil for biodiesel production. In 13 separate experiments published between 1976 and 2002, SBH (fed at up to 20 per cent of the dietary DM) successfully replaced ground corn or high moisture corn in rations for lactating cows. In most of these studies, the dietary starch levels were probably greater than what is recommended today, based upon the proportion of corn in the diets fed in these studies. Therefore the replacement of corn with SBH still left enough starch to support adequate intake and milk production.

One study fed a more conservative amount of corn in the control diet and deserves further attention here. The control diet was 23 per cent high moisture corn, 26 per cent corn silage, and 26 per cent alfalfa silage. Adding SBH to the diet at 14 per cent of the dietary DM reduced the corn to 9 per cent of the diet. This dropped the starch from roughly about 25 per cent to 15 per cent, calculated from average values. Diets were fed starting at 8 days fresh.

Lactating cows ate more DM when offered the ration containing SBH but milk production was not changed. Lactating heifers fed SBH ate the same amount of feed and produced the same amount of milk as those not fed SBH. Concentrations of milk fat and protein were unchanged. Replacing 60 per cent of the corn with SBH, so that SBH made up 14 per cent of the ration, was effective to support high milk production in early lactation. Even though dietary NDF increases when SBH replace corn, feed intake has not been decreased. This is in contrast with the well-documented fact that as traditional forage replaces concentrate in the diet, concentration of dietary NDF increases and feed intake decreases. However the NDF in nonforage fibre byproducts like SBH does not have the same physical characteristics as the NDF in traditional, properly chopped forage. The fibre in SBH is highly digestible, very short and has a specific gravity that allows it to move out of the rumen quickly so that SBH fibre does not have the same rumen fill as forage fibre at similar NDF concentrations. In order to use SBH most efficiently, the diet should contain sufficient "effective" fibre from traditional forages.

Corn Silage

Corn silage hybrids differ in the proportion of grain in the total crop. Selecting corn silage hybrids that contain more starch will reduce the amount of ground corn needed. In a test of 55 corn silage hybrids grown in Gainesville in 2006, starch concentration ranged from 22 per cent to 35 per cent of silage dry matter, with an average of 29.3 per cent. Selecting a corn silage hybrid with high levels of starch and digestible fibre, while maintaining high, could be a good strategy to reduce corn costs in the ration. How much could be saved? If a corn silage hybrid had 33 per cent starch instead of 30 per cent, and corn silage was fed at 33 per cent of the ration dry matter, dietary ground corn could be reduced from 18 per cent to 16.5 per cent, which is about 1 lb/day less ground corn. The cost savings would depend upon the relative price of the feedstuff(s) used.

Corn silage contains much more starch than other forages fed in Florida. Feeding more corn silage in the ration and less sorghum silage, cottonseed hulls, or bermudagrass will increase the starch in the diet and allow for less ground corn to be fed. If corn silage inventory allows, increasing the corn silage from 33 to 38 per cent of the diet DM by replacing another forage like sorghum silage, will allow ground corn to be reduced by 1.5 percentage units.

Glycerol

Glycerol is a byproduct made as plant and animal fats are processed to make diesel fuels. Diesel fuel produced from fat burns cleaner, as there is no soot or particulate matter produced. Starting with animal fat to make diesel fuel is more profitable than starting with vegetable oil because animal fats are cheaper. However, diesel made from animal fat may not work as well in colder climates because it "clouds up." This process is expanding worldwide and therefore the supply of glycerol in the future is likely to increase. Ten pounds of glycerol result from every 100 pounds of biodiesel produced.

The purity of this byproduct can vary widely. The more impure the product, the more water, methanol, phosphorus, and potassium it contains. A product with these contaminants is labeled glycerin. The glycerin fed in a German study contained 2.2 per cent potassium and up to 2.4 per cent phosphorus. Methanol is present due to its use in the manufacture of biodiesel. Although ruminal bacteria can detoxify methanol, it can be problematic if present in large quantities. According to the FDA, glycerin is considered a substance that is GRAS (Generally Recognized as Safe) for general purpose use in animal feed, unless methanol is present at concentrations exceeding 150 ppm. The energy content of glycerol is similar to corn when fed in high starch diets. Dietary glycerol would be converted mainly to propionate and butyrate by ruminal bacteria. As a feed ingredient, it could substitute for corn or molasses.

Little research has been done regarding the feeding of glycerol to dairy cows. Lactating dairy cows fed glycerol (1.3 per cent methanol) at either 0 per cent, 2.7 per cent, or 5.3 per cent of dietary dry matter ate the same amount of feed and produced the same amount of milk. Milk composition was unchanged with the exception that milk urea nitrogen concentrations were lower in cows fed glycerol.

Efficiency of fat-corrected milk production was improved from 1.46 to 1.59 and 1.60 lb of FCM yield per lb of feed dry matter intake. Interestingly, animals of other species consumed more water when they were fed glycerol. This may have a benefit to dairy cows managed under heat stress conditions. German researchers fed glycerol up to 10 per cent of ration DM successfully. Glycerol appears to be a good pelleting agent when added at 5 per cent.

New Technologies to improve Ethanol Production

As the ethanol manufacturing industry works at improving the efficiency of conversion of corn to ethanol, different byproducts will become available. Applegate et al. lists the following potential corn byproducts.

- Called the "quick germ quick fibre method," an enzyme is added to the water used to soak the ground corn, causing the germ and fibre to float before the fermentation process begins. The product from this process contains 28 per cent protein, 5 per cent fat, and 25 per cent NDF.
- Collecting the pericarp fibre and germ prior to fermentation by modifying the dry grinding process (drum degerminator) results in a product that is 24 per cent protein, 8-9 per cent fat, and 28 per cent NDF.

- Removing the fibre by sieving and air aspiration results in a product that is 40+ per cent protein, 15 per cent fat, and 20 per cent NDF.
- Modifying the yeasts used to ferment the sugar to ethanol could allow for an increased concentration of lysine in DDGS, thus giving DDGS a more favourable amino acid profile.

As each mill adopts what they consider the best ethanol-producing technology, the feeding industry will be faced with a variety of byproducts on the market which will need to be properly identified prior to purchase.

With the increased price for corn due to demand for ethanol, farmers will be planting more of their acres to corn. This will reduce the number of acres committed to other crops, such as cottonseed. This shift will likely have a large impact on the market price of a number of feed commodities. The availability and price of each commodity will need to be evaluated for optimal pricing. Some acceptable, alternative feeds may allow some reduction of corn grain in the diet of lactating dairy cows.

NUTRIENT REQUIREMENTS: COW AND RUMEN

REQUIRED NUTRIENTS

All living organisms require essential nutrients to support metabolic processes to keep them alive. General classification of required nutrients include: water, the most essential, energy, protein, minerals, and vitamins. Minerals can be further subdivided into macrominerals, microminerals based on the daily amounts (gm or mg) required. Vitamins are separated into fat or water soluble sources. Daily requirements for these essential nutrients are a function of the cow's body weight and physiologic state (*e.g.*, maintenance, growth, lactation, pregnancy) as modified by environmental conditions. Bacteria have similar requirements for maintenance and growth (*i.e.*, reproduction). Differences between the cow and microbes are seen in where they derive their nutrients.

As a consequence of a pregastric fermentation system, consumed feeds are exposed to microbial fermentation before being available for digestion and absorption by the host. In many instances, this process enhances nutrient availability to the host animal. However, an opposite effect can also occur. Exposure to microbial fermentation can result in nutrient alteration to a form that is either no longer absorbed or has lost its biologic activity. As an anaerobic fermentation system, the rumen is a highly reduced environment with an abundance of reducing equivalents looking for a place to go. Highly oxidized compounds are prime targets. Absorption of selenium (Se) occurs only in its highly oxidized forms as selenite (+4) or selenate (+6) via an intestinal sulfate transport system. Selenium absorption from selenite, is more efficient in non-ruminants than ruminants with a total retention of 77% and 29%, respectively.

Microbial reduction of Se to its elemental (+0) or selenide (-2) forms renders it unavailable thus reducing Se absorption in ruminants. High concentrate diets are associated with increases in insoluble Se complexes, possibly due to changes in rumen pH, redox potential, microbial populations or some combination.

Table. Substances Which Supply Essential Nutrient Needs for the Cow and Rumen Microbial Population.

Nutrient	Cow	Bacteria
Energy	VFA's Glucose	Complex Carbohydrates Sugars, Starches, Amino Acids
Protein	Amino Acids Microbial Protein	Ammonia, Amino Acids, Peptides
Minerals	Dietary	Dietary
Vitamins	Dietary Bacterial	Dietary Synthesized

Rumen interactions can also result in a reduced availability of other minerals, namely copper (Cu) and magnesium (Mg). Best studied relative to copper availability are the interaction of molybdenum (Mo) and sulfate (SO_4) in the rumen. High Mo and SO_4 concentrations in the rumen allow microbial synthesis of thiomolybdates, which can chelate copper making it unavailable for absorption or utilization. Dietary Mg absorption efficiency is low and occurs in the rumen by a sodium (Na)-potassium (K) ATPase system.

Many dietary factors, including excess K, nitrogen, Ca, sulfur and organic acids as well as deficient Mg and phosphorus levels, can predispose an animal to a form of hypomagnesemic tetany. Another factor, a direct consequence of rumen microbial activity, also plays an important role in the pathogenesis of this syndrome. Grasses typically contain significant amounts (>1% of plant dry matter) of sucrose and trans-aconitate, an organic acid, during periods of rapid growth.

In the rumen trans-aconitate is metabolized to acetate or reduced to tricarballyic acid. One of the most abundant rumen bacteria, *Selenomonas ruminantium*, is primarily responsible for the reduction of trans-aconitate. This organism also prefers to grow when sucrose is readily available as substrate. Tricarballylic acid is a very potent chelator of calcium (Ca), Mg and Zn inducing increased urinary excretion and reduced tissue status of these minerals.

MAKING MORE BUGS - MICROBIAL PROTEIN

The cow derives a majority of her energy and protein from microbial end products. In other words, the more we make the bugs grow (reproduce), the less additional, more expensive feedstuffs we need to provide in the cow's diet. Why is production of microbes so important to a ruminant feeding programme?

Microbes contain approximately 62% crude protein, which is 80% true protein and 80% digestible. This is considered high quality protein.

Microbial protein production alone can support up to 25 kg of milk production. The first goal of a ruminant feeding programme should be to maximize microbial protein production and then secondly, meet additional cow's nutrient requirements over-and-above those not met by microbial fermentation end products. This type of feeding approach would theoretically be the most economical and efficient. So how do we get the rumen microbes to abundantly grow without disturbing the rumen ecosystem?

Bacteria require a number of essential nutrients for the synthesis of protein, similar to that of the cow. However unlike the cow, bacteria can use a greater variety of potential nitrogen sources to synthesize amino acids, the building blocks of proteins. In addition, bacteria can synthesize both essential and non-essential amino acids unlike the cow which needs to be supplied with preformed essential amino acids. Figure presents an overview of the processes required to synthesize microbial protein.

Table . Microbial Protein Synthesis Relative to Daily Protein Needs of the Cow.

Efficiency of Microbial Protein Synthesis	Daily Milk Yield		
	25 kg	35 kg	45 kg
gm N/kg OM digested	% of protein from microbes		
20	49	42	39
30	73	64	59
40	98	85	79

The microbial protein production is a function of rumen available substrates, primarily carbohydrates and nitrogen. If any of the required building blocks are in limited supply, microbial protein production will be determined by the availability of the most limiting substrate. Usually this is energy from carbohydrate fermentation. Energy production (generation of ATP) will be dependent upon the available carbohydrate source and its rate of degradation. Ammonia (NH_3) may be provided from non-protein nitrogen sources, amino acids, peptides, or proteins where utilization of a nitrogen source is dependent upon the specific population of bacteria.

For example, cellulolytic bacteria can only use NH_3 as their nitrogen source. Microbial protein production is more complex than just providing the necessary amounts of substrate in the diet. The rumen is a dynamic system that constantly has fermentation end products, liquid, bacteria, and particles being removed via digestion and passage through the rumen as well as new substrate added. So not only do we need to address concepts of total substrate requirements, availability of substrate relative to other substrates needs to be addressed. We must be able to

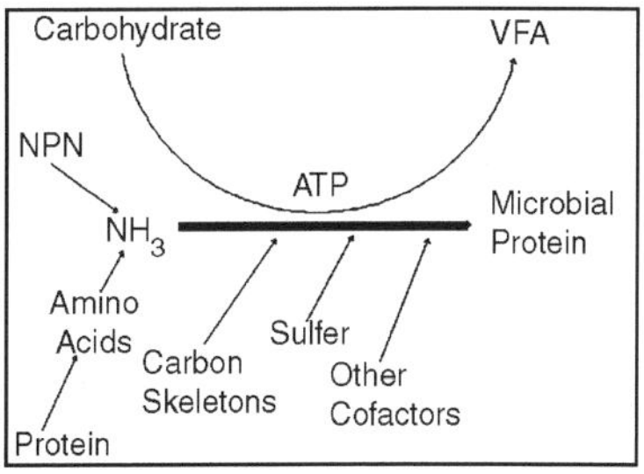

Metabolic processes involved in microbial protein synthesis.

NPN = non-protein nitrogen; NH_3 = ammonia.

Predict rate and extent of carbohydrate and protein degradation takes place in the rumen. This is the critical component of a dynamic modeling system for the rumen and requires more comprehensive and complex feed analysis procedures.

NEW CONCEPTS IN CARBOHYDRATE AND PROTEIN NUTRITION

Amino acids are supplied to the duodenum of ruminants by microbial protein synthesized in the rumen, undegraded dietary protein and endogenous protein. Microbial protein usually accounts for a substantial portion of the total amino acids entering the small intestine. Microbial protein is a high quality protein for the animal that is highly digestible in the small intestine.

Because microbial growth rates affect amino acid supply to the ruminant animal, it is important to maximize microbial protein synthesis in the rumen. Microbial protein synthesis in the rumen provides the majority of protein supplied to the small intestine of ruminants, accounting for 50 to 80% of total absorbable protein. The total amount of microbial protein flowing to the small intestine depends on nutrient availability and efficiency of utilization of these nutrients by ruminal bacteria.

Ruminal degradation of protein from dietary feed ingredients is one of the most important factors influencing intestinal amino acid supply to ruminants. Proteolysis determines the availability of ammonia nitrogen, amino acids, peptides and branched-chain volatile fatty acids, which influence microbial growth rates in the rumen. Rate and extent of ruminal proteolysis not only affect microbial protein synthesis but also the quantity and quality of undegraded dietary protein that reach the duodenum. Although microbial protein alone may be adequate for low producing ruminants, it can be inadequate for supporting higher levels of growth, wool or milk production.

As animal production increases, additional protein must be provided from dietary protein that leaves the rumen undegraded to meet the animal's protein requirement. The use of high rumen undegradable proteins (RUP) in diets fed to ruminants with high protein requirements can improve the amino acid supply to the animal provided that enough degradable protein is included in the diet to maximize microbial protein.

When formulating diets for ruminants, various criteria can be used to select the protein supplement including palatability, ruminal protein degradability, protein quality, intestinal absorption of amino acids, cost per unit of protein, availability and consistency of product, and impact on animal performance. Nutritional models for feeding protein to dairy cattle in the USA have evolved from basic crude protein (NRC, 1978) to more complex systems based on rumen degradable protein (RDP), RUP and intestinal digestion of RUP (NRC, 2001).

The NRC (1989) recognized that intestinal digestion of protein supplements differs; however, empirical data were lacking and as a result, a constant value of 80% was used for all feeds. The main problem was the lack of reliable techniques for estimating intestinal digestion of proteins.

With improved techniques, the NRC (2001) has assigned estimates of intestinal digestion to the RUP fraction of each feedstuff. These estimates were obtained from literature using the mobile bag technique and the three-step in situ/*in vitro* procedure of Calsamiglia and Stern (1995). Values used in the French Protein System were adopted for feeds with limited or no data.

Feeding proteins to ruminants that are resistant to microbial degradation in the rumen can provide a practical way to increase dietary protein and alter the amino acid profile of the protein reaching the small intestine for digestion and absorption.

However, the effects of feeding high RUP on intestinal amino acid supply and animal performance have been inconsistent. Lack of response has been attributed to various factors including depression in ruminal microbial protein synthesis and factors related to quality of the dietary protein such as inadequate or overprotection of protein, reduced intestinal availability of amino acid or inherent amino acid limitations of the dietary protein.

A considerable amount of variation among and within feeds in ruminal degradation and intestinal digestion of protein has been reported (Calsamiglia and Stern, 1995; Howie *et al.*, 1996; Yoon *et al.*, 1996). It is important that this variation be considered when determining the value of feeds as sources of protein for the ruminant animal.

Standardized chemical methods of feed analysis were developed over 150 years ago.

The proximate analysis system that includes crude protein, crude fibre, ether extract and ash has been in use for over 100 years. However, this system is not adequate in characterizing feed composition relative to rumen and cow needs. Newer chemical and biological methods of feed analysis that better relate to nutritional function have been developed over the past 40 years and continue to be developed. The following describes the rationale and methods used to better characterize feed carbohydrates and protein fractions.

UNDERSTANDING CARBOHYDRATE FRACTIONS

For the most part, carbohydrates (CHO) are in the diet to support rumen microbial populations. Various populations of microbes are capable of utilizing any CHO compound. In contrast, the cow has intestinal enzymes capable of only digesting sugars and starch. In fact, the cow like other mammals does not really have a CHO requirement as long as other glucose precursors are present in the diet. Therefore, dietary CHO need to be fractionated according to their capability of supporting rumen microbial growth and potential impact on rumen environment. Carbohydrates are a tremendously diverse group of organic compounds and usually comprise more than 60% of the total diet in most dairy cattle rations. From a nutritional perspective, all CHO provide energy upon oxidation and depending upon their chemical structure may be a precursor for glucose (sugars, starches CHO) or fat (fibre CHO) synthesis.

Due to the diversity and complexity in CHO structure, our ability to chemically characterize important nutritional fractions of CHO has been somewhat limited.

The challenge for nutritionists has been to be able to adequately quantify different CHO fractions relative to their extent and rate of ruminal degradation or intestinal digestion as to determine their impact on animal performance.

Plant CHOs are primarily differentiated on the basis of their association to the cell wall. Carbohydrates that make up the cell wall are termed structural carbohydrates and are quantified as Neutral Detergent Fibre (NDF). Amount of NDF in a plant is determined by plant species and maturity. Structural CHOs have the property of being slowly fermented, if at all.

Therefore, energy yield from these sources would be minimal and slow compared to non-fibre carbohydrates. Due to its slower fermentation rate and need for mastication, NDF is often associated with intake capacity in ruminants. A subset of NDF is acid detergent fibre (ADF), which quantifies the most slowly fermentable or non-fermentable portions of the cell wall.

The ADF portion of feeds is often used to estimate digestibility or energy availability. Non-fibre carbohydrates (NFC) are those compounds not associated with the cell wall, with the exception of neutral detergent soluble fibre (NDSF) compounds (pectic substances, fructosans, beta-glucans).

Again, this is a very heterogeneous group of CHO that include organic acids, sugars, starch, and NDSF. A subset of NFC is non-structural CHO (NSC), which are primarily sugars and starches that are very rapidly fermented in the rumen. In contrast to structural CHO, non-fibre CHO can rapidly provide large amounts of energy for microbial protein production or directly to the cow and their presence in a feed increases its digestibility. The diversity of compounds within this nutritionally defined group makes it difficult to directly measure amounts in a feed. Determination of non-fibre CHO is usually by difference, thus accumulating all errors in laboratory analyses for other feed fractions.

Within this diverse group of compounds there is tremendous nutritional differences in rate and extent of fermentation or digestion. Ideally, we would like to be able to quantitatively subfractionate this group into organic acids, sugars, starch, and neutral detergent soluble fibre based on differences in nutritional responses. However, procedures to identify some of these non-fibre CHO fractions are only being developed.

Carbohydrate Terminology - The following is a summary list of commonly used CHO terms based on either chemical composition analysis or nutritionally-important fractions.

Crude Fibre (CF) - Original proximate analysis procedure of determining indigestible fibre (cell wall) content of feeds. This procedure incompletely accounts for total cell wall contents, due to limited recovery of hemicellulose and lignin, and thus underestimates total cell wall content of forages and roughages. Crude fibre content will always be less than NDF content and equal to or greater than ADF for a given feed.

Nitrogen Free Extract (NFE) - Proximate analysis procedure to determine readily available carbohydrate sources (sugars and starches). Determined by subtraction: (100 - CP - CF - Ash - EE). Due to the underestimate of fibre by crude fibre, NFE overestimates available carbohydrate, especially for forages.

Neutral Detergent Fibre (NDF) - Van Soest detergent methodology to recover all cell wall components excluding pectin. Contains primarily cellulose, hemicellulose and lignin as well as other resistant non-CHO substances. Correlated with dry matter intake in ruminant animals.

Acid Detergent Fibre (ADF) - Van Soest detergent methodology to recover the indigestible cell wall components. Contains cellulose and lignin as well as heat-damaged protein and other resistant plant compounds. Often associated with feed digestibility.

Non-fibre Carbohydrates (NFC) - Readily digestible (non-cell wall) carbohydrates as determined using the detergent system (organic acids, sugars, starches, soluble fibre). Similar to NFE determination, but always less than NFE for any given feed. Calculated by subtraction: (100 - (NDF + CP + EE + Ash)).

Nonstructural Carbohydrates (NSC) - those carbohydrates that can be digested by mammalian enzymes. Includes organic acids, sugars and starches. Newer chemical fractionation methods are allowing better determinations of starch in feedstuffs.

Lignin - Polyphenol compound, not a carbohydrate, but intimately associated with the plant cell wall carbohydrates. Totally unavailable portion of the cell wall. Lignin increases in amount within a plant structure with increasing plant maturity.

Plant species differ in lignin amounts and its impact on carbohydrate availability. Lignin content (per cent DM) times 2.4 is used to estimate the amount of unavailable fibre in a feed.

Effective Fibre or NDF (eNDF, peNDF) - The portion of the total plant cell wall that is effective in increasing rumination and rumen motility. Effectiveness of NDF is based on particle size, degree of lignification, hydration, and density within classes of feeds. Effective fibre will increase rumen pH through its impact on rumination and saliva production.

Factors Affecting Rumen Carbohydrate Availability - A variety of factors beyond chemical composition of CHO can influence rate and extent of ruminal CHO degradation.

1. *Plant Maturity* - As a plant matures, there is an increase in the cell wall content thus diluting out the more digestible components (protein, sugars, minerals; Table). In addition as the plant matures, lignification of cell wall also increases making the cell wall less available for fermentation.
2. *Environmental Conditions* - Rainfall, soil temperature, fertility, cloud cover, location, cutting strategies, etc. all can influence the availability of carbohydrates in the plant. Environmental light, temperature and their interaction have the greatest impact on plant growth. Increased temperature stimulates plant cell wall and lignification reducing plant digestibility. Light exposure will increase soluble CHO content making the plant more digestible.

Table. Typical test value of alfalfa and grass hays harvested at various stages of plant maturity (all values on dry matter basis).

Type of Hay/Stage	CP %	ADF %	NDF %	TDN %
Alfalfa				
Pre-bloom	> 19	< 30	< 35	> 62
Early bloom	17-19	30-35	35-39	57 - 62
Mid bloom	13-16	36-41	41-47	51 - 56
Late bloom	< 13	> 41	> 48	< 51
Grass				
Prehead	17	< 29	< 55	> 54
Early head	12-17	30-35	56-61	47 - 54
Head	8-12	36-44	60-65	44 - 46
Post-head	< 8	> 45	> 65	< 44

Abbreviations: CP = crude protein; ADF = acid detergent fibre; NDF = neutral detergent fibre; TDN = total digestible nutrients.

3. *Processing* - Particle size reduction (grinding) increases surface area for available microbial attachment and degradation and is very beneficial in increasing cell wall digestion. Steam, extrusion, and popping will alter starch configuration to make it more available. Fermentation (ensiling) will make lesser available carbohydrates

more available. Heating can make soluble proteins insoluble and less rumen available.

4. *Plant Species* - Degree of lignification and distribution of lignin within the cell wall will affect rate of digestion of plant carbohydrates. Chemical structure of starch within a feed will dictate rate of fermentation or digestion. Corn starch (amylose) is less digestible than oat or barley starch (amylopectin), but if processed (ensiled, flaked, ground), it can become very available an a potential problem with ruminal acidosis.

RUMINAL NITROGEN METABOLISM

Ruminal Microbial Protein Synthesis The ultimate goal of proper rumen nutrition is to maximize microbial growth and the amount of RDP that is captured into rumen microbial cells. Maximizing the capture of degradable N not only improves the supply of AA to the small intestine, but also decreases N losses. The rumen is a complex environment inhabited by different microbial species, with each species having different nutrient requirements and metabolism. Therefore, considering the nutrient requirements of ruminal microorganisms is crucial to understanding N metabolism in the rumen as well as the factors that may modify it (Bach *et al.*, 2005).

Factors that affect microbial protein synthesis. Bacteria can utilize carbohydrates and proteins as energy sources. Carbohydrates are the main energy source for bacteria, although they can also be used as carbon skeletons for protein synthesis in combination with ammonia.

Ruminal microbial protein synthesis depends on supply of adequate amounts and type of carbohydrate as an energy source for the synthesis of peptide bonds. Readily fermentable carbohydrates such as starch or sugars are more effective than other sources such as cellulose in promoting microbial growth (Stern and Hoover, 1979). Several *in vitro* (Stern *et al.*, 1978; Henning *et al.*, 1991) and *in vivo* (Casper and Schingoethe, 1989; Cameron *et al.*, 1991) studies demonstrated that infusions of increasing amounts of readily fermentable carbohydrate decreased ammonia-N concentrations due to improved N uptake by ruminal microbes.

However, the optimum ratio of non-fibrous carbohydrates to ammonia-N has not yet been determined. In addition to the importance of the amounts of nutrient supply, the synchrony at which nutrients become available is also important. When rate of protein degradation exceeds the rate of carbohydrate fermentation, large quantities of N can be lost as ammonia, and conversely when the rate of carbohydrate fermentation exceeds protein degradation rate, microbial protein synthesis can decrease (Nocek and Russell, 1988). Whereas the concept of synchronous protein and energy supply has a solid theoretical basis, it is likely that in the complex ecosystem of mixed ruminal microorganisms when nutrient supply is synchronized for a specific subpopulation, it might not be synchronized for other populations.

Therefore, average microbial efficiency remains fairly stable. Also, recycled N to the rumen may contribute to stabilize microbial growth even when N supply is not well synchronized (Bach *et al.*, 2005).

Ruminal bacteria are classified as cellulolytic and amylolytic based on their preferential use of energy. Russell *et al.* (1992) proposed a simplified model to describe energy and protein requirements of microbial subpopulations. Microbes that degrade structural carbohydrates (cellulolytic) have low maintenance requirements, grow slowly, and use ammonia-N as their main N source, whereas microorganisms that degrade non-structural carbohydrates (amylolytic) have higher maintenance requirements, grow rapidly, and use ammonia, peptides, and amino acids (AA) as N sources (Russell *et al.*, 1992). However, bacterial growth has been shown to increase with addition of AA and (or) peptides in cellulolytic and amylolytic bacteria (Maeng and Baldwin, 1976; Argyle and Baldwin, 1989; Kernick, 1991). Similarly, fibre digestion was reported to increase with the supply of AA (Griswold *et al.*, 1996; Carro and Miller, 1999) and peptides (Cruz Soto *et al.*, 1994) to pure cellulolytic bacteria.

Atasoglu *et al.* (2001) demonstrated with pure cultures of cellulolytic bacteria, that the incorporation of ammonia-N into microbial cell-N decreased as the proportion of AA increased in the medium, suggesting that cellulolytic bacteria would use AA if available. Similar findings were reported with increasing concentrations of peptides, although Atasoglu *et al.* (2001) reported a greater preference of cellulolytic bacteria for incorporating AA-N compared with peptide-N into their cell-N. However, at typical ruminal peptide and AA concentrations, about 80% of the cell-N is derived from ammonia-N.

Addition of branched-chain AA that will ferment to branched-chain VFA, and addition of peptides to ruminal fluid has increased fibre digestion, microbial protein production and growth efficiencies (Russell and Sniffen, 1984; Thomsen, 1985). The increase in microbial growth observed with addition of AA and (or) peptides, may be due to direct incorporation of AA into microbial protein and (or) to increased availability of carbon skeletons from AA deamination, which can be used for energy production or as carbon skeletons for new microbial AA. Importance of microbial protein to ruminants.

The theoretical contribution of microbial protein to the total protein requirement of the lactating dairy cow, calculated at three efficiencies of microbial protein synthesis, is presented in Table. Contribution of microbial protein to total protein requirement was determined using NRC values (2001) for a 680-kg lactating dairy cow producing 25, 35, or 45 kg/d of 4% FCM.

At these three milk yields, microbial protein would contribute 51, 49 and 48%, respectively, of the total protein required by the cow when microbial synthesis in the rumen is 30 g of N/kg of organic matter truly digested (OMTD).

When milk yield is 45 kg/d, the contribution of microbial protein to total protein required by the cow would increase from 32 to 63% as efficiency of microbial protein synthesis increases from 20 to 40 g of N/kg of OMTD. Stern and Hoover (1979) reviewed the literature and reported that approximately 30 g of N were synthesized per kilogram of OMD in the rumen; values ranged from 10 to 50 g.

Efficiencies of microbial protein synthesis used in Table fall into the range of reported values; therefore, calculated contributions of microbial protein clearly depicted the importance of optimizing microbial protein synthesis in the rumen of high producing dairy cows. In addition, these calculations demonstrate that, as milk yield increases, a substantial quantity of RUP from protein supplements must leave the rumen to meet the protein requirement of the cow.

Contribution of microbial protein to growing beef cattle calculated at three efficiencies of microbial protein synthesis is presented in Table. Contribution of microbial protein to total protein requirement was determined using NRC values (1996) for a 250 kg growing beef animal with average daily gains (ADG) of 300, 1200 and 1700 g/day. At these three ADG, microbial protein would contribute 95, 57 and 40%, respectively, of the total protein required by the animal when microbial synthesis in the rumen is 25 g of N/kg of organic matter truly digested (OMTD). Lower efficiencies of microbial protein synthesis, ranging from 20 to 30 g N/kg OMTD, were used because beef cattle are generally less efficient, most probably due to a high population of amylolytic bacteria. Microbial protein contribution to total protein requirement demonstrated a more dramatic difference in beef cattle compared with dairy cattle.

This response is probably due to the lack of increase in dry matter intake in beef cattle as ADG increases. However, in both situations, it is clear that substantial quantities of RUP from protein supplements must be incorporated into the diet of high producing ruminants. Estimates of rumen undegradable protein. Under ideal circumstances, the ruminant nutritionist would prefer to use "book values" for RUP of various protein supplements, such as those found in the NRC (2001) publication. However, ruminal protein degradation is a complex process influenced by various factors such as solubility, protein structure, microbial proteolytic activity, microbial access to the protein and ruminal retention time of dietary protein.

There is an inherent variability in reported values for RUP of feeds that is related to conditions under which experiments are conducted; however, part of the variation can also be attributed to differences in quality of protein supplements.

UNDERSTANDING PROTEIN FRACTIONS

Dietary crude protein (N% x 6.25) is not a very useful measure of what potentially happens with consumed nitrogen relative to utilization by the rumen or cow. Crude protein can be chemically separated into fractions on the basis of rumen degradability and solubility and have nutritional relevance to meeting the protein needs of the rumen and cow. Proteins that are rumen degradable would be able to provide nitrogen for microbial protein production. Rumen solubility suggests that the protein source would be more rapidly available. For example, urea, a non-protein nitrogen source, is 100% soluble and degradable and therefore would very rapidly provide ammonia for microbial protein production. Rumen insoluble and very slowly degraded is that dietary protein that can bypass to the abomasum to be digested enzymatically by the cow. The amount of unavailable dietary protein is measured as the amount of nitrogen found in the ADF fraction.

In meeting our goal of maximum microbial protein production, we need to match carbohydrate and protein rates of degradation. This allows for somewhat equivalent amounts of rumen energy and nitrogen availability promoting efficient microbial protein yield and overall dietary protein incorporation.

If we can reduce the amount of protein used in the ration and yet maintain or improve milk yield, the dairy cow becomes much more efficient, profitable and environmentally friendly! Protein Terminology - the following is a summary list of commonly used terms relative to protein nutrition of ruminant animals.

- *Crude Protein (CP)*: Total nitrogen per cent of a feed times 6.25 factor. Crude protein analysis does not differentiate between true protein and non-protein nitrogen sources.
- *Digestible Protein (DP)*: Amount of protein absorbed from the intestine from a feed source. Digested protein in the ruminant animal comes from microbial protein and undegraded dietary protein.
- *Soluble Protein (Sol CP, SIP)*: Dietary protein that readily goes into solution in rumen fluid. This is a rapidly available source of NH_3 for the rumen bugs. Composed of NPN and soluble true proteins.
- *Rumen Degradable Protein (RDP, DIP)*: Dietary protein that can be fermented in the rumen by microbes and contribute to the rumen ammonia pool. Rate of passage of digesta through the rumen will determine the extent of which RDP will be degraded.
- *Rumen Undegradable Protein (RUP, UIP), Escape Protein, Bypass Protein:* Protein which is either very slowly degraded in the rumen or which is unavailable to rumen fermentation. This protein is available for digestion in the abomasum to provide amino acids for the cow.
- *Metabolizable Protein (MP)*: Dietary protein (feed and microbial) that has been digested and is available for absorption and utilization.
- *Non-protein Nitrogen (NPN):* Nitrogenous compounds that do not contain linkages of amino acids. Potential protein source for rumen microbes but the utilization is dependent upon dietary CHO

availability. Sources include urea, ammonia, biuret and amino acids.

- *Unavailable Protein, Bound Protein (ADIN, ADF-N)*: Protein that is unavailable to rumen degradation or abomasum digestion. This is the amount of nitrogen present in the ADF residue and is a measure of protein bound to fibre due to heat damage. It is very indigestible and reduces feedstuff quality.

INTESTINAL PROTEIN DIGESTION

Methods for Measuring Intestinal Protein Digestion The total amount of protein available for absorption from the small intestine depends on the flow of microbial and dietary protein to the duodenum and their respective intestinal digestibilities.

Digestion of protein that leaves the rumen starts in the abomasum with acid-pepsin digestion and is completed in the small intestine with pancreatic and intestinal proteases. The NRC (1989) recognized that intestinal digestion of protein supplements may differ; however, empirical data were lacking, and as a result, a constant value of 80% was used for all feeds. The main problem was the lack of reliable techniques for estimating intestinal digestion of proteins (Stern *et al.*, 1997).

In vivo estimation of intestinal protein digestion involves expensive and labor-intensive experiments and requires the use of surgically prepared animals. Apparent digestion of protein is calculated as the disappearance of CP or amino acids between the duodenum and ileum, which is subject to considerable error associated with digesta sampling, use of digesta flow rate markers, and inherent animal variation. Therefore, several alternative procedures for estimation of intestinal protein digestion in ruminants have been developed.

MAKING A DYNAMIC RUMEN MODEL

Over the past few decades much research has been devoted to quantification of total dietary nutrient needs of dairy cows to support all levels of performance. Many computerized feeding programmes a based on comparison between nutrients delivered in the diet and defined daily nutrient requirements. However, this approach is not sensitive to dynamic changes in feed intake and other factors that impact feed degradability and availability, which ultimately impact the amount of metabolizable nutrients available to the cow to support production. Nutrient availability from a feed either by microbial fermentation or intestinal digestion is not constant. Competition between inherent feed properties of degradability or digestibility and rate of passage through the digestive tract can explain the amount of a given nutrient pool will be available. This interaction of rates is mathematically described in equations. This concept is the central component of any dynamic rumen model such as the Cornell Net Carbohydrate and Protein System (CNCPS).

(1) Amount Degraded = Pool size x $K_d / [K_d + K_p]$

(2) Amount Passed = Pool size x $K_p / [K_d + K_p]$

How much of a given carbohydrate or protein fraction within a feed is degraded will depend upon how rapid the rate of degradation is compared to rate of passage. Rate of passage through the rumen depends upon the fraction. Forages or fibre are slowly passed between 4 and 8 per cent/hr whereas the liquid fraction passes more quickly, between 12 and 20 per cent/hr. If a feed fraction has a rate of degradation that greatly exceeds rate of passage, then most or all of the fraction will be degraded in the rumen. If rate of degradation is slower than rate of passage, then only a small fraction will be degraded in the rumen.

The dynamic components to this system are those CHO or protein fractions where their rate of degradation is nearly equal to the rate of passage. For these fractions, *i.e.*, fermentable fibre CHO and insoluble degradable protein, slight changes in either rate of passage or degradation can greatly influence how much of the fraction will be degraded in the rumen. The following describes a variety of factors that can influence ruminal degradation and passage rates.

RATE OF DEGRADATION (K_D)

Degradation of a feed ingredient is inherently related to the chemical and physical properties of all component compounds. Relative to CHO feeds, degradation (ruminal or intestinal) of sugars is extremely rapid compared to starch or fibre. Within starches there is some range of degradation. Fibre is more slowly fermented, while some fibre is not fermented at all. Similarly with proteins, soluble proteins are rapidly degraded while others are more slowly or not at all. It is important then to have an accurate chemical analysis of feed ingredients to determine how much of each of these different fractions comprise a given feed ingredient.

Beyond chemical and physical properties of a feed, degradability can be modified by feed processing, grinding, and rumen conditions. Heating, grinding, ensiling or some combination will increase rumen degradability of feed CHO starch and fibre fractions. For example, corn starch is very crystalline in nature and hydrophobic, thus being fairly resistant to fermentation without processing. If corn grain is ensiled and ground starch availability will be increased dramatically.

This explains why coarsely ground corn may be seen in manure compared to ground high moisture corn causing acidosis. Ensiling will increase protein availability to the rumen through increasing solubility. Heat treatment of protein makes the insoluble degradable fraction less rumen degradable. Proteolytic activity in the rumen will dictate extent of rumen protein breakdown of degradable fraction. A decline in rumen pH will greatly reduce fibre fermentation and may reduce proteolytic activity.

RATE OF PASSAGE (K_p)

Rate of ruminal passage is calculated for forages and concentrates. Rate of passage will always be slower for forages compared to concentrates. Dry matter intake, forage concentration in the diet and body weight are the primary determinants of rate of passage. Rate of passage increases with increasing dry matter intake and dietary forage concentration. Rate of passage declines slightly with increasing body weight. For each feed an adjustment factor to rate of passage is calculated based on effective NDF of the feed. Passage rate will be reduced with higher eNDF content of feeds.

PUTTING THE CONCEPTS TOGETHER

When designing a ruminant feeding programme, concepts of rates need to be addressed in attempting to make sure ruminal availability of energy and nitrogen are coordinated in order to achieve maximal microbial protein production. The goal is to achieve some level of synchrony between ruminal energy and nitrogen sources in order to maximize microbial protein yield. If energy and nitrogen sources are not synchronized, then loss of available energy or nitrogen and reduced microbial activity will occur.

Asynchronous energy and nitrogen sources will result in reduced milk production and efficiency and elevated blood urea nitrogen values. In this scenario diseases such as urea toxicity, grain overload or other rumen dysfunctions may occur. If energy and nitrogen sources are synchronized, we maintain maximal microbial protein synthesis. In this scenario not only will the amount of additional dietary protein be reduced, but the cows will have increased dry matter intake and remain healthier, all contributing to increased milk producing efficiency. How is this synchronization achieved?

Intuitively, synchronization of ruminal energy and nitrogen sources is achieved by utilizing various CHO and protein feed ingredients that have differing degradability properties. Use of a single protein (canola meal or soybean meal) or energy source (corn grain) does not give one this flexibility. Using ensiled feeds as the sole forage programme unbalances protein to the more readily degradable fractions. One can alter degrability properties with grinding, heating or some other processing procedure as previously discussed. This is where the strength of the dynamic formulation programmes comes into place. One can use such software to appropriately balance feed ingredients to best meet ruminal needs. In addition, one can use milk or blood urea nitrogen as a diagnostic aid in evaluating the balanced achieved between dietary CHO and nitrogen sources.

Another more simplified way of achieving relative synchrony to the rumen environment is through the feeding programme. Feeding a total mixed ration (TMR) is the single best method of approximating rumen nutrient synchrony. The needed balance between CHO and protein fractions is achieved with multiple meal feeding. This effect accounts for the observed 5-8% improvement in milk production when shifting a conventional to a TMR feeding programme. More slowly degraded CHO fractions can trap readily available nitrogen at a later meal. When concentrates are fed in the milking parlor, this balance is lost. In a conventional feeding system where feed ingredients are fed separately, one then needs to more closely account for differences in CHO and protein availability within feeds.

Thus, sequence of feeding can have an impact on milk production response. Forages should be fed prior to concentrate meals to maintain slower rate of passage. Large concentrate meals are best divided and fed 3 to 4 times or more per day to limit the effect of acidosis.

FEEDING RECOMMENDATIONS FOR DIETARY CARBOHYDRATES AND PROTEIN

The preceding discussion has focused on understanding the rumen system and presented concepts on how to better feed the microbial populations to improve milk production efficiency. To move into this new era of dairy cattle feeding, more comprehensive methods of feed composition analysis are needed.

With a better understanding of feed composition, we can apply concepts of rates of degradation and passage to a dynamic rumen system to achieve improved feeding practices that will not only improve milk production efficiency, but decrease wasted dietary nitrogen from entering the environment. The following tables provide some practical feeding guidelines in applying these concepts to ration formulation.

NDF AND DRY MATTER INTAKE

As previously described, NDF has been shown to influence dry matter intake in ruminant animals. Mertens had shown optimum intake of NDF as a per cent of body weight (%BW) was 1.2 " 0.1 %BW. Neutral detergent fibre intake capacity is influenced by physiologic state and age. Cows can consume more or less dietary NDF, however, there will be negative consequences on production or performance as a result of reduced intake.

NON-FIBRE CARBOHYDRATES

The NFC portion of the diet is primarily responsible for rapid ruminal fermentation in support of milk production. Milk production will be severely compromised in the face of low dietary NFC content. Conversely, excess dietary NFC content may result in ruminal dysfunction, namely subclinical or clinical acidosis.

The balancing act with NFC is to provide sufficient amounts to support maximal microbial growth, yet not providing excess amounts that will reduce ruminal pH and compromise fermentation activity. Table presents suggested dietary concentrations for total NFC and NSC (sugar and starch) components over differing physiologic states.

PROTEIN FRACTIONS

From a dietary protein formulation perspective, nitrogen needs of the rumen should be met first, then metabolizable protein requirements for the cow should be addressed. Dietary crude protein is a poor method by which to formulate diets in this manner. However, one needs some sophisticated dynamic models to adequately predict dietary metabolizable protein delivery.

If one formulates diets to the specific protein fractions, the amount of metabolizable protein should be reasonably adequate. Again, one can only truly predict the degradability and undegradablility of protein sources with a dynamic modeling system, which accounts for dry matter intake and ingredient interactions on availability and contribution to either microbial protein yield or digestible undegraded protein.

Table. Suggested Dietary Concentrations for Various Protein Fractions in Diets for Dairy Cattle.

	Crude Protein	**Soluble CP**	**RDP[1]**	**RUP[1]**
Dry Cows	% DM	% of CP	% of CP	% of CP
Far Off	12 - 13	35	70 - 72	28 - 30
Close-up	14 - 15	30	63 - 65	35 - 37
Lactating Cows				
Fresh	19	30	60 - 62	38 - 40
22-80 DIM	18	31	64	36
81-200 DIM	16	32	66	34
>200 DIM	14	34	68	32

Abbreviations: RDP = rumen degradable protein; RUP = rumen undegradable protein. In formulating for dietary protein, rumen nitrogen needs must be considered first. Ruminal degrable protein should be between 10.5 and 11% of dry matter to support microbial nitrogen needs. One can never separate rumen nitrogen from CHO needs. To address this interaction between dietary CHO and protein in the rumen, one should formulate for a NFC:RDP ratio between 3.0 - 3.5 to 1. This suggests as NFC content goes up, then more degradable protein can be assimilated.

Rumen protein needs should be met with forage and plant source proteins. If additional bypass protein is necessary, then heat-treated plant proteins or animal proteins (if allowable) can be used to meet the cow's metabolizable protein needs. Use of rumen protected amino acids should be left to only fine tune a diet if necessary, given their cost.

FEEDING GOATS

Goats are browsers, like deer, which means they prefer trees, bushes, and woody weeds; rather than standing still and eating grass down to the roots, they like to stay on the move, eating a bit of this and a bit of that. Goats can learn to graze a pasture, but don't expect it to be "mowed." The grass helps supplement the goats' diet, but low grazing also can spread parasites. Goats have specific nutritional needs, only some of which are met by the plants on your farm that they browse on. You have to provide feed for the needs that can't be met by browsing. Unless you have a lot of property with a variety of browse, feed will be your biggest expense in raising goats. Don't scrimp on goat feed — it will pay dividends in good health, milk production, and lower veterinary bills.

Feeds for goats can be divided into several different categories: forages, grains, protein supplements, minerals and water. Depending on what levels of each category that is fed. It will be found that goats will gain about.25 to.5 a pound each day. Goats will readily eat a variety of forages. These include pasture, hay, haylage, and brush. Any species of forage is acceptable for goats, although be careful of feeding a lot of alfalfa to wethers and billies because the higher calcium levels can lead to urinary calculi or kidney stones. Goats are very useful in many operations for cleaning up unwanted brushy areas. Most brushy areas will only be able to survive about 2 years with intensive defoliation. Forages are an important part of a ruminant's ration because the high fibre is necessary to keep the digestive system healthy. Grains provide the energy portion of a ration for goats. A variety of grains are available, with corn, oats, wheat and barley being the most widely used. Limit the amount of wheat in the ration to no more than 2 per cent to prevent any digestive problems. When starting goats on a feedlot ration, be sure to gradually introduce them to grain.

Feeding high levels of grain suddenly can cause bloating and overeating disease. Disease can be prevented by vaccinating with type C and D perfringens. This is often available in one vaccine, along with tetanus. Any male goats that are castrated should receive a dose of tetanus vaccine. Castration is a personal consideration as many ethnic groups prefer to purchase intact males.

Protein Supplements

Market goats should be fed a grain ration that runs between 14 per cent and 18 per cent protein. To reach these protein levels, we need to include some sort of protein supplement. Typical supplements will include soybean meal, roasted soybeans, cottonseed, brewers or distiller's grains, or flour mill by-products. Beware of using any dairy protein supplements because of the amount of copper in the supplements. Many nutritionists still disagree over the amount of copper required in a goat ration, but care should still be given to overfeeding this nutrient.

Minerals

One of the best ways to provide minerals to a market goat is through a commercially mixed salt and mineral mix. Be sure to purchase a mix formulated specifically for goats. Allow goat's free access to a mineral feeder that contains the mineral mix at all times. Minerals are especially important for helping to ensure that goats remain healthy. These mineral mixes will often include some of the essential vitamins for goats as well.

Water

Water is often one of the most overlooked aspects of livestock feeding operations. Clean, fresh water should be available at all times for all species of livestock. Animals that have adequate amounts of water are more likely to stay healthy as well as have a faster growth rate. Goats will often be able to supply all of their water needs through dew and lush pasture. However, it is still in their best interest to have a supply of water available to them at all times.

FEEDING YOUR GOAT HERD

Feed is the largest cost associated with raising goats. It can affect herd reproduction, milk production and kid growth. Late gestation and lactation are critical periods for doe nutrition. Nutrition level determines kids growth rates. Goats not receiving adequate nutrition are more prone to disease and will fail to reach their full potential. Goats require energy, protein, vitamins, fiber and water. Energy is the most limiting nutrient, while protein is the most expensive. Imbalances of vitamins and minerals can limit animal performance and lead to various health problems. Fiber maintains a healthy rumen enviroment and prevents digestive disturbances. Inadequate water intake can cause various health problems.

WATER - Water is the cheapest feed ingredient and often the most neglected. Goats should have free access to clean fresh water at all times. It is critical that they have an adequate supply of water in the Winter months. A mature goat will consume between 3/4 to 1 1/2 gallons of water per day.

PASTURE AND BROWSE- Pasture and browse are the most economical source of nutrients for goats. Pasture tends to be high in energy and protein in its vegetative state. As the pasture plants mature the palatability and digestibility decline, making it important to rotate pastures to keep plants in a vegetative state. Some of the best pastures for goats are Bahiagrass, millet, sorghum, sudan grass and a mixture of a grain, grass and clover. During the early part of grazing season, browse tends to be higher in protein than ordinary pasture. Goats are natural browsers and select plants at their most nutritious state. Goats that browse have less problems with internal parasites.

HAY- Hay is the primary source of nutrients for goats during the winter months. Hay varies in quality and the only way to know the nutritional content is to have it analyzed by a forage testing laboratory. Legume hays- alfalfa, clover, lespedeza- tend to be higher in protein; this also depends on which cutting it is.

Briers and such - Goats love eating multifloral rose bushes, green briers, poison ivy and just about every type of unwanted brush located on your property. Many herds are used specifally to "clean" up overgrown property, which can be more economical than hiring people to do the job.

VITAMINS AND MINERALS- Many mierals are required by goats. The most important are salt, calcium and phosphorus. Vitamins are needed in small amounts. Goats require vitamins A, D, and E. Offer a pre-mix of loose minerals free choice; goats will consume more if loose is available.

GRAIN- It is often necessary to feed grains when forage alone cannot provide enough nutrients. CREEP FEEDING and supplemental feeding of kids does increase growth weight, but should only be done to the extent that increases profits. There are two types of feed- carbon and protein. Carbon or 'energy' feeds include cereal grains- corn, barley, wheat, oats, milo, and rye- various by products feeds such as fat, soybean hulls and wheat middlings. Protein supplements may be of animal or plant origin and include soybean meal, cottonseed meal and fish meal. 14% CP to 16% CP will usually fill the nutritional requirement for various classes of goats.

OTHER FEED ADDITIVES

For herds that are prone to urinary calculi, ammonium chloride can be added to prevent the disease. Another common practice is to add a coccidiostat to the ration to prevent as well as control coccidiosis. Common additives used to prevent coccidiosis include Rumensin, Bovatec or Deccox (decoquinate). Lastly, no matter what the ration, be sure to provide enough feeder space for all the goats to eat. Plan for at least 1 foot of feeder space for each goat. And, don't forget to clean feeders and waterers on a regular basis.

Sound feeding practices and good housing facilities result in optimum growth and high milk production and contribute to the good health and comfort of goats.

When feeding goats, keep these objectives must be in mind:

- Feed a young animal enough energy for growth, and feed a mature animal enough energy to maintain a fairly constant body weight;
- Provide enough protein, minerals and vitamins in a balanced feeding programme to maintain a healthy animal; and
- Offer does enough extra food during gestation and lactation for fetus development and milk production.

Optimum growth, good health and high milk production are the results of sound feeding practices. Dairy goats are not unique in their body requirements; they will respond to good nutritional practices. Digestible fibre is especially important in dairy goat diets. Too much grain in relation to forage does not foster good ruminant action and is a costly feeding practice. When feeding, keep minerals and trace mineral salt separate. Always feed them separately.

Feed hay in a rack that will not permit wasting. One recommended type of feeder is a keyhole feeder. Do not overlook forage testing as an economical way to feed the correct ration to herd.

Feeding Kids and Yearlings

Kids are born with no natural protection from disease. The first milk (colostrum) from the mother offers protection and gets the digestive system working. To be most effective, it must be fed before disease-producing organisms enter the mouth and digestive tract. Wash the fresh doe's udders and teats with warm water.

Hand milk half a cup of colostrum and feed it to the kid within 15 minutes of birth. This is the best way to ensure that the newborn receives some milk and to provide it with the most protection from organisms present on the skin of the doe. Complete the first milking and store the colostrum for later feedings if hand feed is elected. Otherwise, permit the kid to nurse at its convenience following the first hand feeding.

Clean the feeding utensils immediately after each use. The same cleaning procedures should be followed for washing milk-handling equipment. Milk replacer may be fed from the fourth day; it should contain at least 20 percent protein, 20 percent fat and be free of vegetable products. A lamb or high-quality calf milk replacer is recommended. Provide hay and grain at one to two weeks of age.

Wean from milk when grain intake reaches 1/4-pound daily and kids are readily consuming hay. Make sure the solution is thoroughly mixed. Mix a fresh mixture daily and feed in place of milk. Feed as soon as diarrhea is noticed. Use for 1-1/2 or 2 days, then return to the regular milk diet. After four to six months of age, the kids may be fed a ration similar to that of the milking herd. Good hay and 1/2 pound of grain per day should provide an ample growth rate. Poor hay may require 1 to 1-1/2 pounds of grain daily.

Feeding the Milking Herd

If milk production is important, feed maximum amounts of high quality hay balanced with a grain ration containing enough protein, minerals and vitamins to support production and animal health. Grass or legume hays are equally acceptable. As the percentage of legumes is increased, the need for protein in the grain mix is reduced.

To determine the amount of grain to feed, consider level of milk production, amount and quality of forages consumed, appetite and state of fleshing. Thin, high-producing does should have access to all the hay they can eat plus grain to the limit of their appetite. Does in mid-lactation that are in good flesh should have all the hay they will eat plus 1 pound of grain for each 3 pounds milk produced. Late lactation does may not need more than 1 pound of grain for each 5 pounds of milk.

Feed a grain ration formulated for a milk-producing ruminant (dairy cows). Rolled or cracked grain is more palatable than ground grain. Because of palatability problems, urea is not recommended. Some commercial cow feeds may contain byproduct ingredients unpalatable to goats. Wet molasses is more palatable than dry molasses. Beer or citrus pulp is a valuable source of fibre, especially if the available hay is of low quality.

If the doe is not thin, reduce the amount of grain to 1/2 to 1 pound per day. Feed her all the forage she will eat. Hay fed during the dry period may be of lower quality, but if so, the grain ration should contain additional protein. Browse, leaves and weeds are often useful to recondition the stomach. If the dry ration differs from the milking ration, be sure to change to the milking forage and grain ration two weeks before the doe freshens. Most bucks do not need more than a pound of grain per day plus forages. Don't let them grow fat. Adjust grain upward or downward accordingly. Always feed full forage.

FEEDING AND NUTRITION OF PIG

Feeding guinea pig is a fairly simple manner, though there are a few special needs that we should be aware of. Guinea pigs are strict vegetarians, and in addition to their pellets, they will eat fruits and vegetables, bread, and will even nibble on vegetarian dog biscuits. Before getting in to the details, it is important to point out that guinea pigs like a routine, predictable life. Sudden changes in their diet can cause stress, so be careful. If the brands of pellets is changed, gradually by adding a few of the new ones to their old pellets, and slowly switching the mixture over. Feeding excessive amounts of fruits and vegetables at one time can cause diarrhea, particularly if new foods are introduced.

All guinea pigs need water to survive. Though they can obtain a large portion of their water from fresh fruits, vegetables and greens, they will still require water from a water bottle to stay healthy. This water should be changed daily, and the water bottles themselves washed periodically to keep them clean. To prevent algae growth, make sure that the sun does not directly shine on the water bottle for long periods of time. If there is arrangement of tap water bottled water or a water filter system should be considered. Distilled water is not recommended for regular drinking, as it can disturb the osmotic pressure in the body cells, which can potentially lead to osmotic shock.

Be aware that some guinea pigs like to "play" with their water bottle, leaking out large amounts of water by holding the steel ball up in the tube. Others like to try and blow water back up into the bottle, which can contaminate the entire supply. Hay is the mainstay of every guinea pig's diet. Though guinea pig pellets do contain hay, it is not in sufficient quantity to keep guinea pig healthy. Guinca pig must havc a frcsh supply of hay every day in order to keep its digestive system regular, and as a general rule. Timothy hay is the best type of hay to feed, though any grass hay will do. If grass hay is not available, a legume hay can be fed, such as alfalfa, though in general this should be avoided as much as possible. Legume hays are high in calcium, and this can lead to bladder stones (uroliths) in some animals. Most pet stores carry dried timothy hay in large and small bags, but it is far better to purchase hay fresh from a feed and garden store, or directly from a farmer. Buying hay by the bale is especially economical, as a bale of hay will literally last for months, and will stay fresh as long as it is stored properly. Some tips on hay storage can be found here.

Guinea Pig pellets

Guinea pig pellets will provide guinea pig with the proper balance of vitamins (save for Vitamin C), minerals and other nutrients. Although it's not necessary to feed pellets to guinea pigs, it is certainly recommended that they be primary feed, after hay. Guinea pig pellets are, in particular, a prime source of protein; obtaining the necessary amounts of this and other nutrients will require careful dietary planning if pellets are not chosen. As a general rule of thumb, adult guinea pigs will eat between one and two ounces of pellets each day. More active animals will eat less, and animals without much stimulation will eat more out of boredom. Nursing and pregnant sows may demand a little more than this, and animals that are given plenty of hay and some fresh vegetables each day might eat a little less. It will be easy to tell if they are fed too much or too little: if there are leftover pellets. If the bowl is empty and they seem to be foraging for food, then they aren't fed enough.

Try to feed at the same hour each day, as guinea pigs like a routine and predictable life. Remove any leftover pellets after an hour or so. Guinea pig pellets are a prime source of protein: a diet that is heavy on pellets will lead to a fat guinea pig. Hay is the only food that they should have access to 24 hours a day. When choosing pellets, be sure to *avoid* the guinea pig "mixes" that contain nuts, seeds and dried fruits. These mixes are high in fats and oils, which can lead to excessive weight gain. Additionally, many of these mixes contain sunflower seeds in their shells. Guinea pigs should *never* be fed nuts or seeds that are still in their shells (peanuts, sunflower seeds, etc.): dozens of guinea pigs in the Pacific Northwest alone die each year from choking on these shell fragments.

One thing that should be noted before we move on: guinea pigs should be fed guinea pig pellets, not rabbit pellets or pellets for another species of animal. Guinea pig pellets are nutritionally balanced for guinea pigs, and feeding them pellets for other animals can result in serious health problems. Though some people may tell that their guinea pigs survive well on rabbit pellets, for instance, it is not trusted that all rabbit food is safe for guinea pigs. Their particular brand may work for guinea pigs, but another brand may lack necessary vitamins or minerals, or have them in insufficient or excessive quantities.

There is no regulation in the pet food industry, so pellets and feeds vary considerably between manufacturers. Don't play Russian roulette with cavies: only feed them pellets that are explicitly and *exclusively* for guinea pigs. It is *extremely important* that guinea pigs receive enough vitamin C each day to prevent scurvy. Guinea pigs cannot manufacture or store vitamin C, so they must obtain it from their diet on a daily basis. Adult guinea pigs will require 10mg of vitamin C each day, and nursing and pregnant sows will require twice that amount (20mg). The caveat: vitamin C, also known as ascorbic acid, breaks down very quickly, so it must be supplied fresh. Although many pellet manufacturers will claim that their feed contains the required amount of vitamin C.

Vitamin C in pellets breaks down during storage, and after about 90 days, there won't be enough vitamin C in the feed to keep guinea pigs healthy. Also note that the food starts breaking down on the shelf, not when the bag of pellets are opened. Unless the manufacturer dates the bag, showing when the feed was packaged, we really have no idea how long pellets have been on the shelf at the local pet store. Since pellets are generally an inadequate source of vitamin C, most owners will either feed sufficient quantities of fruits and vegetables rich in vitamin C each day, or provide vitamin C in the form of supplements added to the guinea pigs' food or drinking water.

Vitamin supplements come in many forms: pre-packaged vitamin supplements for guinea pigs can be bought from pet store, or self grind up vitamin C tablets. In either case, most people prefer to add the vitamins to the animal's drinking water, as that is the only way to guarantee that it will be ingested daily. If the supplement routes are chosen, here are a few pointers:

- Add the appropriate amount of supplement to the amount of water that the guinea pig will drink in a day. Most water bottles work best if they are filled halfway or more, so end up wasting a lot of vitamins. Fortunately, they aren't very expensive. If guinea pig drinks 4 fluid ounces of water a day, and fill water bottle with 8 ounces, then need to add 20mg of vitamin C supplement. This will lead to 4 ounces of wasted water and 10 mg of wasted vitamin C.
- Vitamin C (ascorbic acid) breaks down very quickly in water, so need to add it on a daily basis.
- The chlorine in tap water can actually inactivate ascorbic acid. If vitamin supplements in water is opted, then use tap water that has been standing for at least 24 hours, or filtered tap water from a filter system that reduces chlorine content.

Being vegetarians, guinea pigs will eat many kinds of fruits, vegetables and fresh greens. And, many fruits and vegetables are sources of vitamin C, providing a natural way to meet guinea pig's daily requirement. Calcium and Phosphorus totals for the given food amounts are presented as well. Although these are necessary minerals, and a part of every guinea pig's diet, they are getting them from their pellets, as well. The nutrients guinea pig is getting along with their vitamin C. Too much calcium can cause bladder sludge, a precursor to bladder stones. The proper balance of phosphorus, calcium and vitamin C prevents guinea pigs from needing vitamin D supplements.

Although most guinea pig pellets contain vitamin D as a precaution, there's no guarantee that particular brand of feed does. In addition to providing vitamin C and variance in cavy's diet, fresh greens and fruits contain a significant amount of water. Feeding "wet foods" will aid nursing sows in lactating, and will reduce the amount of water that the average guinea pig will drink from the water bottle. It is especially useful to take lettuce, green peppers and other watery greens on trips to provide moisture for guinea pigs while in the car, since water bottles will leak heavily as the bumps in the road rattle the ball in the drinking tube.

To reduce or completely eliminate pellets from guinea pig's diet, they can survive on water, hay and fresh fruits and vegetables. However, this will require very careful dietary planning on part, to make sure that they receive the necessary nutrients in the proper amounts. Also, be aware that many fresh greens are laxative in action, which means that run the risk of giving guinea pigs loose bowels, or even diarrhea, if the are fed too much at one time. If runny droppings are noticed immediately cut fresh greens out of their diet and feed dry foods until the feces returns to normal. Vegetables that are not laxative (such as carrots) may still be fed. Guinea pigs will not eat what they don't like: some cavies have very discriminating tastes, while others will eat anything that put in front of them. Each pig has its own preferences, and it may take some time to figure out what he or she likes best. Given below is a list of foods that guinea pigs can/will eat. This list is neither comprehensive nor complete.

Apples, bananas, bread (slightly stale & crunchy, but not moldy), broccoli, carrot greens, carrots and baby carrots, celery (cut into small pieces first), cilantro, cucumber, dandelion greens, grass, green & red bell peppers, green leaf & romaine lettuce, kale, kiwi, mustard greens, oats, oranges, parsley, raspberries, spinach, tomatoes. When feeding "wild" greens, such as grass and dandelion greens, make sure they have not been sprayed with chemicals, or contaminated by droppings or urine from other animals, such as cats, dogs and birds. The dangers of pesticides are obvious, and feces can carry any number of parasites which can be transmitted to cavies eating contaminated greens. Even if dogs and cats are dewormed regularly, their feces can contain protozoa/ bacteria which can cause debilitating diarrhea in a guinea pig.

Some foods to avoid are listed below:

- Long celery stalks
- Iceberg lettuce
- Any shelled nuts or seeds

- Raw beans
- Rhubarb

There is some confusion as to whether or not potato peelings are good or bad: some books indicate that they are poisonous to guinea pigs, while others say that they are okay in small amounts. The truth is that potatoes are *okay*, however, any *green* in a potato *is poisoinous*. If potato peelings are given, make sure there are *no green spots* anywhere in the portions. Also, some people may recommend yogurt in small amounts, for it's bacteria-growing properties (to aid the digestive system). Instead, look into acidophilus powders or liquids: acidophilus is also a bacteria growing culture, and it';s likely to give better results since less of it needs to be fed.

There are several commercially available treats on the market that are aimed at guinea pigs. Berry flavoured "crunchies" and flavoured chew sticks are popular among some owners and guinea pigs, and are generally safe. Some owners also have experimented with vegetarian dog biscuits and dry cereals such as Cheerio's. However, the guinea pig "treat sticks", which are also available commercially. These sticks are essentially seeds and nuts that are held together with honey. Most of the treat sticks on the market contain sunflower seeds that are still in their shells, which is a big no-no for guinea pigs due to the choking hazards.

FISH FEEDING

Although fresh animal by-products such as liver and blood are available in limited quantities and could be used by a small-scale trout farmer (< 5 tonnes fish/year) within an in-house produced moist diet, a dry diet formulation and feeding regime is recommended for the Project fish farms.

Three dietary formulations are recommended for use within three distinct feed lines, namely starter, fingerling and production diets. Table 3 and Table 4 shows the recommended nutrient levels and dietary formulations for the three feed lines, respectively.

The formulations are conservative in that emphasis has been placed on using high dietary inclusion levels of known quality ingredients (ie. such as fishmeal; Table 5) rather than experimenting with unknown cheaper alternatives such as rapeseed meal. For example, although oilseeds (ie. soybean meal, rapeseed meal) and animal by-product meals (ie. meat and bone meal, blood meal) can and have been successfully used at high dietary inclusion levels within practical salmonid rations (Table 6), their nutritional success is dependent upon the individual manufacturing process used to produce them. Thus, since the level of technology available in Qinghai Province to manufacture these ingredient sources is still in its infancy, and the proximate composition of these ingredient sources is highly variable, their inclusion level within the proposed diet lines has been kept to a minimum.

Providing fish with the right type of food in the right proportions is a very important task. This can be much more of a challenge in a saltwater tank than in a freshwater tank, because most marine fish are still wild-caught, and their natural diet can be difficult to maintain in captivity. With the right information, however, most marine fish will learn to eagerly anticipate their feeding time and will voraciously devour their food.

Fig. Fish Food

A good rule of thumb to follow is that fish should be fed no more than they can consume within a five minute period. Any food that is left behind after feeding.

DIET AND NUTRITION OF SHEEP

Mostly sheep eat grass, clover, forbs, and other pasture plants. They especially love forbs. It is usually their first choice of food in a pasture. A forb is a broad-leaf plant other than grass. It is a flowering plant. Forbs are often very nutritious. As compared to cattle, sheep eat a greater variety of plants and select a more nutritious diet, but less so than goats.

Sheep are exclusively herbivorous mammals. Like all ruminants, sheep have a complex digestive system composed of four chambers that allows them to break down otherwise inedible cellulose from stems, leaves, and seed hulls in to complex carbohydrates. When sheep graze, vegetation is chewed in to a mass called a bolus, which is then passed in to the first chamber: the rumen. The rumen is a 5-10 gallon (18.9-37.8 liter) organ which ferments feed via a symbiotic relationship with the millions of bacterium, protozoa, and yeasts which are collectively called flora.

The bolus is periodically regurgitated back in to the mouth as cud for additional chewing and salivation, this process usually takes around six hours. Cud chewing is an evolutionary adaptation that allows ruminants to graze more quickly in the morning, and then fully chew and digest feed later in the day. This is beneficial because grazing, which requires lowering the head, leaves sheep more vulnerable to predators, while cud chewing does not. During fermentation, the rumen produces excess gas which must be expelled; disturbances of the organ, such as sudden changes in a sheep's diet, can cause potentially fatal conditions such as bloat. After fermentation in the rumen, feed passes in to the reticulum and the omasum; special feeds such as grains may bypass the rumen altogether.

Following the first three chambers, food moves in to the abomasum for final digestion before processing by the intestines. The abomasum is the only one of the three chambers directly equatable with the usual mammalian stomach, and is accordingly sometimes called the "true stomach".

Sheep follow a diurnal pattern of activity, beginning to feed at dawn and throughout the day until dusk, stopping sporadically to rest and chew their cud. Ideal pasture for sheep is not lawn-like grass, but an array of grasses, legumes and forbs. Types of land where sheep are raised vary widely, from pastures that are seeded and improved intentionally to rough, native lands. Common plants toxic to sheep are present in most of the world, and include (but are not limited to) oak and acorns, tomato, yew, rhubarb, potato, and rhododendron. Sheep also may perish from consuming man-made substances.

Sheep are largely grazing herbivores, unlike browsing animals such as goats and deer that prefer taller foliage. With a much narrower face, they can crop plants very close to the ground and can overgraze a pasture much faster than cattle. For this reason, many shepherds utilize some type of managed grazing, whereby a flock is rotated through multiple pastures, giving plants time to recover.

Paradoxically, sheep can be both the cause of and solution to one of the major ill effects of overgrazing: the spread of invasive plant species. By disturbing the natural make-up of pasture, sheep and other livestock can pave the way for invasive plants. But sheep also prefer to eat invasives such as cheatgrass, leafy spurge, kudzu and spotted knapweed over native species such as sagebrush—thereby making grazing sheep an effective measure in restoring native pastures. Other than forage, the only other common staple feed for sheep is hay, most often during the winter months. The ability to survive solely on pasture (even without hay) varies with breed, but all sheep are capable of doing so. Also included in most sheep's diets are minerals, either in a trace mix or in licks.

Naturally, a constant source of potable water is also a fundamental requirement for sheep. The amount of water needed by sheep fluctuates with the season and the type and quality of the food they consume. When sheep feed on large amounts of new growth and there is precipitation (including dew, as sheep are dawn feeders), sheep need less water. When sheep are confined or are eating large amounts of cured hay, more water is typically needed. Sheep also require clean water, and may refuse to drink water that is covered in scum or algae.

COOLING PONDS FOR DAIRY CATTLE

Cattle have used water for cooling probably for as long as there have been cattle. The last survey done in Florida reported that 30 per cent of the dairies had cooling ponds. The reader should be aware that cooling ponds, if not maintained properly, could harbor infectious diseases such as Leptospirosis and many mastitis organisms. The water usage is very high to fill and keep full.

Environmental concerns that must be addressed are the run-off from ponds and ground seepage. Every state seems to have many regulations that must be addressed before building cooling ponds. The benefits of cooling ponds and some suggestions on building and maintaining them. It is the dairymen's responsibility to use them properly and to be aware that ponds can harbor diseases and the pond water must be dealt with according to state and local laws.

What is a cooling pond for dairy cattle? For this discussion they are a man made hole in the ground with a constant inflow of clean fresh water. Usually the ponds drain or overflow from the top.

If a series of ponds is used, the overflow water may be run into a retention pond to be irrigated on crops or enter the waste management system. Cooling ponds may also be fed from springs or artesian wells. So cooling ponds for dairy cattle should have constant fresh water in and a constant outflow of water to some place for proper disposal.

Cement ponds may also be built. These may drain from the bottom and can be used to flush feed lanes or barns. Bottom draining will help remove some of the solids that have settled to the bottom of the pond. Care must be taken on the entrance and exit to cement ponds.

These may be very slippery if the cement is not cross-grooved. A stagnant pond without fresh water entering, a lagoon, a wallow, and a mud hole do not classify as a cooling pond for dairy cattle.

Site Selection

A site near the feed and drinking water, ideally with some form of shade for cows, is best. The idea is for cows to cool, leave the pond and goes eat and drink, lay down in the shade and then do it all over again. If cows do not have feed and drinking water near, they will stay in the pond and not eat and we have cool wrinkled cows that do not give much milk.

Cooling ponds should be located close to a water supply as fresh water needs to enter the ponds constantly. There must be a way to dispose of runoff water properly. This will vary by areas of the country.

Some slope from the ponds to the waste management area would be helpful, so pumps need not to be used. Cooling ponds should not be too far from the milking parlor as long walks in the sun increase heat stress. Exit lane sprinklers help make the long walk more comfortable.

Concrete Ponds

Concrete ponds will work and have been used but they are expensive to build. However, they are easier to maintain, as the entrances and exits do not have to be rebuilt. The finish of the concrete on exit and entrance slopes should be cross-grooved to prevent cows from falling. A 1:8 slope would be a safe slope if concrete ponds are constructed.

Sizing Cooling Ponds

There is no official listing for cooling pond size. If a holding area requires $15/ft^2$ / cow and cows are driven into a pond and taken out as a group then $15/ft^2$ / cow might be adequate. If $50/ft^2$ / cow is recommended for most general animal space, then this might be the correct figure for a cooling pond.

This amount of space would allow cows to enter and exit the pond at all times when all cows were not using the pond at the same time. If $50/ft^2$ / cow for a pond is used, then $5000/ft^2$ is then needed for the 100 cow group.

Pond Shapes and Depths

Using the sample of $5000/ft^2$ /100 cows a pond could be 50' wide × 100' long, or a circle 80' in diameter (A= (Pi) r^2; A= 3.14 × 40 = $5024ft^2$). If a 50' × 100' pond is used and the sides are fenced so the cows can only enter or exit at each end of the pond, the amount of dirt falling into the pond will be reduced. If a circular pond is made, cow carpet should be used to eliminate the problem of dirt falling in the pond. As cows can swim, pond depth is not a problem.

Deeper ponds may allow the settling of organic matter on the bottom of the pond. This organic matter may not be disturbed by cows walking through it as they will swim over the top of it. Preference should be given to depths of 3' or 4' as it seems to cool the cows and let them stand and walk in and out at will. While deep ponds allow cows to submerge, they do not seem to like to float too long so they go to where they can stand and may block cow traffic. A circular pond that is deep in the middle lets cows dunk themselves while others enter and exit the pond at will.

Water Usage

A 100 cow group with $50/ft^2$ / cow could have a 50' wide × 100' long × 3' deep cooling pond (it may be 4' in the middle and slope in and out). A cooling pond 50' wide × 100' long × an average 3' deep=$15000/ft^3$. There are 7.48 gallons/ft^3, thus the pond requires 112,200 gallons of water.

The pond must be filled and then kept fresh with running water at all times while in use. A 1" pipe 300 ft. long and water pressure at 35psi will result in a flow rate of about 23.5 gpm. At this rate it would then take 80 hours to fill the pond initially. The dairy water usage would then be 1,410 gph × 24 hours or 33,840 gallons of water per day. This fresh water must be pumped into the pond and there is going to be close to that amount of run off water that will exit the pond for disposal.

Water Run-off

Disposal of run-off water must be in accordance with whatever local, state, and federal laws that applies. Many dairies run the water into the waste management system, while others add it to retention ponds to be used for irrigation.

Shading over Ponds

Experience has led to discontinued use of shade over ponds as the cows never leave the ponds. Remember, the object of this exercise is to increase dry matter intake and cows will not eat much soaking in the pond.

Pond Maintenance

At least once a year, usually during the winter, the pond should be fenced and the water removed by pumping. The mud and manure on the bottom of the pond should be removed and the pond left empty for the sun to shine on it.

In the spring, the pond should be filled again and the fences removed. Some people just fence the pond in the winter, dig a new pond and fill in the old one with new pond's dirt.

Mycoplasma mastitis is often cultured in water from cooling ponds. This is a real problem if pulsators do not work well resulting in damaged teat ends. Many herds with mycoplasma positive ponds, have no mycoplasma in the bulk tank.

If the water is filthy, cows should not be permitted in the ponds and the ponds should be pumped dry. The sludge should be cleaned and the ponds filled with water again. Both the sludge and the water shoud be removed from cow areas.

Cooling Ponds to Reduce Body Temperature

In the summer of 1986, an experiment was conducted at a North Florida dairy that was using man-made cooling ponds. Ten early lactation Holstein cows were fitted with radio transmitters that transmitted inner ear temperatures every five minutes to a radio receiver and data logger.

Data were collected over a four-day period in mid-August. The cows' activities were also monitored during this time. Entering and exiting the ponds, eating, drinking and laying were recorded.

The cooling ponds lowered the cows' temperature by 1-2°F depending on the time of day they entered the cooling ponds. The average length per stay in the pond was 18 minutes for events from midnight to noon and 12 minutes per visit from noon to midnight. It was obvious that cooling ponds cooled cows.

Pond Water Quality

In 1987, samples were taken from a dairy's man-made earthen ponds. The ponds were sampled weekly from May 19,1987 to September 28, 1987. Six to ten ponds were sampled each week.

Total bacteria counts (TBC), varied greatly from week to week. This may have been due to no water entering the ponds. There did not seem to be any great increase in total bacteria counts or coliform counts as time progressed. The total bacteria count averaged 3,133,700 CFU/ml for all ponds for the 20 weeks. The coliform counts averaged 14,340 CFU/ml for the same period.

Cooling Ponds and Their Effect on Milk Quality

In a previous study on a dairy in Florida with man-made cooling ponds, it was reported that cows exposed to cooling ponds did not experience more clinical mastitis than cows that did not have access to cooling ponds. In fact, cows exposed to cooling ponds during the trial period were only half as likely to develop a case of clinical mastitis as cows not exposed to cooling ponds.

Clinical Mastitis Organisms

The same dairy had provided its clinical mastitis records for several years. The total number of clinical mastitis cases was high in the first quarter of the year and then declined sharply about the time hot weather came and the cows started using the ponds. The incidence stayed low, even in the last quarter of the year when the ponds were not in use. This can be explained many ways. The first quarter of the year was an extremely wet period.

The cows were quite dirty and the cow wash system was available to one-half of the herd. This may explain some of the variation, even though one-half of the herd had never had a cow wash. Another confounding variable was that this herd switched to Clorox for pre- post-teat dipping in April of that year. It had previously pre- and post-dipped with a Chlorhexidine product. From this it could be concluded that Clorox caused the reduction of clinical mastitis. However, there were no controls in this study, one could conclude nothing except that the ponds did not increase clinical mastitis.

This herd was on a lactation treatment study during the summer of 1987. A total of 40 cows were sampled and treated. There were no unusual organisms treated. These 40 cows were not all the cows treated for clinical mastitis during the period the cows had access to the ponds, but the results were encouraging.

Statewide Effects of Ponds on Milk Quality

Regulatory samples for all the herds in Florida were obtained from the Division of Dairy Industry for the year of 1987 and were analysed for the effects of no ponds, man-made ponds and natural ponds. The numbers presented are least squares estimates of Standard Plate Counts (SPC) and Direct Microscope

Somatic Cell Counts (DMSCC) for each month. The data are not biased by different numbers of dairies with no ponds, natural, or man-made ponds. The data were analysed by herds whose milking cows had access to ponds as indicated on the survey.

The data indicated that milking cows with access to man-made cooling ponds had lower SPC and DMSCC counts than other herds in the state with no access to ponds and those herds with natural ponds were higher in both categories. From these data it could be argued that people who built man-made ponds were better managers and thus had higher milk quality.

4

Reproduction System

Reproduction is the main theme of life. Obviously, without reproductive success an individual (or species) is evolutionarily doomed. All of the organs of the body and their functions may be viewed as structures and processes that enable us to survive in order to reproduce. In this sense each adult individual is an elaborate device for producing eggs or sperm and for bringing them together in the process of fertilization.

Sexuality (fertilization) has evolved because (1) it permits a substantial reshuffling of genes (via random segregation and recombination), and (2) it provides a mechanism by which the diploid state can be reproduced after the genetic diversity has been created. This allows an increased genetic variability in populations, and is thus advantageous in evolutionary terms.

The concept of the germ line holds that only germ cells are passed from one generation to the next. They do not originate from a germ layer (in the embryonic sense). This idea arose from work in lower forms, but was demonstrated in vertebrates when Bounoure (1934) showed that the vegetal region of fertilized frog eggs contains a material with special staining properties. He was able to trace this cytoplasm, called germ plasm, through cleavage and gastrulation to a few large cells, the primordial germ cells. These cells were located initially in the prospective endoderm, and migrated to the genital ridge. When those cells were removed or destroyed by irradiation with ultraviolet light, the frogs became sterile. They could produce no germ cells, though they were otherwise normal. If fragments of endoderm from an unirradiated gastrula that contained some of these cells were transplanted to an irradiated host embryo, the resulting frogs were again fertile. The importance of the germ plasm was confirmed when it was shown that ultraviolet irradiation of the vegetal region of the zygote also resulted in sterile frogs. But when vegetal cytoplasm that contained the germ plasm was transferred with a micropipette to an irradiated zygote, it counteracted the effects of the irradiation. The frogs were fertile. Thus, the germ plasm in frog zygotes contains a determinant for germ cell formation that is sensitive to ultraviolet light and can be transferred with the fluid portion of the cytoplasm.

There has been considerable controversy concerning whether primordial germ cells, PGCs, arise in the gonads during embryogenesis or are derived from extragonadal origin. Once it was established that germ cell lines in mammals were derived from an extragonadal origin, investigations focused on two questions:

Where and when does the germ line arise? Two possibilities exist: (1) germ cells are set aside before embryonic cells become committed to specific pathways, (2) germ cell lines appear de novo from other cells at some point in development.

All the evidence (based on following Green Fluorescent Protein-tagged cells) leads to the conclusion that mammalian germ cells are not morphologically distinct during early development and are induced to form by surrounding cells. There is only fragmentary data for humans, but in mouse the story is coming clear and so far human data are in line with the mouse data. During primitive streak formation, extraembryonic ectoderm cells near the posterior end of the epiblast begin secreting BMPs. This induces some nearby cells to differentiate and express the transmembrane protein *fragilis*. A subset of these cells then begins to express the protein *stella* which is thought to encode an RNA splicing factor (though that is yet unproven). It is this last set of cells that give rise the the PGCs.

It was long thought that PGCs migrate out of the epiblast, into the extraembryonic mesoderm, then back into the embryo via the allantois. That appears not to be true. Based on visualizing fluorescent live cells, the PGCs migrate directly into the embryonic endoderm from the posterior region of the primitive streak.

There they find themselves in the developing hindgut. Migration then proceeds from the posterior end of the gut to the genital ridge where they take up residence and complete differentiation. They continue to grow and divide during the migration such that the 10-100 cells that begin the process are ~5000 cell by the time they reach the ridge.

Two major questions concerning the migration of PCGs in mammals remain unanswered: 1) What cues PCGs to migrate specifically to the gonadal anlage (i.e. the developing gonadal ridge), and 2) what mechanisms do PCGs use to accomplish this migration?

Chemotaxis, with movement being directed by a substance emitted by the developing gonads, is the long suggested cue for PCG migration, but evidence is incomplete. In vitro TGF-b-like proteins secreted by the mouse gonadal ridge are able to induce migration of mouse PGCs. A similar story exists in zebrafish where the known lymphocyte chemoattractant, SDF-1, seems to be used by PGCs. The in vivo significance of these observations is still unknown. Alternative mechanisms for directed cell movement include contact guidance, and differential adhesion using molecules likefibronectin. A theory based on contact guidance proposes that motile PCGs move along paths determined by endodermal and mesodermal tissues. The differential adhesion theory proposes the PCGs move to maximize the strength of their adhesive contacts. Most evidence indicates that PCGs move by ameboid motion during the time of active migration. The cell propels itself using an actomyosin-based force generating mechanism with small membrane specializations (filopodia) on the PCG cell membrane acting as attachment points for the cells during migration.

The number of PCGs, as defined by alkaline phosphatase staining, changes greatly during the migratory period. After completing migration, PCGs undergo divisional arrest. Regulation of the cell division (both mitotic and meiotic) in PCGs is poorly understood and a major area of research.

The development of the gonads, either testes or ovary, is directly influenced by the localization of the PGCs. And the chromosome complement of a given set of PGCs results in the production of specific gene products that directly effect gonadal differentiation. During embryonic weeks 5-7, in prospective females (chromosomal XX), germ cells localize in the cortex of the indifferent gonad and participate in the formation of an ovary, while the medulla regresses. In males, primordial germ cells (chromosomally XY) migrate deeper to the medullary region of the gonad and are incorporated into the forming testis, and stimulate the primary sex cords to form seminiferous tubules.

Reproduction is a marvelous culmination of individual transcendence. Individual organisms come and go, but, to a certain extent, organisms "transcend" time by reproducing offspring. Let's take a look at reproduction in animals.

WHAT IS REPRODUCTION?

In a nutshell, reproduction is the creation of a new individual or individuals from previously existing individuals. In animals, this can occur in two primary ways: through asexual reproduction and through sexual reproduction. Let's look at asexual reproduction.

Asexual Reproduction

In asexual reproduction, one individual produces offspring that are genetically identical to itself. These offspring are produced by mitosis. There are many invertebrates, including sea stars and sea anemones for example, that produce by asexual reproduction. Common forms of asexual reproduction include:

Budding

- In this form of asexual reproduction, an offspring grows out of the body of the parent.
- Hydras exhibit this type of reproduction.

Gemmules (Internal Buds)

- In this form of asexual reproduction, a parent releases a specialized mass of cells that can develop into offspring.
- Sponges exhibit this type of reproduction.

Fragmentation

- In this type of reproduction, the body of the parent breaks into distinct pieces, each of which can produce an offspring.
- Planarians exhibit this type of reproduction.

Regeneration

- In regeneration, if a piece of a parent is detached, it can grow and develop into a completely new individual.
- Echinoderms exhibit this type of reproduction.

Parthenogenesis

This type of reproduction involves the development of an egg that has not been fertilized into an individual.

Animals like most kinds of wasps, bees, and ants that have no sex chromosomes reproduce by this process. Some reptiles and fish are also capable of reproducing in this manner.

Advantages and Disadvantages of Asexual Reproduction

Asexual reproduction can be very advantageous to certain animals. Animals that remain in one particular place and are unable to look for mates would need to reproduce asexually. Another advantage of asexual reproduction is that numerous offspring can be produced without "costing" the parent a great amount of energy or time. Environments that are stable and experience very little change are the best places for organisms that reproduce asexually. A disadvantage of this type of reproduction is the lack of genetic variation. All of the organisms are genetically identical and therefore share the same weaknesses.

If the stable environment changes, the consequences could be deadly to all of the individuals.

Asexual Reproduction in Other Organisms

Animals are not the only organisms that reproduce asexually. Yeasts, plants and bacteria are capable of asexual reproduction as well. Bacterial asexual reproduction most commonly occurs by a kind of cell division called binary fission. Since the cells produced through this type of reproduction are identical, they are all susceptible to the same types of antibiotics.

MECHANISMS OF REPRODUCTION

Death by accident, disease, or aging awaits all organisms. So, if a population is to continue, those that die must be replaced. Reproductive mechanisms are as diverse as species, but even though we usually associate reproduction and sex, some species can reproduce asexually. In *asexual* reproduction a single parent produces offspring with characteristics identical to itself. This method has important advantages. For one, it isn't dependent on the presence of other individuals. One organism can start a whole new population. Furthermore, asexual reproduction may be simpler and more efficient. Asexual reproduction does not occur in vertebrates, however, it occurs in some chordates specially the Urochordates. By comparison, sexual reproduction may seem needlessly complicated.

Sexual reproduction is almost universal in vertebrates; only a very few lizards normally reproduce *parthenogenetically*—without needing (or having) mates. In vertebrates, the sexes are functionally separate in the overwhelming majority of cases. The sexual reproduction consists of ovaries and testes, which produce and nurture the gametes, and ducts that transport the eggs or sperms.

Fertilization may take place either internally or externally. If fertilization is external, the female deposits her eggs in the water (*spawns*) and the male fertilizes them during the process. Although the sperm remains viable for several hours, the egg soon loses its ability to become fertilized, perhaps because of the absorption of water by its jelly coats, which then act as a barrier to sperm penetration. *Intermittent organs* develop in those animals (mainly terrestrial) with internal fertilization and aid in guiding the sperm to the female reproductive trace. *Accessory sexual structures* present in the male compensate for the lack of an aquatic environment by secreting a fluid medium for the sperms. Since the sperms are hardly more than sacks of chromosomes, this material, the semen, also provides nourishment for them and induces their motility. The type of fertilization practiced by a few vertebrates, such as newts and salamanders, does not fall into either of the above categories. The male of these animals deposits immobile sperms in small gelatinous packets called *spermatophores,* on twigs and debris near the female, and she picks them up with her cloacal lips and stores them. Fertilization occurs internally as the eggs leave the oviducts.

The number of eggs released by the vertebrate animal during a breeding period shows a definite relationship to the type of fertilization practiced. Fishes and amphibians ovulate large number of eggs. Since the adults abandon the eggs after they are spawned and show no further interest in their fate, the prodigious numbers ensure species survival. For example, it is estimated that a population of Pacific herring releases from 8 to 75 billion eggs each year. Most of them hatch (95%), but only 0.1% survive long enough to reach maturity. On the other hand, in those species in which the young ones are nurtured within the body of the mother until the time of birth, only a few eggs are usually released during a reproductive cycle. Normally, a human female ovulates only one.

The vertebrate reproductive system, like all systems in the body, is under the control of chemical agents or *hormones.* Ductless or endocrine glands of the body generally produce the hormones, which basically function by modifying cellular metabolism. As a result, they coordinate and integrate the adjustments made by the body to living conditions. Changes in structure, function, or development of the individual organs reflect hormonal control and result in the maintenance of a stable state. The vertebrate reproductive system is an excellent example of a system operating under endocrine control. The cyclic changes in structure, function, and behaviour associated with the female reproductive cycle can be correlated with fluctuations in hormone secretion.

PHENOMENON OF SEXUAL CYCLE

Sexual cycle is a common phenomenon among animals but in mammals, the period of sexual cycle is far more marked in the female than in the male. In some animals, such as those of the deer family, there is a brief period of intense sexual activity at one particular season of the year and then a long period during which there is no sexual importance with the cessation of spermatogenesis. A brief period of pronounced sexual activity, when it occurs in male is known as *rutting season.*

This period always corresponds in time with the female period of strong mating impulse which breeder called the *"period of heat"* and physiologist as *estrus.* The female of most species will accept the male in copulation only during the estrus periods. In cats and cows, the period estrus is accompanied by a period of restlessness and *sexual receptivity.*

The estrus cycle is marked not only by the change in the intensity of the sex urge, but by changes in the lining of the vagina and uterus, which prepare the latter to receive a fertilized egg. The uterine lining becomes softer and thicker and its blood vessels and glands increase in number and size. These processes reach the height a short time after ovulation and if fertilization and the embedding of a fertilized of fertilized egg do not occur, the lining gradually reverts to its original condition. The females repeat the cycle at intervals that vary species to species. Most of the wild animals have one estrus cycle in a year and condition is called the *mono estrus.* Cats and dogs have about two estrus periods each year thereby exhibit *diestrus condition. Poly estrus conditions* repeats the cycle more than twice a year. Female rats and mice have them as frequently as every five days; the guinea pig, every fifteen days; the cow every three weeks.

Human and anthropoid ape females do not experience any distinct period of estrus; instead, the cycle is marked by period of bleeding, known as *menstruation,* which occurs about every twenty-eight days and lasts about four days.

Non primates cycle

In non-primate animals like pig the estrus cycle is of 21 days duration and if pregnancy does not occur then it continues throughout the year. The estrus cycle is divided in several well marked phases as given below:

(a) Pro-estrus

The first phase (pro-estrus) of the estrus cycle is referred to as the "*building up*" phase. It is during this phase the ovarian follicle with its enclosed ovum increases in size primarily by increasing the follicular fluid, which contains estrogenic hormones.

Estrogen absorbed from the follicles into the blood stream stimulates increased vascularity and cell growth of the tubular genitalia. Uterus and vagina become congested and secrete a clear sanguineous fluid, however, during this period copulation is not permitted.

(b) Estrus (heat period)

This is the *period of heat,* and copulation is allowed only at this time. During this period the vaginal epithelium thickens further and the superficial layers are fully keratinised. Examination of the ovaries shows that the *Graafian follicles* come to maturity at about the middle of this period, and at that point ovulation occurs. Then after the ovum is expelled from the follicle to pas into the upper part of Fallopian tube, follicular rupture occurs spontaneously in most animal species. However, in the cat, rabbit, mink, ferret and a few other animals rupture is possible only if coitus occurs. Apparently some nervous reaction at this time initiates the follicular rupture. If coitus does not occur in these animals, the follicle with enclosed ovum regresses. A sterile mating frequently is followed by *pseudopregnancy.*

The duration of this period varies considerably in different groups of animals, for example, in rats it lasts for 9 to 15 hours while in gods it lasts for nine days.

(c) Met-estrus

Met-estrus is the period of functional development of corpus luteum. This occurs shortly after ovulation and is also said to be the *post ovulatory phase.* During this period there is a decrease in *estrogen* and increase in *progesterone* formed by the ovary. The lining of the ruptured follicle begins to grow inward as the blood vascular supply increases within the cavity. The cells lining the cavity which have not been expelled increase in size, multiply, and become laden with fat droplets. The newly re organized structure is called the *corpus luteum* or *yellow body.* In the absence of pregnancy, the corpus luteum degenerates and the changes of the generative organs subside.

(d) Diestrus and Anestrus

Diestrus (resting internal) is a relatively short period of quiescence between estrus cycle in polyestrus animals. It shows preparatory changes for the initiation of second estrus. There is a functional regression of the corpus luteum. In rat it lasts for 4-5 days.

Anestrus is the resting asexual period. In mono-estrus animals, it lasts up to the next mating season.

When another breeding season comes about, the ovary is again activated, and a new cycle is started. Pathologic anestrus results from disease or malfunctions involving the hormonal apparatus of the female reproductive organs.

ENDOMETRIUM OF PRIMATE CYCLE

In most mammals the *endometrium* (epithelium of the uterus) undergoes rhythmic cycle along with ovulatory and estrus cycles. In man and the higher primates the periodic sloughing of the lining of the uterus in the absence of pregnancy is known as *menstruation*. The menstrual cycle is characteristic of primates. It does not occur in other vertebrates. In mammals except primates, the female receives the male only during a definite period (*heat period*) in their sexual cycle. But in primates (menstruating mammals), copulation may be performed at any time during the sexual cycle. In temperate climate, menstruation usually commences at the age of 13 to 14 years and marks the onset of puberty in human females. The menstruation continues until the time of the *menopause* (45 years). Menstruation has a duration of 3 to 5 days. Ovulation occurs about midway between the menstrual period following the commencement of the bleeding. The menstrual cycle begins with the casting of endometrium and bleeding. The casting of endometrium can be considered as periodic preparation for fertilization of the ovum and pregnancy. If pregnancy takes place, the proliferated mucosa becomes converted into placenta. If pregnancy does not occur, the hypertrophied mucosa breaks down and is discharged as *menstruation.* Menstruation, therefore, may be described as the *'funeral of the unfertilized ovum'*. The disappointed uterus weeps and the weeping is swept out as the menstrual flow. The menstrual cycle involves the following phases:

Proliferative Phase or Estrogen Phase or Follicular Phase

When the damage resulting the menstrual period has been fully repaired the *proliferative phase* begins. The *estrogen* is secreted in increasing quantities by the ovary. The endometrium is initially thin and consists of a cilliated columnar epithelium, dipping down into a loose stroma to form simple tubular glands. During the next 8 days or so (i.e., 6th-4th day) the mucosa thickens, becomes more vascular, and the glands elongate and become dilated in their deeper part. Upto the end of this phase the endometrium increases 2 to 3 mm in thickness in case of human female. The phase continues up to the time of ovulation (10 days from the end of menstruation).

Ovulatory phase

No remarkable changes have been reported in the endometrium. During this phase the ovum is expelled out and the process is called *ovulation.* The temperature of the body is comparatively high and remains high until the onset of next menstrual period. Corpus luteum develops in ruptured follicle.

Secretory phase or luteal phase or progravid phase or premen strual phase

During this phase both *estrogen* and *progesterone* are secreted in large quantities by the corpus luteum. The endometrium progressively increases in thickness, e.g., to 4-6 mm.

The outstanding features are the increase in the length and diameter and the change in the outline of the glands; they are greatly distended with mucous and the lining is thrown into folds which project into the lumen giving the gland wall a saw-edge, tufted appearance. The lipid and glycogen deposits of the stroma cells increase.

The whole purpose of all these endometrial changes is to produce and store, highly secretory endometrium and large amount of nutrients that can provide appropriate conditions for implantation of a fertilized ovum respectively. This phase lasts for about 13 to 14 days.

(iv) Menstruation phase

This phase is characterized by bleeding and shedding of the superficial part of the endometrium, leaving the basal layer intact. The mechanism is obscure. It is suggested that the spiral arteries close down for hours and the walls of the contained capillaries are weakened. When the spasm passes away and the circulation is restored, blood leaks out through the destroyed areas of the capillary wall into the stroma, under the superficial epithelium, and into the lamina of the glands. The necrotic endometrium, together with exuded blood and much mucin, is cast off into the lumen of the uterus, whence it passes to the exterior.

The destruction of the endometrium occurs partly resulting the gradual loss of blood. During menstruation, approximately 35 ml of blood and an additional 35 ml of serous fluid are lost. Within approximately five days after menstruation starts, the loss of blood ceases, and by this time the endometrium has become completely re-epithelialized.

MALE SEXUAL CYCLE

Sexual cycle is of most common occurrence in females. Males also show a tendency towards activity of sexual cycle. Testicular activity in the majority of mammals is a seasonal event, as is associated period of sexual excitement known as *rutting period.* At such time the males may develop very special secondary sexual characters such as the antler of the buck-deer, as well as great irritability and desire for combat with other males. In seasonal breeders an annual period of spermatogenesis is followed by an intervals of inactivity. Man, who is polybreeder, does not show rutt cycle.

THE GESTATION PERIOD

The *gestation period* or *pregnancy period* is the time from fertilization to parturition. When the ovum is fertilized the pregnancy starts, the corpus luteum persists instead of degenerating at the end of the menstrual and menstruation does not occur. Failure to menstruate at the expected date is thus one of the first presumptive signs of pregnancy in woman. In all mammals, the ovary is necessary in the early stages of pregnancy since the hormone from the corpus luteum (*progesterone*) is responsible for preparing the endometrium for the reception of a fertilized ovum but after this preliminary stage, the ovary can be removed in several species, including man, without disturbing the pregnancy. During the period the pregnant female contributes vitally to offspring development. It is studied under two sub-heads.

(a) The luteal phase of gestation

The ovum is fertilized somewhere in the middle segment of the Fallopian tube. Cell division or cleavage begins at once to form the blastocyst which at first floats free in the Fallopian tube and then in the uterus. The developing embryo reaches the uterus when the endometrium is in the premenstrual phase and sinks into it about the seventh day be digesting away the superficial layers, cells and cell debris taken up by two trophoblast presumably provide food material for the growth of the blastocyst. Thus the embryo is implanted in the wall of the uterus and the process is called *implantation.* The presence of an embryo in the uterine wall is signalled via the brain to the pituitary which continues to produce LH as a result.

The corpus luteum consequently continues to secrete *progesterone.* Thus, the luteal phase of the menstrual cycle in which fertilization has occurred now does not terminate as usual but continues directly as a luteal phase of gestation.During the luteal phase, the *progesterone* output increases and the uterine lining becomes even more glandular and vascularized. The placenta apparently provides all the hormonal material necessary for the continuation of pregnancy. The amount of *progesterone* increases as pregnancy advances. At term the placenta produces about 250 mg progesterone per 24 hrs. So when the corpus luteum stops its endocrine function, the luteal phase is changed over to a placental phase.

(b) The placental phase

In a little more than 20 per cent of all known pregnancies the infant is born before it is prepared to carry on an independent existence. When this occurs, the birth is called *abortion* or *miscarriage.* Such preventive births may be caused by improper implantation of the embryo, by faulty functioning of the placenta or by injury to the mother. If the corpus luteum should degenerate a little too soon and if the placenta should reach full development a little late, then the amount of available *progesterone* during this gap is likely to be inadequate. In the absence of *progesterone,* placental tissue could not be maintained. Just at the end of a menstrual cycle, the uterine, lining would disintegrate and the embryo, no longer anchored securely, would be aborted. This may happen near the end of 3rd month of pregnancy. Injection of *progesterone* is the remedy to check the abortion or miscarriage. The placental phase continues upto ninth month during which almost all systems are properly developed.

As the pregnancy advances the products of conception increase in size and eventually occupy the whole of the uterine cavity. The growing foetus attached by the umbilical cord to the placenta is bathed by the *amniotic fluid* which is contained within the bag of membranes consisting of the *amnion* and *chorion.* The volume of the amniotic fluid can be estimated by the dye solution technique. The maximum volume, about 1100 ml, is reached at the 38th week of gestation, it decreases by 145 ml per week thereafter. Amniotic fluid is fairly freely exchanged with the maternal fluids and its protein and amino acid content approach that of blood plasma; it is mainly an excretory product of the amniotic epithelium but it also contains urea and bile pigments (08 Ug per ml) derived from the foetus. The foetus swallows amniotic fluid and absorbs water and solid constituents in its intestine, the water absorbed is excreted by the foetal kidney into the amniotic cavity and is again swallowed by the foetus which in this way regulates the volume of the amniotic fluid. Amniotic fluid always contains a number of foetal cells and the sex of the foetus can be discovered by their sex chromatin content.

REPRODUCTIVE CYCLES OF SEXUALLY-REPRODUCING ANIMALS

The reproductive cycles of sexually-reproducing animals are influenced by a number of hormones, including follicle-stimulating hormone, leutinizing hormone, estrogen, testosterone, and progesterone, produced by the animals' bodies. Additionally, reproductive cycles are influenced by environmental factors such as day length. Since the hormone melatonin is only secreted in the dark, animals' bodies produce more melatonin in winter months when there is less daylight. Melatonin (or lack thereof) produces a number of physiological changes, including regulation of fertility in many species. In humans, melatonin levels drop 75% just before puberty, and it has been noted that blind girls whose brains/pineal glands do not receive light from their eyes have delayed onset of puberty. In many other species, melatonin controls sexual receptivity. For example, ewes only are receptive in fall/winter so the lambs will be born in spring and have all summer to grow. Typically, special sex pheromones are emitted to convey information about receptivity to members of the opposite sex or to regulate sexual cycles of other members of the same species.

Sperm production in males is under the control of follicle-stimulating hormone (FSH) and leutinizing hormone (LH). LH stimulates production of androgens (including testosterone) by the testes, while FSH stimulates the testes to produce sperm. As in females, production of FSH and LH are under the control of gonadotropin releasing hormone (GnRH) produced by the hypothalamus. Via a negative feedback loop, androgens control levels of GnRH, FSH, and LH. Additionally, androgens are responsible for development of the male reproductive organs during embryonic development, development of secondary sexual characteristics (beard, deeper voice) and behaviors (singing in birds), and production of sperm. Also via negative feedback loops, FSH and LH control production of GnRH. In human males, it appears that blood levels of all these hormones remain fairly constant, but in males of many other mammal species, blood levels of these hormones vary seasonally (hinting at involvement of melatonin in some way).

Thus, in a number of species of mammals, such as goats, deer, camels, and sheep, due to environmental and hormonal cues, the male periodically goes into rut, a periodic sexual excitement in the male that is analogous to estrous in the female. Billy goats (bucks) in rut stink worse that they do at other times. Buck deer are more likely to use their antlers to engage in sexual competition with other males. People who work with camels know that male camels in rut become more unpredictable, meaner than usual, and harder to work with and control.

In most species of mammals, the female has either an estrous cycle or a menstrual cycle. An estrous cycle is also called "being in heat" because the female's body temperature rises a little. However, more pronounced than changes in temperature are noticable behavioral changes. If the female does not become pregnant, the endometrium is reabsorbed and (usually) there is no "bleeding". In species in which the female has a menstrual cycle, the unused endometrium is shed and voided from the body as the menses or period or menstrual flow. Note that while this term was derived in the context of human females who happen to cycle in about the same length of time as a lunar month, it is also applied to other species whose cycles are not one month long. Most other species have cycles that are shorter or longer than a month, and in some species, the timing of ovulation and female receptivity is so tied to environmental cues (light, temperature, etc.) that they only cycle and mate at one or two specific times of the year. Human females are somewhat unusual because the females of most species are only receptive around the time of ovulation (release of a fertile egg). A human female is more receptive around the time of ovulation, but that is not the only time she is receptive. Human females generally are receptive to sexual activity throughout their cycles.

The average menstrual cycle in humans ranges about 20 to 40 days in length, with a statistical average of about 28 to 29 days. Some women experience regular cycles, while others are quite irregular. Much is understood about the hormones involved in the female cycle, but we are only beginning to understand the role of pheromones. It is well-known that the cycles of college women living in a dorm situation frequently adjust such that many of the women "have their periods" at the same time as each other. It has been shown experimentally that when perspiration from one woman is dissolved in alcohol and applied above the upper lip of other females (several times a week), their cycles all adjust to coincide with hers. There is also experimental evidence that frequent exposure (several times a week) to male perspiration (therefore, pheromones) can regulate a female's cycle, causing irregular, short, or long cycles to become more regular and closer to the 28 to 29 day average. This would make sense evolutionarily because it would increase a woman's chances of conceiving if she was in a situation where she had frequent contact with a stable, supportive mate who could provide her and her child with enough food to survive (at a time when she would be more vulnerable and less able to gather food herself, yet need more than usual).

The first half of a woman's cycle is the proliferative phase (follicular phase), during which the endometrium starts to thicken. The pituitary secretes FSH which causes (usually one) follicle to mature and the ovaries to secrete estrogen. The ovarian estrogen secretion gradually increases until just prior to ovulation. This gradually supresses secretion of FSH and stimulates the hypothalamus to secrete a larger amount of GnRH which, in turn, triggers the pituitary to secrete a burst of LH, causing ovulation. During the proliferative phase, a woman's body temperature is low, perhaps as low as 98.0° F (36.7° C) or less. Sometimes there is a slight rise near the end of the phase during the pre-ovulatory burst of LH before it dips again at ovulation. Throughout this phase, the cervical mucus becomes progressively clearer and thinner.

On about day 14 of an average 28-day cycle, ovulation, the rupture of the follicle and release of the egg, occurs in response to the surge of LH. LH stimulates the remaining follicle cells to form a corpus luteum after ovulation. Often at ovulation there is a sharp drop in the woman's body temperature and her cervical mucus becomes very thin and clear and forms "threads".

Interestingly, both the low body temperature and the condition of the mucus are designed to prolong the life of and aid in the mobility of sperm to increase the chances that the egg being released will be fertilized.

The second half of the cycle is called the secretory phase in the uterus and the luteal phase in the ovaries. The endometrium continues to build up due to the secretion of progesterone (and estrogen) by the corpus luteum, and this prepares the uterus for a possible pregnancy. In case an egg has been fertilized, the growing embryo needs a warmer environment to do well, so the body temperature rises to 98.6° F (37.0° C) or higher. The cervical mucus becomes thick and paste-like, blocking any more sperm from entering the uterus.

If an egg is not fertilized, as the corpus luteum deteriorates at the end of a cycle, both estrogen and progesterone levels go very low. In a pregnancy, implantation triggers hormone production that keeps the corpus luteum secreting estrogen and progesterone for a while, but without that, the corpus luteum disintegrates and stops secreting these hormones. At this time, all of the hormone levels are low and decreasing. This frequently causes the mood swings and other symptoms of premenstrual syndrome (PMS), but PMS is also affected by diet, and large amounts of salt, sugar, yeast, and/or caffeine can make things worse.

Throughout this discussion, the roles of the five hormones that influence a woman's menstrual cycle have been mentioned, but their interactions are quite complex and involve both positive and negative feedback loops. Reiterating the previous discussion specifically from the point of view of the female hormone cycle will hopefully aid in better understanding the roles played by these hormones.

Copulation in most species of animals is preceded by various courtship rituals which help insure that the most genetically fit individuals pass on their genes. In humans, various activities including "flirting" and "foreplay" lead up to and physiologically stimulate and prepare the couple's bodies for coitus, and in general, most women need considerably more stimulation over a longer period of time than many men, to physiologically prepare their bodies for sexual intercourse. Human mating physiology includes four phases in the sexual response cycle.

- The excitement phase prepares the vagina and penis for coitus or sexual intercourse. Vasoconstriction, filling of the erectile tissue with blood occurs because of increased flow in the arteries and constriction of the veins, especially in the penis and clitoris, but also in the testes, ovaries, female labia minora, and female breasts. Myotonia, increased muscle tension, causes erection of the nipples; tension in the arms, legs, and other skeletal muscles; and sustained or rhythmic contractions of the smooth (and skeletal) muscles.

- Many of these characteristics continue into the plateau phase. In the female, the outer third of the vagina becomes vasoconstricted, while the inner two thirds becomes slightly expanded, and the uterus becomes elevated, all in preparation for receiving sperm. Breathing and heart rates increase, not because of physical activity, but because of stimualtion from the autonomic (sympathetic) nervous system.
- Orgasm is rhythmic, involuntary contractions of the reproductive structures/organs in both sexes. In the male, this occurs in two stages: emission, in which the semen is injected into the urethra; and ejaculation, in which the semen is propelled into the woman's vagina. A male normally has only one orgasm (and some men may fall asleep during/after the resolution phase which follows). In the female, orgasm involves the uterus and outer vagina but not the upper two thirds of the vagina. Typically it is harder to stimulate a woman to reach orgasm, and the man needs to take care to make sure the woman is ready when he is. Women, however, can have multiple orgasms, and once aroused, are often not ready to "quit" when, physiologically, the man's body is progressing to the resolution phase. Many women appreciate continued, gentle stimulation after coitus.
- The resolution phase involves a return to normal. During the resolution phase, vasoconstriction returns to normal, the muscles relax, etc. While the changes that occur in most structures occur within the first 5 min. after orgasm, loss of the male erection often takes a little longer.

If the woman is just pre-ovulation or at ovulation, conception, the start of a new person, of pregnancy, may occur. Interestingly, there are some things a couple can do to increase their chances of having a baby of a given sex. It has been found that sperm which contain an X chromosome tend to live longer than Y sperm, so having intercourse a couple days prior to ovulation increases the chances of a girl baby. On the other hand, Y sperm swim faster than X sperm, so having intercourse right at ovulation can increase the chances of a boy baby. As previously mentioned, at the time of ovulation, a woman's body temperature and cervical mucus are "right" to help sperm survive. The contractions caused by orgasm and the prostaglandins in semen help propel sperm up into the uterus, and from there, up the Fallopian tubes. An egg is fertilized near the end of the Fallopian tube, then finishes meiosis, the nuclei unite, and the embryo starts dividing as it begins to travel to the uterus.

Recent work on human pheromones has turned up some interesting effects they may have on pregnancy. It is not unusual in our busy society, for a couple who desire a baby to keep track of where the woman is in her cycle, and plan to have intercourse "at the right time," yet, despite no physical problems, many of these couples still fail to conceive. Recent work on human pheromones suggests that this is not enough. Apparently the woman's body may need continued exposure to the man's pheromones, such as would be acheived by the close contact during coitus, several times a week during the last two weeks of her cycle to make implantation more likely. Again, evolutionarily, this would make sense in terms of signaling the presence of a pair-bonded male who could provide her and her developing child with enough food to survive the pregnancy. It would, then, make sense that lack of continued pheromonal stimulation (as in when the couple go back to their busy schedule until next month at the appointed time or in a "one night stand") would signal the absence of a male that could mean increased hardship and lower chances of survival if pregnancy were to result. However, in the interest of passing on one's genes, evolutionarily, the possibility of pregnancy is not totally ruled out, just diminished.

If an egg is fertilized, fertilization occurs in the far end of the Fallopian tube near the ovary, and as the embryo travels down the tube, cell division starts. By the time it reaches the uterus (about a week later), it is a "hollow ball" or blastocyst, which has about 100 cells.

The embryo, specifically the placenta, starts producing hormones which prevent the corpus luteum from disintegrating, and hormones from the corpus luteum maintain the endometrium until the placenta is large enough to secrete more hormones on its own. Note that whereas contraceptives, as their name suggests, prevent conception (union of egg and sperm to form a new person), intrauterine devices (IUDs) as well as "morning after" pills and some other drugs, including Depo-Provera, work to prevent implantation of the week-old, genetically-unique, growing embryo thereby causing it to die, and thus are abortifacients.

When the embryo implants in the endometrium, it starts to form the placenta. A human placenta attains the size of a dinner plate. The placenta serves as the site of transfer of nutrients and wastes between the maternal and fetal blood, and when large enough, it secretes hormones to maintain the endometrium, thus the pregnancy. Keep in mind that the two blood supplies do not come into contact with each other, but chemicals can be passed back and forth.

In the first trimester (the first three months) of pregnancy, most organogenesis, development of body organs occurs. The baby's heart starts beating at about four weeks (when the mother's first period is about two weeks late and she's beginning to suspect she might be pregnant). Because so much critical growth and development takes place in the first few weeks before she even knows she's pregnant, it is very important that she be well nourished (including vitamin supplements if needed, especially folic acid to prevent neural tube defects) before she even tries to get pregnant. By the end of the eighth week (when her second period is two weeks overdue) all of the major body structures and organs of an adult are present in at least rudimentary form, so the embryo is now called a fetus. The rest of the time during pregnancy is mostly just growth. Note that because of organogenesis, the first trimester is the time when the growing baby is the most sensitive to adverse effects of drugs like alcohol, tobacco, and caffeine, viruses like German measles, and lack of necessary nutrients such as folacin. During the first trimester, the placenta secretes HCG to maintain the corpus luteum which continues to secrete the estrogen and especially progesterone needed to maintain the endometrium. The cervical plug forms to protect the growing baby from "foreign invaders", and the woman's breasts begin to enlarge in preparation for lactation, or nursing. The level of HCG is so high in the blood that some is excreted in the urine. Most pregnancy tests look for the presence of HCG in urine. As the sex organs begin to develop, the embryo starts with rudimentary forms of both male and female systems, and only one set continues to develop while the other atrophies. Also the initial development of external genitalia is similar and it can be difficult to tell the sex of a baby early on.

In the second trimester growth occurs. The baby is very active, and eventually these movements can be felt by the mother. The baby can hear by the fourth month, and when born recognizes not only mom's voice, but also dad's voice (if he's around) and/or any other frequently-heard, familiar sounds, perhaps including things like any musical instruments the mother plays, etc. During this trimester, HCG production declines and the corpus luteum stops producing progesterone as it deteriorates. As this is happening, the placenta takes over and begins to secretes progesterone itself to maintain the endometrium.

The third trimester is a time of rapid growth. The baby tends to move less just because the uterus has become so crowded. Estrogen levels in the mother's blood reach their highest in the last weeks of pregnancy, and as this increases, it triggers the formation of oxytocin receptors in the uterus. Late in pregnancy, fetal cells produce increasing amounts of oxytocin, another hormone, as does the pituitary.

Oxytocin is a powerful smooth muscle stimulant, and due to the receptors which have formed in the uterus, causes uterine contractions. The high estrogen levels also stimulate the placenta to make prostaglandins (similar to those that cause "cramps") that also stimulate contractions. The hormonal induction of labor is a positive feedback loop. Oxytocin and the prostaglandins cause uterine contractions that, in turn, stimulate the release of more oxytocin and prostaglandins.

This is partially due to pressure of the baby's head against the cervix that both dilates the cervix and acts on cervical nerve endings to stimulate the production of oxytocin, which causes stronger contractions that cause the baby's head to be pressed more strongly against the cervix, etc.

REPRODUCTIVE SYSTEMS OF INVERTEBRATES

GONADS, ASSOCIATED STRUCTURES, AND PRODUCTS

Although asexual reproduction occurs in many invertebrate species, most reproduce sexually. The basic unit of sexual reproduction is a gamete (sperm or egg), produced by specialized tissues or organs called gonads. Sexual reproduction does not necessarily imply copulation or even a union of gametes. As might be expected of such a large and diverse group as the invertebrates, many variations have evolved to ensure survival of species. In many lower invertebrates, gonads are temporary organs; in higher forms, however, they are permanent. Some invertebrates have coexistent female and male gonads; in others the same gonad produces both sperm and eggs. Animals in which both sperm and eggs are produced by the same individual (hermaphroditism) are termed monoecious. In dioecious species, the sexes are separate. Generally, the male gonads ripen first in hermaphroditic animals (protandry); this tends to ensure cross-fertilization. Self-fertilization is normal, however, in many species, and some species undergo sex reversal.

SPONGES, COELENTERATES, FLATWORMS, AND ASCHELMINTHS

Sponges are at a cellular level of organization and thus do not have organs or even well-developed tissues; nevertheless, they produce sperm and eggs and also reproduce asexually. Some species of sponge are monoecious, others are dioecious. Sperm and eggs are formed by aggregations of cells called amoebocytes in the body wall; these are not considered gonads because of their origin and transitory nature.

In hydrozoan coelenterates, temporary gonads are formed by groups of cells in either the epidermis (outer cell layer) or gastrodermis (gut lining), depending on the species; scyphozoan and anthozoan coelenterates generally have gonads in the gastrodermis. The origin and development of gonads in coelenterates, particularly freshwater species, are often associated with the seasons. Freshwater hydrozoans, for example, reproduce asexually until the onset of cold weather, which stimulates them to form testes and ovaries. Colonial hydrozoans asexually produce individuals known as polyps. Polyps, in turn, give rise to free swimming stages (medusae), in which gonads develop. The body organization of sponges and coelenterates is such that most of their cells are in intimate contact with the environment; consequently, gametes are shed into the water, and no ducts are necessary to convey them to the outside.

In contrast to sponges and coelenterates, platyhelminths generally have well-developed organ systems of a permanent nature and, in addition, have evolved secondary reproductive structures to convey sex products. One exception is the acoels, a group of primitive turbellarians; they lack permanent gonads, and germinal cells develop from amoebocytes in much the same manner as in sponges. The majority of flatworms, however, are monoecious, the primary sex organs consisting of one or more ovaries and testes. The tube from the ovary to the outside is called the oviduct; it often has an outpocketing (seminal receptacle) for the storage of sperm received during copulation. In many species the oviduct receives a duct from yolk (vitelline) glands, whose cells nourish the fertilized egg. Beyond the entrance of the duct from the yolk glands the oviduct may be modified to secrete a protective capsule around the egg before it is discharged to the outside. The male organs consist of testes, from which extend numerous tubules (vasa efferentia) that unite to form a sperm duct (vas deferens); the latter becomes an ejaculatory duct through which sperm are released to the outside. The sperm duct may exhibit expanded areas that store sperm (seminal vesicles), and it may be surrounded by prostatic cells that contribute to the seminal fluid. The sperm duct eventually passes through a copulatory organ. The same basic structural pattern, somewhat modified, is found in most higher invertebrates.

Aschelminthes (roundworms) are mostly dioecious; frequently there are external differences between males and females (sexual dimorphism). The males are generally smaller and often have copulatory spicules. Nematodes have relatively simple reproductive organs, a tubular testis or ovary being located at the end of a twisted tube. The portion of the female tract nearest the ovary forms a uterus for temporary storage of fertilized eggs. Some species lay eggs, but others retain the egg in the uterus until the larva hatches. The sperm are released into a cavity called the cloaca. A number of free-living nematodes are capable of sex reversal—if the sex ratio in a given population is not optimal or if environmental conditions are not ideal, the ratio of males to females can be altered. This sometimes results in intersexes; i.e., females with some male characteristics. Hermaphroditism occurs in nematodes, and self-fertilization in such species is common.

ANNELIDS AND MOLLUSKS

Annelids have a well-developed body cavity (coelom), a part of the lining of which gives rise to gonads. In some annelids, gonads occur in several successive body segments. This is true, for example, in polychaetes, most of which are dioecious. Testes and ovaries usually develop, though not invariably, in many body segments; and the sperm and eggs, often in enormous numbers, are stored in the coelom. Fertilization is external. In oligochaetes (all of which are monoecious) on the other hand, the gonads develop in a few specific segments. Sperm are stored in a seminal vesicle and eggs in an egg sac, rather than in the coelom. A portion of the peritoneum, the membrane lining the coelom, becomes a saclike seminal receptacle that stores sperm received from the mate. The earthworm, Lumbricus terrestris, is an example of a specialized annelid reproductive system. Female organs consist of a pair of ovaries in segment 13; a pair of oviducts that open via a ciliated funnel (i.e., with hairlike structures) into segment 13 but open to the exterior in segment 14; an egg sac near each funnel; and a pair of seminal receptacles in segment 9 and another in segment 10. Male organs consist of two pairs of minute testes in segments 10 and 11, each associated with a ciliated sperm funnel leading to a tiny duct, the vas efferens. The two ducts on each side lead to a vas deferens that opens in segment 15. Testes and funnels are contained within two of three pairs of large seminal vesicles that occupy six body segments. Leeches (Hirudinea), also monoecious, have one pair of ovaries and a segmentally arranged series of testes with duct systems basically similar to those of earthworms.

Although mollusks have a close evolutionary kinship to annelids, they have reduced or lost many structures characteristic of segmented worms. The coelom persists only as three regional cavities: gonadal, nephridial (kidney), and pericardial (heart). In ancestral forms these were interconnected so that gametes from the gonad passed through the pericardial cavity, the nephridial cavity, then to the outside through a nephridial pore. The various groups of mollusks have tended to modify this arrangement, with the result that gonads have their own pore; among amphineurans, for example, the sexes are usually separate, and there is one gonad with an associated pore. Gastropods show considerable variability, but generally one gonad (ovary, testis, or ovotestis—a structure combining the functional gonads of both sexes) is located in the visceral hump and connected to the outside by a remnant of the right kidney. In hermaphroditic forms, one duct carries sperm as well as eggs. The gonadal ducts of gastropod females often secrete a protective capsule around the fertilized eggs; in males, the terminal portion of the duct is sometimes contained in a copulatory organ. Pelecypods may be either monoecious or dioecious, but the gonads are usually paired. In mussels and oysters, the gonads open through the nephridial pore, but in clams the reproductive system opens independently. The cephalopods are all dioecious. The single testis or ovary releases its products into the pericardial cavity and this, in turn, leads to a gonopore, the external opening. The oviduct of the squid is terminally modified to form a shell gland. The male system is more complex—the gonoduct leads into a seminal vesicle where a complicated torpedo-shaped sperm case (spermatophore) is secreted and contains the sperm. Spermatophores are then stored in a special structure (Needham's sac) until copulation occurs.

A remarkable characteristic of some mollusks is the ability to alter their sex. Some species are clearly dioecious; however, among the monoecious species there is considerable variability in their hermaphroditic condition. In some species, male and female gonads, although in the same individual, are independent functionally and structurally. In others, an ovotestis produces both sperm and eggs. Oysters display a third condition; young oysters have a tendency toward maleness, but, if water temperature or food availability is altered, some individuals develop into females. Later, a reversal to the male condition may occur. The sexual makeup of an entire oyster population also has a seasonal aspect; in harmony with the group, an individual may undergo several alterations in the course of a year. A similar phenomenon, called consecutive sexuality, occurs in limpets. These gastropods stack themselves in piles, with the younger animals on top. The animals on top are males with well-developed testes and copulatory organs; those in the middle are hermaphroditic; those on the bottom are females, having lost the testes and copulatory organ (penis) by degeneration. A decrease in the number of females in a stack induces males to assume female characteristics, but the transition is retarded when an excess of females is present. The degree of maleness or femaleness is probably controlled in part by environmental and internal factors.

ARTHROPODS

The phylum Arthropoda includes a vast number of organisms of great diversity. Most arthropods are dioecious, but many are hermaphroditic, and some reproduce parthenogenetically (i.e., without fertilization). The primary reproductive organs are much the same as in other higher invertebrates, but the secondary structures are often greatly modified. Such modifications depend on whether fertilization is internal or external, whether the egg or zygote (i.e., the fertilized egg) is retained or immediately released, and whether eggs are provided some means of protection after they have left the body of the female. The mandibulate arthropods (e.g., crustaceans, insects) include more species than any other group and have invaded most habitats, a fact reflected in their reproductive processes.

Crustaceans (e.g., crabs, crayfish, barnacles) are for the most part dioecious. The primary reproductive organs generally consist of paired gonads that open through paired ventral (bottom side) gonopores. Females often have a seminal receptacle (spermatheca) in the form of an outpocketing of the lower part of the female tract or as an invagination (inpocketing) of the body near the gonopore. Males have appendages modified for clasping the female during copulation or for guiding sperm. A number of groups have members that reproduce parthenogenetically. Branchiopods (e.g., water fleas, fairy shrimp) have simple paired gonads. The female gonopore often opens dorsally (on the back side) into a brood chamber; the male gonopore opens near the anus. Males have appendages for clasping females during copulation. Ostracods, or seed shrimp, have paired, tubular gonads. The eggs may be brooded by the female, or they may be released into the water via a gonoduct and gonopore. The terminal portion of the male gonoduct is enclosed in a single or paired penis. Many species reproduce parthenogenetically. Some experts contend that this is the only method employed, even though functional males may be present in the population. Copepods (e.g., Cyclops) have paired ovaries and an unpaired testis. The terminal portion of the oviduct constitutes an ovisac for storage of eggs. The male deposits sperm in a spermatophore that is transferred to the female. Sexual dimorphism is particularly evident among parasitic copepods. Frequently, parasitic females can hardly be recognized as copepods except for the distinctive ovisacs. Males, on the other hand, are free-living and are recognizable as copepods.

The hermaphroditic Cirripedia (e.g., barnacles) are among the exceptions to the generalization that crustaceans are dioecious. It has been suggested that hermaphroditism in barnacles is an adaptation to their sessile, or stationary, existence, but cross-fertilization is more common than self-fertilization. The ovaries lie either in the base or in the stalk of the animal, and the female gonopore is near the base of the first pair of middle appendages (cirri). The testes empty into a seminal vesicle through a series of ducts; from the vesicle extends a long sperm duct within a penis that may be extended to deposit sperm in the mantle cavity of an adjacent barnacle. The terminal portion of the oviduct secretes a substance that forms a kind of ovisac within the mantle cavity, where fertilized eggs undergo early development. Although most barnacles are hermaphrodites, some display a peculiar adaptation in that they contain parasitic dwarf or accessory males. Dwarf males are much smaller than the host barnacle in which they live and are degenerate, except for the testes. In some species they live in the mantle cavity of hermaphroditic forms and produce accessory sperm; in other species only the female organs persist in the host animal, and the accessory male is a necessity.

Amphipods and isopods (e.g., pill bugs, sow bugs), like most crustaceans, are dioecious and have paired gonads. Females of both groups have a ventral brood chamber (marsupium) formed by a series of medially directed (i.e., toward the body midline) plates (oostegites) in the region of the thorax, the region between head and abdomen. Many isopods are parasitic and have developed unusual sex-related activities. Certain species are parasitic on other crustaceans. After a series of molts (i.e., shedding of the body covering) a parasitic larval (immature) isopod attaches to the shell of a crab. If it is the only larva to do so, it increases in size and develops into an adult female. If another larva subsequently attaches, the new arrival becomes a male. It has been demonstrated that the testes of the functioning male larva will change to ovaries if the larva is removed to a new, uninfected host. Thus, the larvae of these species apparently are intersexual and can develop into either sex. This phenomenon, reminiscent of that in mollusks, demonstrates the way in which similar adaptations have evolved in diverse groups of organisms.

The gonads of crabs and lobsters are paired, as are the gonopores. The females of many species have external seminal receptacles on the ventral part of the thorax; those of other species have internal receptacles in the same region. In some species, seminal receptacles are absent, and the male simply attaches a spermatophore to the female. Thus, males either have appendages (gonopods) by which sperm are inserted in the body of the female or produce spermatophores for sperm transfer. The sexual dimorphism of many decapods can be altered by parasitism. An example of this is the crab that is parasitized by a barnacle. A barnacle infection in male crabs induces the secondary sex characters of the crab to resemble those of a female; however, masculinization does not occur in parasitized females. At each molt a parasitized male crab increasingly resembles a female even though the testes may be completely unaffected. Feminization results from a hormonal alteration of the parasitized crab.

Insects are rarely hermaphroditic, but many species reproduce parthenogenetically (without fertilization). The insect ovary is composed of clusters of tubules (ovarioles) with no lumen, or cavity. The upper portion of each ovariole gives rise to oocytes (immature eggs) that mature and are nourished by yolk from the lower portion. The oviduct leads to a genital chamber (copulatory bursa, or vagina), with which are often associated accessory glands and a seminal receptacle. Some accessory glands form secretions by which eggs become attached to a hard surface; others secrete a protective envelope around the egg. The eighth and ninth body segments are often modified for egg-laying. The paired testes consist of a series of seminal tubules that form primary spermatogonia (immature spermatozoa) at their upper ends. As the spermatogonia mature a covering is secreted around them. Eventually they enter a storage area (seminal vesicle). The terminal portion of the male system is an ejaculatory duct that passes through a copulatory organ. A pair of accessory glands, often associated with the ejaculatory duct, contributes to the semen (fluid containing sperm) or participates in spermatophore formation. The ninth body segment and sometimes the tenth bear appendages for sperm transfer. Scorpions and spiders have tubular or saclike gonads; the female system is equipped to receive and store sperm, and, in some species, the female retains the eggs long after fertilization has occurred. Male spiders may have a cluster of accessory glands associated with the terminal portion of the reproductive system for the manufacture of spermatophores, or they may have expanded seminal vesicles for the retention of sperm until copulation takes place. Often specific appendages are adapted for sperm transfer.

REPRODUCTIVE SYSTEMS OF VERTEBRATES

GONADS, ASSOCIATED STRUCTURES, AND PRODUCTS

The reproductive organs of vertebrates consist of gonads and associated ducts and glands. In addition, some vertebrates, including some of the more primitive fishes, have organs for sperm transfer or ovipository (egg-laying) organs. Gonads produce the gametes and hormones essential for reproduction.

Associated ducts and glands store and transport the gametes and secrete necessary substances. In addition to these structures, most male and female vertebrates have a cloaca, a cavity that serves as a common terminal chamber for the digestive, urinary, and reproductive tracts and empties to the outside. In lampreys and most ray-finned fishes in which the cloaca is small or absent, the alimentary canal has a separate external opening, the anus. In some teleosts the alimentary, genital, and urinary tracts open independently. Hagfishes, which are closely related to the lampreys, have a short cloaca.

In many vertebrates other than mammals, especially reptiles and birds, the cephalic, or head, end of the cloaca is partitioned by folds into a urinogenital chamber (urodeum) and an alimentary chamber (coprodeum) that open into a common terminal chamber (proctodeum). Above monotremes (e.g., platypus, echidna) the embryonic cloaca becomes completely partitioned into a urinogenital sinus conveying urine and the products of the gonads, and an alimentary pathway; the two open independently to the exterior.

Gonads arise as a pair of longitudinal thickenings of the coelomic epithelium and underlying mesenchyme (unspecialized tissue) on either side of the attachment of a supporting membrane, the dorsal mesentery, to the body wall. At first, gonadal ridges bulge into the coelom and are continuous with the embryonic kidney.

The germinal epithelium covering the gonadal ridges gives rise to primary sex cords (medullary cords) that invade the underlying mesenchyme. These cords establish within the gonadal blastema (a tissue mass that gives rise to an organ) a potentially male component, the medulla. Secondary sex cords grow inward, spreading just beneath the germinal epithelium to form a cortex. If the gonad is to become a testis, only the medullary component differentiates. If the gonad is to become an ovary, only the cortex differentiates.

The length of an adult gonad depends, in part, upon the extent of gonadal-ridge differentiation. In cyclostomes (lampreys and hagfish), elasmobranchs (sharks, skates, and rays), and teleosts most of it differentiates, and the gonads extend nearly the length of the body trunk. In tetrapods (amphibians, reptiles, birds, and mammals), the cranial portion, at the anterior end, generally does not differentiate; in toads only the more caudal, or posterior, portion does so. The middle segment in toads of both sexes gives rise to a Bidder's organ containing immature eggs.

In anurans (frogs and toads) and some lizards of both sexes, one segment of the gonadal ridge gives rise to yellow fat bodies that, especially in anurans, diminish in size just prior to the breeding season. In mammals, only the middle portion of the gonadal ridge differentiates.

Some vertebrate species have only one gonad, which may lie in the midline or on one side; the condition is more common among females. Adult cyclostomes of both sexes have one gonad. In lampreys it is in the middle of the body; in hagfishes it is on the right side. Birds are the only other major group of vertebrates in which most females have one gonad, the right ovary being typically absent. Male birds have a pair of testes, however. Exceptions to the condition of single ovaries among birds include members of the falcon family, in which more than 50 percent of mature hawks have two well-developed ovaries. In all bird species a small percentage of females probably have two ovaries; reported instances include owls, parrots, sparrows, and doves, with estimates for doves ranging from 5 percent to 25 percent. A few teleosts and viviparous elasmobranchs have only one ovary; in sharks the right one is usually present, in rays, the left.

In amniotes (i.e., reptiles, birds, and mammals) unpaired gonads are unusual. Some lizards have one testis, and some female crocodiles have one ovary. Among mammals, the platypus usually has only a left ovary, and some bat species (family Vespertilionidae) have only the right.

One of two explanations may account for unpaired gonads: the paired embryonic gonadal ridges may fuse to form a median gonad—as in lampreys and the perch—or only one gonadal ridge may receive immigrating primordial germ cells (immature sperm or eggs), with the result that the opposite gonad does not develop—as in chickens and ducks. Both gonadal ridges have been reported to exhibit an equal number of primordial germ cells in embryonic hawks, and these typically have two ovaries.

Among lower vertebrates, mature gonads sometimes produce both sperm and eggs. Hermaphroditism is more general in cyclostomes and teleosts than in other fishes. A teleost may function as a male during the early part of its sexual life and as a female later. In some teleost families sperm and eggs mature simultaneously but in different regions of the gonad.

These fish normally function as males during one season and as females the next. Cyclostomes generally are ambisexual during juvenile life—i.e., immature male and female sex cells exist side by side, or, as in Myxine, the anterior part of the immature gonad may be ovary and the caudal part, the testis. It is thought that cyclostomes normally become unisexual at maturity. Hermaphroditism is uncommon among amphibians, although it frequently occurs as an anomaly. In vertebrates above amphibians, true hermaphroditism probably does not exist.

Both male and female duct systems are occasionally absent. In cyclostomes, a few elasmobranchs, and some teleosts, such as salmon, trout, and eels, the gametes are propelled toward the posterior within the coelom, often by cilia (minute hairlike structures), and exit via a pair of funnel-like genital pores near the base of the tail. In cyclostomes, the pores lead to a sinus, or cavity, within a median papilla (i.e., a fingerlike structure) and are open only during breeding seasons.

MALE SYSTEMS

Tests

In anurans, amniotes (reptiles, birds, and mammals), and even some teleosts, testes are composed largely of seminiferous tubules—coiled tubes, the walls of which contain cells that produce sperm—and are surrounded by a capsule, the tunica albuginea. Seminiferous tubules may constitute up to 90 percent of the testis. The tubule walls consist of a multilayered germinal epithelium containing spermatogenic cells and Sertoli cells, nutritive cells that have the heads of maturing sperm embedded in them. Seminiferous tubules may begin blindly at the tunic, or outermost tissue layer, and pass toward the centre, becoming tortuous before emptying into a system of collecting tubules, the rete testis. Such an arrangement is characteristic of frogs. In certain amniotes—the rat, for example—the tubules may be open ended, running a zigzag course from the rete to the periphery and back again. The average length of such tubules is 30 centimetres (12 inches), and they seldom communicate with each other. In many mammals the tubules are grouped into lobules separated by connective-tissue septa, or walls.

The arrangement permits the packing of an extensive amount of germinal epithelium into a small space. In immature males and in adult males between breeding seasons, the tubules are inconspicuous and the epithelium is inactive; in some species, however, spermatogenesis, or production of sperm, proceeds at a variable pace throughout the year. An active epithelium may exhibit all stages of developing sperm. The lumen, or tubule cavity, contains the tails of many sperm (the heads of which are embedded in Sertoli cells), free sperm, and fluid that is probably resorbed. In mammals, in any single zone along a tubule, all sperm are at the same stage of maturation; adjacent zones contain different generations of sperm, and a period of sperm formation and discharge is followed by an interval of inactivity.

In cyclostomes, most fishes, and tailed amphibians the germinal epithelium is arranged differently. Instead of seminiferous tubules there are large numbers of spermatogonial cysts (also called spermatocysts, sperm follicles, ampullae, crypts, sacs, acini, and capsules) in which sperm develop, but in which the epithelium is not germinal. Spermatogenic cells migrate into the cysts from a permanent germinal layer, which, depending on the species, may lie among cysts at the periphery of the testes or in a ridge along one margin of the testis. After invading the thin nongerminal epithelium of a cyst, spermatogenic cells multiply, producing enormous numbers of sperm. The cysts become greatly swollen and whitish in colour; the entire testis also swells and has a granular appearance. As sperm mature, they separate from the epithelium and move freely in the cystic fluid. Finally, the cysts burst, and the sperm are shed into ducts. In the case of cyclostomes and a few teleosts the sperm are shed into the coelom. The cysts, totally emptied, collapse. Then either they are replaced by new ones, or they become repopulated by additional spermatogenic cells. It is not yet known which of these processes occurs.

Testicular stroma, which fills the spaces between seminiferous tubules or spermatogenic cysts, consists chiefly of connective tissue, blood and lymphatic vessels, and nerves; it is more abundant in some vertebrates than in others. Glandular Leydig (interstitial) cells are also present in most, if not all, vertebrates. Thought to be a primary source of androgens, or male hormones, Leydig cells are not always readily distinguishable, and, in some bird species, they may be seen only with the electron microscope. The capillary system of the rat testis, and probably that of many other vertebrates, is such that blood that has bathed the Leydig cells flows to the tubules; it is thus probable that Leydig cell hormones have an immediate effect on the germinal epithelium.

Testes in vertebrates below mammals lie within the body. This is also true of many, sometimes all, members of the mammalian orders Monotremata, Insectivora, Hyracoidea, Edentata, Sirenia, Cetacea, and Proboscidea. Some male mammals—most marsupials, ungulates, carnivores, and primates after infancy—have a special pouch (scrotum) that the testes occupy permanently. A few mammals have a pouch into which the testes descend and from which they can be retracted by muscular action. These include a few rodents such as ground squirrels; most, if not all, bats; and some primitive primates (loris, potto). The scrotum consists of two scrotal sacs, each connected to the abdominal cavity by an inguinal canal lined with the peritoneal membrane.

The canals are the path of descent (and retraction) of the testes to the sacs. In descending, the testes carry along a spermatic duct, blood and lymphatic vessels, and a nerve supply wrapped in peritoneum and constituting, collectively, the spermatic cord. Rabbits, most rodents, and some insectivores, which lack scrotal sacs, have instead a wide inguinal canal into which the testes may be drawn and from which they are retracted when in danger of injury. In these mammals, descended testes cause a temporary bulge in the perineal region (i.e., between the anus and the urinogenital opening). In a small number of mammals, the testes permanently occupy the perineal location.

The scrotum is a temperature-regulating device. Warm blood approaching the testis comes close to the vessels carrying cool blood leaving the testis, so that the blood approaching the testis is cooled; the vessels form an intricate vascular network (pampiniform plexus) within the spermatic cord. Failure of both testes to enter the scrotal sacs (cryptorchidism) results in permanent sterility. In cold weather two sets of muscles, the dartos and cremasteric, pull the testes close to the body. The dartos lies between the two scrotal sacs and is attached to the scrotal skin. The cremaster, wrapped around the spermatic cord, is an extension of the abdominal wall musculature. It retracts the testis. Birds, like mammals, are homoiothermic (warm-blooded), and their testes are near air sacs (extensions of incurrent respiratory tubes). Air in the sacs may help regulate the temperature of the testes.

Ducts

The male duct system begins as the rete testis, a network within the testis of thin-walled ductules, or minute ducts, that collects sperm from the seminiferous tubules. The rete is drained by a number of small ducts—usually fewer than ten—called the vasa efferentia, which are modified kidney tubules. In some fishes and amphibians the vasa efferentia connect the testes with the cranial (anterior) end of the kidneys. In anamniotes (e.g., fish and amphibians), therefore, except teleosts, the ducts that drain the kidneys usually drain the testes also. In most amphibians these ducts pass caudad, or posteriorly, to empty independently into the cloaca; in some fishes they pass through a median urinogenital papilla.

Although drainage of the testis and the kidney by the same duct is a basic pattern, there has been a tendency in many vertebrates toward separate spermatic and urinary ducts. This tendency is manifested in one of two ways among anamniotes. In many sharks and in some amphibians (Plethodontidae, Salamandridae, Ambystomatidae), the embryonic kidney duct ultimately drains the testis, and one or more new ducts (ureters) drain the adult kidney. On the other hand, in the primitive fish Polypterus and in most teleosts, the embryonic kidney duct drains the adult kidney, and a new duct arises to drain the testis. Many degrees of separation of the two ducts occur in anamniotes, from the condition of the sturgeon, in which the spermatic duct unites with the urinary duct far toward the head, to the condition in Esox (a pike), in which spermatic and urinary ducts empty independently to the exterior.

In amniotes, the mesonephric kidney is a temporary structure confined to the embryo, but the mesonephric duct persists in the adult male as a sperm duct. A separate ureter drains the adult kidney. The spermatic and urinary ducts empty independently into the cloaca except in mammals above monotremes, in which they are confluent with the urethra. The epididymis of amniotes, a highly tortuous duct draining the vasa efferentia, usually serves as a temporary storage place for sperm; it is small in birds and large in turtles. In mammals, the first part of the epididymis consists of a head, body, and tail that wrap around the testis; it gradually straightens to become the spermatic duct. The epididymis secretes substances that prolong the life of stored sperm and increase their capacity for motility.

In all vertebrates certain regions of the spermatic duct are lined by cilia and a variety of secretory epithelial cells. One end may enlarge to form a sperm reservoir or secrete seminal fluid. In the catfish Trachycorystes mirabilis secretions of the spermatic duct form a gelatinous plug in the female similar to the vaginal plug of mammals. A seminal glomulus in birds functions as a sperm reservoir. In some mammals an enlargement of the spermatic duct called the ampulla contributes to the seminal fluid and stores sperm. Small mucous glands (of Littré) and other glandular structures open into the urethra along its length. Cloacal glands in basking sharks and many salamanders form a jelly that encloses sperm in a spermatophore. Cloacal glands of some lizards produce secretions called pheromones. The siphon sac of elasmobranchs is one of the few accessory sex glands that is a separate organ in animals below mammals. It extends as an elongated pocket into the pelvic fin and secretes a nutritive mucus that enters the female reproductive tract with sperm.

ACCESSORY GLANDS

Accessory sex glands that are conspicuous outgrowths of the genital tract are almost uniquely mammalian. The major mammalian sex glands include the prostate, the bulbourethral, and the ampullary glands, and the seminal vesicles. All are outgrowths of the spermatic duct or of the urethra and all four occur in elephants and horses and in most moles, bats, rodents, rabbits, cattle, and primates. A few members of these groups lack ampullary glands, or ampullary glands and seminal vesicles. Cetaceans (whales, porpoises) have only the prostate, as do some carnivores, including dogs, weasels, ferrets, and bears.

The prostate, the most widely distributed mammalian accessory sex gland, is absent only in Echidna (a marsupial) and a few carnivores. It empties into the urethra by multiple ducts. Many rodents, insectivores, and lagomorphs have three separate prostatic lobes; in a few mammals (some primates and carnivores) the prostate is a single mass with lobules and encircles the urethra at the base of the bladder. In a few mammals (e.g., opossum), the prostate is not a compact mass but a partly diffuse gland. In many rodents (e.g., rat, guinea pig, mouse, hamster) and some other mammals, the semen coagulates quickly after ejaculation as a result of a secretion from a male coagulating gland, which is usually considered part of the prostatic mass. Coagulated semen forms a vaginal plug that temporarily prevents copulation.

Bulbourethral (Cowper's) glands arise from the urethra near the penis and are surrounded by the muscle of the urethra or penis. Typically, there is one pair, but as many as three (marsupials) may be found. The glands, small in man, large in rodents, elephants, and some ungulates including pigs, camels, and horses, are absent in cetaceans, mustelids (e.g., mink, weasel), sirenians (manatees, dugongs), pholidotans (pangolins), some edentates, and carnivores such as walrus, sea lion, bear, and dog.

Although many mammals have an ampullary swelling on the spermatic duct near the urethra, only a small number form a separate ampullary gland as an outgrowth of the duct. It is very large in some bats, absent in many mammalian orders, and variable in the rest. Although common in rodents, it is absent in guinea pigs and some strains of mice.

Seminal vesicles are paired, typically elongated and coiled fibromuscular sacs that empty into either the spermatic duct or the urethra. Absent in monotremes, marsupials, carnivores, cetaceans, and in some insectivores, chiropterans, and primates, seminal vesicles are exceptionally large in rhesus monkeys and small in man. They are absent in domesticated rabbits, small or rudimentary in cottontails, large in armadillos, and variable in sloths. They contribute the sugar fructose and citric acid to the semen but do not serve as sperm reservoirs.

FEMALE SYSTEMS

OVARIES

Ovaries lie within the body cavity and are suspended by a dorsal mesentery (mesovarium), through which pass blood and lymph vessels and nerves. Primitive vertebrate ovaries occur in the hagfish, in which a mesentery-like fold of gonadal tissue stretches nearly the length of the body cavity. Unique in the hagfish is the fact that functional ovarian tissue occupies only the forward half of the gonadal mass, the rear part containing rudimentary testicular tissue. In most fishes except very primitive forms, the ovaries are similarly elongated. In tetrapods other than mammals, the ovaries are usually confined to the middle third or half of the body cavity, particularly during nonbreeding seasons. The ovaries of mammals undergo moderate caudal displacement, finally coming to lie between the kidney and the pelvis.

The appearance of an ovary depends on many factors—e.g., whether one egg or thousands are discharged (ovulated); whether the eggs are immature or ripe; whether mature eggs are small or large; or whether pigments occur in the egg cytoplasm, such as those responsible for yellow yolk. Other factors also affect the appearance of the ovary: the season of the year in seasonal breeders (the ovary enlarges during breeding seasons, diminishes in size between seasons); the age of the animal (whether juvenile, reproductively active, or senile, particularly in birds and mammals); and the fate of ovulated, or discharged, egg follicles, or sacs.

The ovaries are covered with a germinal epithelium that is continuous with the peritoneum lining the body cavity. The term germinal epithelium is inappropriate because in most adults it contains no germ cells, these having moved deeper into the ovary. In hagfishes and amphibians, cells that give rise to eggs are known to occur in the germinal epithelium, and it may be that the germinal epithelium in a few other vertebrates contains similar cells. The germinal epithelium undergoes cell division, however. This is particularly true of species in which enormous expansion of the ovary occurs each breeding season. Beneath the epithelium is a layer of connective tissue, the tunica albuginea, which is much thinner than that surrounding the testes.

A typical vertebrate ovary consists of cortex and medulla. The cortex, immediately internal to the tunica albuginea, contains future eggs and, at one time or another, eggs in ovarian follicles (i.e., developing eggs); it undergoes fluctuations in size and appearance that correlate with stages of the reproductive cycle. The cortex also contains remnants of ovulated follicles and, in mammals, clusters of interstitial cells that, in some species, are glandular. The cortical components are embedded in a supportive framework of connective, vascular, and neural tissue constituting the stroma. Internal to the cortex is the medulla, consisting of blood and lymph vessels, nerves, and connective tissue. The medulla, which contains no germinal elements, exhibits no significant cyclical activity, is usually inconspicuous, is continuous with the dorsal mesentery, and, in cyclostomes, is hardly distinguishable from the latter. The mammalian medulla, on the contrary, is almost completely surrounded by cortex and converges on the mesovarium (i.e., the part of the peritoneum that supports the ovary) at a narrow hilus, at which nerves and vessels enter the ovary. In the medulla of the mammalian ovary near the hilus are small masses of blind tubules or solid cords—the rete ovarii—which are homologous (i.e., of the same embryonic origin) with the rete testis in the male. The microscopic right ovary of birds usually consists only of medullary tissue.

Ovaries are characterized as saccular, hollow, lacunate (i.e., compartmented), or compact. The ovary of many teleosts, especially viviparous ones, contains a permanent cavity, which is formed during ovarian development when an invagination of the ovarian surface traps a portion of the coelom. The cavity is therefore unique in that it is lined by germinal epithelium. The lining develops numerous ovigerous folds that project into the lumen and greatly increase the surface area for proliferation of eggs. In most other teleosts, a temporary ovarian cavity develops after each ovulation, when the shrinking cortex withdraws from the outside ovarian wall along one side of the ovary. The resulting cavity is obliterated as eggs of the next generation enlarge. The permanent and temporary cavities of teleost ovaries and a similar cavity in garfish ovaries are continuous with the lumen of the oviduct, and eggs are shed into them.

The ovaries of other fishes lack cavities and are characterized as compact. The amphibian ovary, which contains six or more central, hollow sacs that give it a lobed appearance, is characterized as saccular. The sacs are formed when the embryonic medullary and rete cords become hollow and coalesce. Maturing eggs bulge into the sacs but are not shed into them. The ovaries of reptiles, birds, and monotremes have cavities homologous to those in amphibians; the number of medullary spaces in the adults is considerably larger, however, so that the ovaries contain an extensive network of fluid-filled cavities (lacunae). Such ovaries are characterized as lacunate. The ovaries of mammals above monotremes are compact, having no medullary cavities.

An ovarian follicle consists of an oocyte, or immature egg, surrounded by an epithelium, the cells of which are referred to variously as follicular, nurse, or granulosa cells. In cyclostomes, teleosts, and amphibians, the epithelium is one layer thick. In the hagfish and those vertebrates in which the oocyte receives heavy deposits of yolk (elasmobranchs, reptiles, birds, and monotremes), the epithelium appears to be two cells thick, apparently the result of layering of nuclei in a simple columnar epithelium (i.e., epithelium consisting of relatively "tall" cells). Above monotremes the follicular epithelium appears to be many cells thick; in at least one species, however, this is considered an artifact, and all granulosa cells are said to extend between the outer boundary of the epithelium and the oocyte.

The follicular epithelium originates as a few flattened cells derived from the germinal epithelium. Primary follicles are usually situated just under the tunica albuginea; secondary follicles lie deeper in the cortex. The primitive role of the follicular cells appears to be the secretion of the yolk-forming material onto or into the oocyte. Evidence from mammals indicates that the follicular cells may also have a role in converting substances produced elsewhere into female hormones, or estrogens. In some hibernating bats the granulosa cells are filled with glycogen, or animal starch, which may be a source of energy. Mammalian follicles above monotremes are unique in that they develop a fluid-filled cavity (antrum) within the granulosa layer. During antrum formation cell division of the granulosa cells increases, and fluid-filled spaces develop among the cells.

The spaces coalesce to form the antrum. Under the influence of pituitary gonadotropic hormones, many antral follicles thereafter continue to grow, forming large so-called Graafian follicles—less than 400 microns, or 0.4 millimetre (0.16 inch), in diameter in large mammals, 150–200 microns, or 0.15–0.2 millimetre (0.006–0.008 inch), in small ones. Graafian follicles contain mature eggs and appear as large blisters on the ovary. At this stage the ovum, suspended within the fluid of the antrum (liquor folliculi) by a slender stalk of granulosa cells, is surrounded by a cluster of these cells, the cumulus oophorus, or discus proligerus. The remaining follicular cells form a thin wall surrounding the antrum. Antra are lacking in a few insectivores (Hemicentetes, Euriculus) because the granulosa cells swell and multiply to form corpora lutea, masses of yellow tissue. In the bat Myotis the antrum is likewise compressed and disappears just before discharge of the egg, or ovulation.

In all vertebrates, oocytes that have begun to grow and mature may, at any time until just before ovulation, cease development and undergo atresia, or degeneration. This is a normal process that reduces the number of eggs ovulated. In small laboratory rodents, atresia takes place in 50 percent of the Graafian follicles in each ovary one or two days before ovulation, thus reducing the number of ovulatable eggs by 50 percent. A similar reduction takes place in hagfish prior to ovulation. Atretic follicles eventually become lost in the stroma of the cortex of the ovary. In mammals especially, follicles lacking oocytes and antra, called anovular follicles, as well as polyovular follicles (i.e., containing more than one oocyte), occasionally occur.

The ovarian follicle of vertebrates, commencing with hagfish, is surrounded by a theca, or sheath, composed of two concentric layers of stromal cells. The outer layer (theca externa) is chiefly connective tissue but may contain smooth muscle fibres. The inner layer (theca interna) has more blood vessels and, in vertebrates that produce heavily yolked eggs, the largest vessels carry venous blood. In these species the cell membranes of the oocyte and granulosa cells have many microvilli (i.e., fingerlike projections), which probably facilitate transport of substances important in yolk formation from the blood vessels to the egg. Mature follicles in the marsupial Dasyatus are said to lack theca, and in some bats only one thecal layer has been described.

During the growth phase, eggs in species with massive amounts of yolk may increase in size 106 (1,000,000) or more times as a result of vitellogenesis (deposit of yolk). In goldfish, on the other hand, when vitellogenesis commences, the egg has a diameter of 150 microns (0.15 millimetre [0.006 inch]); that of the mature egg is only 500 microns (0.5 millimetre [0.02 inch]). Mammalian eggs contain little yolk and vary little in size. Oogonia (i.e., cells that form oocytes) of the golden hamster average 15 microns (0.015 millimetre [0.0006 inch]) in diameter, and eggs in Graafian follicles average 70 microns (0.07 millimetre [0.003 inch]). The mature eggs of horses and humans are approximately the same size—somewhat less than 150 microns. In seasonally breeding oviparous fishes and amphibians, all eggs are usually in the same stage of development, and the ovary grows to a mature state quite rapidly as a result of growth of the eggs, which frequently number more than 1,000,000. Such ovaries distend the body wall when mature; following spawning, the ovaries shrink rapidly to inconspicuous bodies consisting mainly of oogonia, immature oocytes, and a few stromal cells. In reptiles and birds, ovarian weight also is high in proportion to body weight during egg-laying seasons. The weight of the ovary of the starling, for example, may increase from eight milligrams in early winter to 1,400 milligrams immediately before ovulation. The mature eggs of reptiles and birds are unique in that they are suspended from the ovary by a short stalk (pedicle). The stalk contains a cortex with additional oocytes in various stages of development and extensions of vessels and nerves. Full growth of the follicle in reptiles and birds requires only a few days or weeks (nine days in the domestic hen). In mammals, the ratio of ovarian weight to body weight varies insignificantly throughout the reproductive life of the female, and follicles in many stages of development are constantly present.

Vertebrate eggs are almost universally shed into the coelom or into a subdivision thereof, from which they enter the female reproductive tract. Even in those teleosts in which the eggs are shed into an ovarian cavity, the latter is often of coelomic origin. In many mammals a membranous sac of peritoneum, the ovarian bursa, traps part of the coelom in a chamber along with the ovary. The bursal cavity (periovarian space) may be broadly open to the main coelom, completely closed off from the coelom, or in communication with the coelom by a narrow, slitlike passage. The bursa, moderately developed in lower primates and catarrhines (Old World monkeys), is poorly developed in man. In horses, one edge of the ovary contains a long groove (ovulation fossa) into which all eggs are shed; the groove is found in a cleftlike ovarian bursa. The ovarian bursa increases the probability that all ovulated eggs will enter the oviduct.

The process of ovulation has been described for all vertebrate classes. Elasmobranchs, reptiles, and birds have massively yolked eggs. As ovulation approaches, the fimbria (i.e., frills, or fringes) of the membranous and muscular funnel surrounding the entrance to the oviduct wave in a gentle, undulating motion. An egg that is nearly free of the ovary is grasped and partially encompassed by the fimbria; when the egg is freed, the fimbria draw the egg into the funnel. At this time, the egg has little shape and is partly squirted and partly flows into the oviduct; never completely free in the coelom, its chances of not entering the oviduct are small. In the case of moderately or poorly yolked eggs cilia help to sweep the eggs into the ostium, or opening, of the oviduct. During ovulation in Japanese rice fish, Oryzias latipes, a tiny papilla, or fingerlike process, develops on the surface of a bulging mature follicle in the centre (stigma) of the follicle. The follicle becomes thin at the stigma, an aperture appears, and the egg rolls out. In rabbits this process differs only in detail. During the final 20 minutes before ovulation in rabbits, some of the tiny blood vessels surrounding the stigma rupture, and a small pool of blood forms under the apex of the cone-shaped papilla. The follicular wall shortly gives way at the apex, and follicular fluid oozes from the opening, followed soon after by the egg. The ovulated mammalian egg typically is surrounded by a layer of columnar follicular cells, the corona radiata; but it is naked in some insectivores and some marsupials. Following ovulation in all vertebrates, the ovary may become smaller, become modified for maintenance of pregnancy, or proceed to form additional eggs.

The process of ovulation in vertebrates has been documented, but the immediate causes remain to be clarified. It is almost certain that an ovulatory hormone is secreted by the pituitary gland (i.e., the so-called master endocrine gland) of all vertebrates. It is highly probable that breakdown of very small fibres that bind the follicular cells together may occur at the stigma, weakening the follicular wall at that location. Hormones from the ovary and other sources may play a role, as may neurohormones, which are hormones released at nerve endings. Rhythmic contractions of the entire ovary occur at ovulation in many vertebrates and have been described in rabbits. The role of mechanical pressure within the follicle, however, is not understood. Ovulation in most mammals (spontaneous ovulators) occurs cyclically as a result of the spontaneous release of the ovulatory hormone. In a few mammals (reflex ovulators) the stimulus of copulation is essential for release of the ovulatory hormone.

Striking postovulatory changes take place in the follicles of mammals and, to lesser degrees, of lower vertebrates. Blood vessels from the theca interna invade the ovulated follicles; the granulosa cells divide, enlarge, accumulate fats, and obliterate any remnants of the collapsed antra. Thereafter, they are known as lutein cells. Theca interna cells undergo changes identical to those of the granulosa cells. The result in mammals is the formation of solid masses called corpora lutea, recognizable as prominent reddish-yellow bulges on the ovary. Corpora lutea produce the hormone progesterone, which is essential for the maintenance of pregnancy. The conversion of postovulatory follicles into structures more or less resembling mammalian corpora lutea has been demonstrated in numerous viviparous reptiles, amphibians, and elasmobranchs; in certain other fishes, including cyclostomes; and in some oviparous amphibians and reptiles. In birds, the postovulatory follicle shrinks, and identifiable corpora lutea do not develop, although some granulosa cells accumulate lipids of unknown significance.

TRACTS

The female reproductive tract consists of a pair of tubes (gonoducts) extending from anterior, funnel-like openings (ostia) to the cloaca, except as noted below. The gonoducts are specialized along their length for secretion of substances added to the eggs; for transport, storage, nutrition, and expulsion of eggs or the products of conception; and, in species with internal fertilization, for receipt, transport, storage, and nutrition of inseminated sperm. The predominately muscular tracts are lined by a secretory epithelium and ciliated over at least part of their length. Fusion of the caudal (tail) ends of the paired ducts may occur. Gonoducts are absent in cyclostomes and a few gnathostome fishes that have abdominal pores. A few vertebrates have only one functional gonoduct.

Gonoducts in lungfishes and amphibians are coiled muscular tubes that are ciliated over most of their length. Only occasionally do they unite caudally in a genital papilla before opening into the cloaca. During breeding seasons their diameter increases severalfold because of the highly active secretory epithelium. Between breeding seasons they are small. In some anurans (frogs, toads), such as Rana, the lower end of each gonoduct is expanded to form an ovisac, in which ovulated eggs are stored until spawning; the tube between the ostium (funnel-like opening) and ovisac is the oviduct. In viviparous amphibians the young develop in the ovisac. In amphibians, numerous multicellular glands extend deep into the lining of the female tract. Six successive glandular zones have been described in some urodeles, and these secrete six different gelatinous substances upon the egg. Female urodeles often have convoluted tubular outpocketings of the cloaca called spermatheca; they temporarily store sperm liberated from the male spermatophore.

The two gonoducts of elasmobranchs share a single ostium, a trait found only in Chondrichthyes. The ostium is a wide caudally directed funnel supported in the falciform ligament, which is attached to the liver. The role of the fimbria of the ostium at ovulation has been described (see above Ovaries). Two oviducts pass forward from the ostium to the septum transversum (i.e., between the heart and abdominal cavities), curve around one end of the liver, then pass posteriorly on each side. Approximately midway between ostium and uterus each oviduct has a shell (nidamental) gland. Fertilization takes place above the shell gland, which may be immense or almost undifferentiated. Half of the shell gland secretes a substance high in protein content (albumen), and the other half secretes the shell—delicate in viviparous forms, thick and horny in most oviparous species. Horny shells may have spiral ridges and many long tendrils, which entwine about an appropriate surface after the egg is deposited. In the viviparous shark Squalus acanthias several eggs pass one after the other through the shell gland, where they are enclosed in one long delicate membranous shell that soon disintegrates. Beyond the shell gland the oviducts terminate in an enlargement, which, in viviparous species, serves as a uterus. An oviducal valve may be found at the junction of oviduct and uterus. Although the two uteri usually open independently into the cloaca, they occasionally unite to form a bicornuate (two-horned) structure. In immature females, the uterus may be separated from the cloaca by a hymen, or membrane. The tract enlarges enormously during the first pregnancy and does not thereafter fully regress to its original size.

The gonoducts of most lower ray-finned fishes resemble those of lungfish, but those of gars and teleosts are exceptional in that the oviducts are usually continuous with the ovarian cavities. A median genital papilla receives the oviducts in teleosts, and the papilla is sometimes elongated to form an ovipositor. European bitterlings deposit their eggs in a mussel by means of the ovipositor, and female pipefish and sea horses deposit them in the brood pouch of a male.

With certain modifications, the gonoducts of reptiles and birds are comparable to those of lower vertebrates. Crocodilians, some lizards, and nearly all birds have one gonoduct; the other is not well developed. Even in birds of prey having two functional ovaries, the right oviduct is sometimes undeveloped. The tracts of reptiles generally show less regional differentiation than do those of birds. The oviduct funnel (ostium) in birds forms the chalazae—two coiled, springlike cords extending from the yolk to the ends of the egg. In both reptiles and birds, much of the length of the female tract is oviduct. This region, called the magnum in birds, secretes albumen; lizards and snakes do not form albumen. Behind the albumen-secreting region is a shell gland. In lizards, the gland is midway along the tract. In birds, the shell gland is at the posterior end, has thick muscular walls, and is often inappropriately called a uterus. It is preceded by a narrow region, or isthmus, which secretes the noncalcareous, or soft, membranes of the shell. The shell gland leads to a narrow muscular vagina that empties into the cloaca. The vagina secretes mucus that seals the pores of the shell before the egg is expelled. Special vaginal tubules (spermatheca) store sperm over winter in some snakes and lizards; seminal receptacles have been described in the oviduct funnel in some snakes. In birds, sperm storage glands (sperm nests) often occur in the funnel and at the uterovaginal junction. In lizards and birds, ovulation does not usually occur into a tract already containing an egg. Some lizards shed very few eggs per season; the gecko, for example, sheds only two.

The female reproductive tracts of monotremes, the egg-laying mammals, consist of two oviducts, the lower ends of which are shell glands. These open into a urinogenital sinus, which, in turn, empties into a cloaca. Marsupials have two oviducts, two uteri (duplex uterus), and two vaginas. The upper parts of the vaginas unite to form a median vagina that may or may not be paired internally. Beyond the median vagina, the vaginas are again paired (lateral vaginas) and lead to a urinogenital sinus. The posterior end of the pouchlike median vagina is separated from the forward end of the urinogenital sinus by a partition. When the female is delivering young, the fetuses are usually forced through the partition and into the urinogenital sinus, bypassing the lateral vaginas. The ruptured partition may remain open thereafter, resulting in a pseudovagina. It closes in opossums, and in kangaroos both the median and lateral routes may serve as birth canals. The lateral vaginas in marsupials receive the forked tips of the male penis. Fertilization in all mammals takes place in the oviducts (Fallopian tubes).

In eutherian mammals (i.e., all mammals except monotremes and marsupials), with exceptions noted below, female reproductive tracts beyond the ostia (oviduct funnels) consist of two narrow and somewhat tortuous Fallopian tubes, two large uterine horns (each of which receives a Fallopian tube), a uterine body, and one vagina. Fallopian tubes often have a short dilated ampulla, or saclike swelling, just beyond the ostium. Implantation of the egg occurs only in the uterine horns; the embryos become spaced equidistant from one another in both horns even if only one ovary has ovulated. In some species one horn is rudimentary—the left in the impala (an African antelope)—and the embryos become implanted in the other horn, even though both ovaries ovulate. The body of the uterus in some mammals (e.g., rabbits, elephants, aardvarks; some rodents, bats, insectivores) contains two separate canals (bipartite uterus). In other mammals (ungulates, many cetaceans, most carnivores and bats) the body of the uterus has one chamber into which the two horns empty (bicornuate uterus). There are numerous intermediate conditions between the bipartite and bicornuate condition. Apes, monkeys, and man have no horns, and the Fallopian tubes empty directly into the body of the uterus (simplex uterus). In all mammals, the uterine body tapers to a narrow neck (cervix). The opening (os uteri) into the vagina is guarded by fleshy folds (lips of the cervix). The vagina in eutherian mammals other than rodents and primates terminates in a urinogenital sinus that opens to the exterior by a urinogenital aperture. In some rodents and in higher primates the vagina opens directly to the exterior. In the young of many species a membrane, the hymen, closes the vaginal opening. In guinea pigs the hymen reseals the opening after each reproductive period. Sperm are stored over winter in the uterus of some bats and in vaginal pouches in others.

ANIMAL REPRODUCTIVE AND ORGAN SYSTEM

Animal reproductive system, any of the organ systems by which animals reproduce. The role of reproduction is to provide for the continued existence of a species; it is the process by which living organisms duplicate themselves. Animals compete with other individuals in the environment to maintain themselves for a period of time sufficient to enable them to produce tissue nonessential to their own survival, but indispensable to the maintenance of the species. The additional tissue, reproductive tissue, usually becomes separated from the individual to form a new, independent organism.

The reproductive systems in metazoans (multicelled animals) from sponges to mammals, exclusive of humans. It focuses on the gonads (sex organs), associated ducts and glands, and adaptations that aid in the union of gametes—i.e., reproductive cells, male or female, that are capable of producing a new individual by union with a gamete of the opposite sex. Brief mention is made of how the organism provides for the development of embryos and of the regulatory role of gonads in vertebrate cycles.

Unlike most other organ systems, the reproductive systems of higher animals have not generally become more complex than those of lower forms. Asexual reproduction (i.e., reproduction not involving the union of gametes), however, occurs only in the invertebrates, in which it is common, occurring in animals as highly evolved as the sea squirts, which are closely related to the vertebrates. Temporary gonads are common among lower animals; in higher animals, however, gonads are permanent organs. Hermaphroditism, in which one individual contains functional reproductive organs of both sexes, is common among lower invertebrates; yet separate sexes occur in such primitive animals as sponges, and hermaphroditism occurs in animals more highly evolved—e.g., the lower fishes. Gonads located on or near the animal surface are common in the lowest invertebrates, but in higher animals they tend to be more deeply situated and often involve intricate duct systems. In echinoderms, which are among the highest invertebrates, the gonads hang directly into the sea and spill their gametes into the water. In protochordates, gametes are released into a stream of respiratory water that passes directly into the sea. Duct systems of the invertebrate flatworms (Platyhelminthes) are relatively complex, and those of specialized arthropods (e.g., insects, spiders, crabs) are more complex than those of any vertebrate. Copulatory organs occur in flatworms, but copulatory organs are not ubiquitous among vertebrates other than reptiles and mammals. The trend toward fewer eggs and increased parental care in higher animals may account for the relative lack of complexity in the reproductive systems of some advanced forms. Whereas trends toward increasing structural complexity have often been reversed during evolution, reproductive behaviour patterns in many phylogenetic (i.e., evolutionary) lines have become more complicated in order to enhance the opportunity for fertilization of eggs and maximum survival of offspring.

A direct relationship exists between behaviour and the functional state of gonads. Reproductive behaviour induced principally but not exclusively by organic substances called hormones promotes the union of sperm (spermatozoa) and eggs, as well as any parental care accorded the young. There are a number of reasons why behaviour must be synchronized with gonadal activity. Chief among these are the following:

Individuals of a species must congregate at the time the gonads contain mature gametes. This often entails migration, and some members of all major vertebrate groups migrate long distances to gather at spawning grounds or rookeries.

Individuals with gametes ready to be shed must recognize members of the opposite sex. Recognition is sometimes by external appearance or by chemical substances (pheromones), but sex-linked behaviour is often the only signal.

Geographical territories frequently must be established and aggressively defended.

The building of nests, however simple, is essential reproductive behaviour in many species.

When fertilization of aquatic forms is external, sperm and eggs must be discharged at approximately the same time into the water, since gametes may be quickly dispersed by currents. Courtship, often involving highly intricate behaviour patterns, serves to release the gametes of both mating individuals simultaneously.

When fertilization is internal, willingness of the female to mate is often essential. Female mammals not in a state of willingness to mate not only will not mate but may injure or even kill an aggressive male. The unwillingness of a female mammal to mate when mature eggs are not present prevents loss of sperm needed to preserve the species. Parental care of fertilized eggs by one parent or the other has evolved in many species. Parental behaviour includes fanning the water or air around the eggs, thereby maintaining appropriate temperature and oxygen levels; secretion of oxygen from a parent's gills; transport of eggs on or in the parental body (including the mouth of some male parents); and brooding, or incubation, of eggs.

Some species extend parental care into the postnatal period, feeding and protecting the offspring. Such behaviour patterns are adaptations for survival and thus are essential; all are induced by the nervous and endocrine systems and are typically cyclical, because gonadal activity is cyclical.

MALE REPRODUCTIVE ORGANS

The reproductive system in most males consists of a pair of *testes, sperm-conducting ducts,* and certain *accessory tissues.* In all vertebrates the testes perform two essential functions necessary for reproduction. They act both as an exocrine organ and an endocrine gland, since they produce sperms and also secrete the male hormone. The testes of most vertebrates are located in the upper part of the abdominal cavity lateral to or just anterior to the kidneys. A double fold of the peritoneal mesentery, the *mesorchium,* usually suspends then from the dorsal wall. In the majority of mammals, however, the testes do not remain in this position within the abdominal cavity, but descend posteriorly into the *scrotal sacs.*

Each testis is covered by a this layer of peritoneal epithelium under which lies a tough connective tissue capsule, the *tunica albuginea.* In some animals septa extend radially to the capsule, thus partitioning the gland into smaller areas; in other cases the testis is lobular or segmented. *Seminiferous tubules,* which produce the sex cells, and *interstitial tissue,* which secretes *androgen* (the general term of the male hormone), compose the testis regardless of its shape, size, or location.

The tubules may be short (ampulla) or long and tortuous, but in either case their walls usually consist of an outer connective tissue capsule, a thin basement membrane, and a lining of stratified epithelium. *Spermatogenesis* occurs in these seminiferous tubules. By a series of rapid cell divisions the spermatogonia, undifferentiated sex cells located beneath the basement membrane, transform into mature spermatozoa and pass into the lumen of the tubules. A cross section through the tubules reveals many stages of differentiating sperm, usually in more or less distinct layers, located between the basement membrane and the cavity of the tubule. It is possible by repeated cell division of a small number of spermatogonial cells to produce several million sperms each reproductive cycle.

Interspersed among the spermatogonia of the seminiferous tubule is a second type of cell, the sustentacular or *Sertoli cell.* These cells, wedged between the many layers of cells undergoing division, extend from the basement membrane to the lumen. They help provide structure to the tubule and function as nurse cells for the sperms. The mature sperm has no way or manufacturing its food and must depend upon some external source for its nutritive materials. The heads of several mature sperms embed in membrane folds of the *Sertoli cells,* where they apparently receive nourishment and remain until released into the genital tract of the male.

In general, most male vertebrates have mature sperms only during breeding periods, whereas during periods of sexual quiescence the tubules undergo changes in which large amounts of lipid material accumulate. There are exceptions to this general statement, however, since some vertebrates have mature sperms the year round. Amphibia and certain male mammals, such as man, remain spermatogenically active. Not all spermatogonia within a tubule mature at one time, since some must remain as reserve cells for future sperms. In some anuran testes, for example, one lobe of the testis may contain active sperms, whereas in an adjacent lobe, all the spermatogonia remain in a quiescent state. In man, one area of seminiferous tubules may be actively producing mature sperms whereas another area may be inactive.

The seminiferous tubules usually anastomose (join together) at their distal ends and empty into straight tubules. These tubules empty into a network of sinuses of channels known as the *rete tubules.* In man, the rete tubules converge to form 12 to 14 *efferent ductules,* which transport the sperms from the testis to the sperms duct. The efferent ductules are located in the mesorchium of lower animals. The ductules appear as this threads leading from the testis to the anterior end of the sperm duct and may be seen even in preserved animal if the mesentery is held up to the light.

The *interstitial cells,* or *Leydig cells,* of the testes produce *andorgen,* the masculinizing hormone. These cells, located between the seminiferous tubules, are obvious in the mammal but may exist in less typical forms in other vertebrates. Although the testes are the main androgen-secreting organs, other structures contain the necessary enzymatic machinery for producing the hormone. For example, the ovary, adrenal cortex, and probably the placenta also are capable of synthesizing the male steroid androgen. *Testosterone* and *androstenedione* are the main androgens secreted by the testes. These compounds contain 19 carbon atoms and possess methyl groups at C-10 and C-13.

FEMALE REPRODUCTIVE ORGANS

The vertebrate *ovary* also functions both as an exocrine and as an endocrine organ. It produces ova and the steroid sex hormones *estrogen* and *progesterone,* which regulate the reproductive trace, secondary sex characteristics, and mating behaviour of the female. The ovary is suspended from the dorsal body wall just behind the kidneys by a section of the peritoneum called the *mesovarium* and may be classified morphologically into either a solid or saccular type. Most vertebrates have a pair of ovaries, but a single structure is found in many fishes, probably representing a fusing of the two original ovaries since it has a typical vertebrate structure. In some animals, birds for example, only one side of the reproductive tract develops and the ovary on the other side remains inactive or degenerates.

The external shape of the ovary varies according to the species and to the reproductive state of the individual. In the frog, during the breeding months the multilobed organ fills the body cavity, actually stretching the body wall. After ovulation, the ovary involutes the looks like a ruffled piece of mesentery. The ovary of birds is an enormous structure because of both the accumulation of yolk in individual eggs and of the fact that the female bird ovulates one egg per day for several months.

The saccular type of ovary, represented by the frog consists primarily of an outer area where the eggs are located and a lymph-filled central space, connecting with the cavities of other lobes. The solid ovary can be divided into two areas: an outer region, the *cortex,* containing many ova in various stages of development, and an inner *medulla,* consisting of irregularly arranged connective tissue, blood, and lymph vessels. *Interstitial cells,* similar to those found in the male, are embedded in the framework or *stroma* of the ovary, and are prominent in some species. A dense connective tissue layer, the *tunica albuginea,* covers the cortex. External to the runic, the peritoneal epithelium surrounds the ovary. This epithelium was unfortunately named the germinal epithelium by the early biologists who believed that these cells were the source of the primordial germ cells, the future eggs and sperms.

Regardless of the morphology of the ovary (saccular or solid) possessed by the vertebrate, the germ cells take up their position within the cortical region where clusters of epithelial cells surround each ovum and form a *follicle.* The follicular cells are believed to arise from the germinal epithelium. In electron micrographs protoplasmic connections extend from the follicle cells to the surface of the egg and evidently aid in providing nutrition for the growing oocyte. For this reason, they are often referred to as "*nurse cells.*" These follicular cells (along possibly with the *theca interna cells*) synthesize the hormone *estrogens.*

A young follicle, one whose wall is only one cell thick, is initially located in the cortex of the ovary embedded in the connective tissue stroma. The stroma contributes additional connective tissue cells to the follicle in the form of sheaths around the follicular epithelium. These sheaths differentiate into an external fibrous layer, the *theca externa,* and an inner vascular layer, the *theca interna.* As the number of connective tissue sheaths increase, the follicles move away from the periphery of the ovary into the deeper areas of the cortex.

When the egg and follicle mature, surface of the ovary thins because of the increased internal pressure. The wall becomes necrotic, ruptures, and releases the egg into the body cavity. This process of freeing the ovum from the ovary is called *ovulation* and is under the control of pituitary, hypothalamic, and ovarian hormones.

The number of eggs released from the ovary at any one time is determined genetically for each species. In those vertebrates that lay their eggs and then desert them, species survival depends on the production of large numbers of eggs, since the mortality is high. Certain fishes spawn several million eggs each season. Frogs shed as many as 2000 to 3000 eggs at one time.

MALE REPRODUCTIVE ORGANS IN MAMMALS

The male reproductive system consists of two testes contained in the scrotum, accessary organs and the penis, which is sometimes referred to as the *external genitalia.* The scrotum provides the favourable environment of a lower temperature for the production of spermatozoa. The remaining structures assist the spermatozoa to reach their ultimate goal, the ovum of the female.

Testes

Each testis is smooth and somewhat oval shaped structure about the size of a walnut. The wall of each testis consists largely of connective tissue which is continuous with wall like partitions or *trabeculae* dividing the interior of the testis into approximately 250 small compartments called *lobules.* In all domestic animals except the horse, these trabeculae unite near the centre of the glans to form a fibrous cord, the *mediastinum testes.* Two to three seminiferous tubules lie within each lobule. In man a single tubule when straightened out measures 21 to inches in length. It has been estimated that their combined length in the human being is 750 ft. of more.

The seminiferous tubules are lined by a thin basement membrane, internal to which are the *spermatogonia,* which divide to form *spermatocytes.* These in turn give rise to more medially situated smaller cells with deeply stained nuclei, the *spermatids,* and finally to *spermatozoa.* The spermatocytes undergo a reduction division by which their chromosome number is reduced by half. In man, no spermatogenesis occurs before puberty, but subsequently the process takes place continuously until it stops in elderly people.

Between the spermatogonia are the elongated supporting cells or *Sertoli cells,* also called *sustentacular cells* or *nurse cells.* The *Sertoli cells* are rich in glycogen and project into the lumen of the tubule; the spermatids and sperms are attached to the inner margin of these cells and may be nourished by them.

Epididymis

It is also known as the main store house of sperms. The sperms in the seminiferous tubules, in spite of their fully formed tails, are non-motile and are believed to be pushed passively onwards by the pressure exerted by the steadily formed fresh cells into the straight tubules and *ductuli (vasa) efferentia.* The latter channels are lined by a ciliated epithelium and a thin muscle coat, which propel the sperms and the thin fluid in which they are suspended into the epididymis. The *epididymis* is a very long convoluted tube and is lined by a muscle coat and a secretory columnar epithelium. While the sperms are within it become motile, mature in other ways and develop a greater fertilizing power.

The fertilizing power of sperms from the distal end of the epididymis is twice as high as that of sperms from the proximal end. It seems, therefor, that the secretion of the epididymis normally exerts a 'spermatotrophic' action.

Vans deferens

Also known as *ductus deferens.* It is lined by a powerful muscle coat and a ejaculation, it propels the spermatozoa from the epididymis to the ejaculatory duct in the *prostatic urethra.*

The scrotum

The scrotum is a cutaneous sac having thin, pliable, and relatively hairless skin. A layer of fibro elastic tissue mixed with smooth muscle fiber, called the *tunica dartos*, is immediately deep to the skin, and in cold weather the muscle fibers of the dartos contract and help in holding the testes against the abdominal wall. In contrast, when the considerable amount of heat is generated within the body then the muscle fibers loosen so that the testes hang down away from the body. In this way a fairly constant testicular temperature is maintained.

The scrotum actually represents a fusion of two *scrotal pouches* which develop during embryonic life. The two halves are separated internally by a partition, the *septum scroti.* Externally the line of fusion is represented by a scarlike *raphe. The* cavity of the scrotum connects the abdominal cavity by means of *inguinal canal.* The canal remains open in those species in which the testes periodically withdrawn into the body cavity. It closes in most forms in which the testes lie permanently in the scrotum. Sometimes, a loop of intestine may pass through the canal into the scrotum and be caught there by the contradiction of the abdominal muscle which cuts off its blood supply producing an *inguinal hernia*. Unless an operation is quickly performed the tissue of that part of intestine dies.

The testes descend into the scrotum by birth or shortly thereafter. Whenever a testis fails to descend down into the scrotum it is called *cryptorchid testis* and the phenomenon is called the *cryptochidism.* An animal in which the testes descend into the inguinal canal but not into the scrotum is called a *high flanker.* A cryptorchid with both testes retained in the abdominal cavity is very likely to be sterile, since spermatogenesis does not occur normally unless the testis is cooler than body temperature a condition provided by the scrotum.

Generally a temperature 3° C lower than that of the body is said to be a optimal temperature for the development and maintenance of sperms.

Sl. No.	*Intra abdominal Testes*	*Periodic withdrawal into the body cavity*	*Permanently in scrotum*
1.	Monotremata	Some Insectivores	Metatheria
2.	Several Insectivores	Tubulidentata	Some Insectivores
3.	Most Edentata	Chiroptera	Some edentata
4.	Proboscidea	Most Rodentia	Carnivores
5.	Hyracoidea		Pinnepedia
6.	Sirenia		Artiodactyla
7.	Cetacea		Perissodactyla
			Primates

Seminal vesicles or vesiculae seminalis

Each vesicle is a muscular convoluted tube lined by an epithelium which secretes an alkaline yellow viscid fluid which forms much of the volume of the ejaculated semen. The secretion of the seminal vesicle contains glucose and fructose which are required by the sperms for their development.

Prostate gland

It is found around the urethera at a point where the ejaculatory ducts enter. If consists of many follicle like spaces leading into ducts. The epithelium of the follicles secretes the *prostatic fluid,* which is thin and opalescent and gives the semen a characteristic odour. The gland discharges its secretion directly into the urethra. In increases the motility of the sperms and neutralizes the acid condition of the urethra and the female reproductive tract. The prostatic fluid in man is milky in nature. It is rich in calcium, citrate, fibrinolysin and *acid phosphatase* and contributes 20 to 30% of the total volume of the semen.

The position of the gland is unfortunate because in older males it has a tendency to enlarge and obstructs the flow of urine. Surgical removal of the gland if often used as a remedy for this condition.

Bulbo urethral (Cowper's) glands

A very small pair of glands, *Cowper's glands,* open in the urethra just below the prostate gland. The mucoid secretion has a lubricating and acid neutralizing function. The glands are not found in the dog.

The Penis

The male organ of copulation, the *penis,* may be divided into three general areas,; the *glans* is the enlarged tip of penis which is normally covered by a fold of skin, the *prepuce;* the *main body;* and the *two crura* or *roots,* which attach to the ischial arch of the penis.Histologically penis shows *erectile tissue,* the *corpora cavernosa* consisting of blood sinusoids separated by sheets of connective tissue called *septa.* The seminal fluid is ejaculated into the vagina of the female by the penis.

Erection of penis

The penis, which is the resting state is small, flabby and covered with wrinkled skin, becomes thickened, elongated and rigid and thus well adapted for introduction into the vagina. The erection is associated with sexual stimulation. During sexual excitement the sponge like space of the erectile tissue become engorged with blood. The arteries dilate resulting in the collection of blood within the these spaces. As the tissues becomes distended they compress the veins (*vasoconstriction*) and so inhibit the flow of blood out of the tissues. The increased blood volume enlarges the penis (from 9.5 to 17 cm) and makes it turgid, hard and erect. In some mammals, notably the carnivores, such as dogs and seals, the penis is further strengthened by a special bone.

The caudal part of the glans of the penis of the dog, the *bulbus glandis,* usually become engorged with blood after the penis enter the vagina of the female. This arrangement makes separation difficult until erection has subsided.

Ejaculation

In sexual intercourse stimulation of sensory nerve endings in the penis as a result of friction against the lubricated walls of the vagina culminates in a short period of heightened paroxysmal sexual simultaneously by *ejaculation,* the forceful coming out of semen from the penis. The process shows involuntary muscular activities like the contraction of the scrotum and raising the testes close to the body resulting the propulsion of sperms from the epididymis through the ducts into the urethra. The heart rate increases and breathing becomes deeper.

Semen

The volume of semen is 3.5 ml per ejaculation, which contains on an average 350,000,000 sperms. The semen consists of the products of the seminiferous tubules, the seminal tract and the related glands. Semen is slightly alkaline with a pH varies from 3-4. Sperms are not optimally active until pH is about 6.5. Human semen is liquid when ejaculated, but soon coagulates in vitro; after 15 minutes it undergoes secondary liquefaction which is due to the specific enzyme *fibrolysin* (*plasmin*), present in prostatic fluid. At 37° C, 2 ml of prostatic fluid can liquify 100 to 1000 ml of clotted human plasma in 18 hours.

Fertility

Male fertility is reduced clinically when (a) the total sperm count falls below 60 million, (b) more than 20% of the sperms have abnormal heads, or (c) the motility of the sperms in a fresh specimen of semen is impaired. Male infertility could be due to failure of sperm production in the testes, or to failure of the accessory glands to secrete a sufficiently nutritive medium in which the ejaculated sperms could thrive and reach the ovum.

Viability of the sperms

Normal sperms are motile and can move at a rate of 1-3 cm per minute. The survival time of sperms introduced into the vagina, though a matter of great importance, has not yet been determined with certainty, it may be a day or two. It is said that sperms become nonmotile in the vagina in less than 60 minutes after coitus, but remain motile in the cervix and body of the uterus for 25 to 40 hours.

FEMALE REPRODUCTIVE ORGANS

The female reproductive system consists of a pair of ovaries that produce egg cells or ova and fallopian tubes where fertilisation occurs and which carry the fertilised ovum to the uterus. Growth of the foetus takes place here. The cervix separates the uterus from the vagina or birth canal, where the sperm are deposited. Anatomically the female reproductive apparatus is even simpler than males. It has a pair of ovaries, a pair of oviducts (Fallopian tubes), the uterus and the vagina.

The ovaries

The ovaries are the primary organs of reproduction in the female and may be considered to be both *endocrine* and the *cytogenic* (cell producing) in nature. The mammalian ovaries are solid structures lying on the sides of vertebral column behind the kidney in the pelvic cavity enclosed in a continuation of the peritoneal epithelium which lines the coelom. In human being each mature ovary is about 3.5 x 2 cm length and weights 4 gm. Over the ovaries this layer of cells is known as the *germinal epithelium* having cuboidal epithelial cells. The epithelial layer penetrates deeply as columns of cells into the underlying connective tissue. From these columns, cells multiply and grow down into the substance of the ovary; some cells enlarge and differentiate to become *ova,* and become surrounded by a single layer of cells forming the *membrane granulosa;* each ovum plus its surrounding membrane granulosa, is called a *priomordial follicle.* It is asserted that the ovary of a new born baby girl may contain 25,000-500,000 ova. Before puberty many of the follicles enlarge because of proliferation of the surrounding granulosa cells; but the whole structure subsequently degenerates, dies, and is absorbed by phagocytic cells. The process called*follicular atresia.* At puberty the ovary contains 100,000-200,000, immature primordial follicles; of these only 300-400 will mature and give rise to the ova which are discharged from the ovary monthly during thirty or more years of active reproduction life; there has thus been a vast over production of ova in relation to adult needs. Histologically the ovary shows an outer *cortex* and inner *medulla.* In mature ovary the cortex contains follicles and corpora lutea in various stages of differentiation and disintegration. The medulla contains only the large blood vessels of the organ.

Between the ages of about 11 and 16 years, the ovaries become active and begin to produce mature eggs. The beginning of the period of life is called *puberty.* At puberty full follicular growth sets in till the mature ovum attains the size of 10 to 12 mm. The ovary enlarges and becomes surrounded by a thick tough membrane, the *zona pellucida.* The granulosa proliferates to form many layers of small cells. Fluid then appears in these cells, first as droplets, which coalesce to form a cavity containing the *liquor folliculi.* The liquid separates the granulosa into two layers:

(1) A thin layer which lines the inner surface of the follicle.

(2) Cells which surround the ovum and form a mass (*discus proligerus*) at one pole, thus attaching the ovum to the margin of the follicle. After the accumulation of liquor the follicle is termed *vesicular, follicle* or *Graafian follicle.*

In human being during each monthly cycle, perhaps 20 or more primordial follicles develop *antra* and begin the vesicular phase of follicular growth, but ordinarily one of the follicles outgrows all the others and reaches a size of approximately 1 to 1.5 cm. This follicle expels its ovum during ovulation.

Ovulation

When the *Graafian follicle* is ripe it is about the size of a pea and projects from the surface oft he ovary. This region of the ovary thins out and finally splits, allowing the follicular contents to be extruded into the body cavity, a process knows as *ovulation.* At this stage the ovum is spherical mass of protoplasm about one-fifth of a millimeter in diameter, with a central nucleus and some of the cells from the follicle still adhering to the outside. Ovulation usually occurs between the thirteen to seventeenth day after the first day of the menstrual period in women but may occur earlier or later. The tiny ovum is carried out with the fluid and in subsequently drawn into the funnel like terminal end of the Fallopian tube. Occasionally, the eggs gets 'lost' in the coelom and never reaches the oviduct. If, however, it should be fertilized while outside the confines to the oviduct, which sometimes happens, the embryo may become attached to any convenient organ like liver and develops for some time. It may cause severe internal haemorrhages that may be fatal if immediate care is not given. In *monotocous animals* (horse and cow) at each heat period one follicle usually develops more rapidly than others, so that when it ruptures one ovum is released. In *polytocous animals* (carnivores) at each heat period several follicles develop and rupture during ovulation.

The time of ovulation in women can be simply determined by recording the body temperature before getting up in the morning. In the pre-ovulatory phase the temperature is 36.6-36.8° C and after ovulation the waking temperature increases by 0.3-0.5°C. The timing of ovulation is of value when it is desired to promote or avoid conception.

Formation of corpus luteum

Just after the discharge of the ovum, the blood is filled in the ruptured follicle which then clots there. The blood filled cavity is called *corpus haemorrhagicum.* Then after a yellow coloured *corpus luteum* appears replacing the clotted blood. The life span of corpus luteum is conditioned upon the establishment of pregnancy. In non-gravid cycles, corpus luteum (*corpus luteum spurium*) lasts only about 14 days. If pregnancy occurs the corpus luteum (*corpus luteum verum* or *corpus luteum of pregnancy*) continues to grow for several months, attains a large size and begins to degenerates at about the sixth month. Finally the corpus luteum degenerates and becomes while coloured *corpus albicans.*

Fallopian tube

In contrast to the male reproductive tract the female tract does not consist of a series of connected ducts. Instead, the egg is released into the abdominal cavity at the time of ovulation, whence it passes into one of the two tubes, the *Fallopian tubes* or *oviducts.* When the eggs has been discharged into the body cavity, it is normally directed into the funnel of oviduct by currents set up by the beating of the cilia lining the oviducts. The activity of the tube is influenced by hormones; its motility is increased by *oestradiol* and depressed by *progesterone.*

Uterus

The two oviducts empty directly into the upper corners of a pear shaped organ, the *uterus* or *womb,* which houses the developing embryo until the time of birth. This organ lies in the lower part of the abdominal cavity just behind the urinary bladder. It is about the size of a clenched fist, and has thick muscular walls and a mucous lining richly supplied with blood vessels.

Vagina

The vagina is that portion of the birth canal that is located, within the pelvis between the uterus cranially and the vulva caudally. The vagina is larger than any other tube in the reproductive tract; it serves both as receptacle for the sperms and so must accommodate the penis, and as the birth canal when pre-natal development is complete. The uterus terminates in a muscular ring, the *cervix,* which projects a short distance into the vagina.

Vulva

The *vulva* is the external portion of the genitalia of the female that extends from the vagina to the exterior. At the lowest extremity of the torso, in front, is a mass of fatty tissue forming a slight elevation known as the *mons veneris.* From this point, two folds of skin covered with hair, the *labia majora,* extend back and down, to enclose the opening of the vagina and grow together into a single tissue behind it. Inside the labia majora are two other folds of skin, free of hair, the *labia minora.* At the junction of these two in front is a sensitive, erectile organ, about the size and shape of a small pea. This is the *clitoris.* Which has the same embryonic origin as the *penis* in the male. Like the penis, it contains spongy tissue which becomes engorged with blood during sexual excitement and nerve endings that responds to erotic stimulation. Before the first sexual intercourse, the vaginal opening its usually more or less occluded by a thing membrane, called the *hymen.* Occasionally the hymen may be complete enough to interfere with copulation.

5

Animal Breeding and Artificial Insemination

The science of Animal breeding is defined as the application of the principles of Genetics and biometry to improve the efficiency of production in farm animals. These principles were applied to change animal populations thousands of years before the sciences of genetics and biometry were formally established. The practice of animal breeding dates back to the Neolithic period, when people attempted to domesticate wild species such as reindeer, goats, hogs and dogs. Domestication was performed through controlled mating and reproduction of captive animals which were selected and mated based on their behaviour and temperament. Judging from cave paintings that have survived, selection was also applied to some qualitative traits such as coat colour and the absence or presence of horns.

Without written records, there is no certain knowledge of the evolution of animal breeding practices, but written documents dating back more than 4000 years indicate that humans appreciated the significance of family resemblance in mating systems, recognized the dangers of intense inbreeding, and used castration to prevent the reproduction of undesirable males. Progress in the performance of domesticated animals through these selection practices was very slow; improvements were mainly due to animals adapting better to their environments.

Robert Bakewell, an English animal breeder of the 18th century, is considered the founder of systematized animal breeding. He was the first to emphasize the importance of accurate breeding records, introduced the concept of progeny testing to evaluate the genetic potentials of young sires, and applied inbreeding to stabilize desired qualitative traits. He also promoted concepts such as "like begets like," "prepotency is associated with inbreeding" and "breed the best to the best." Bakewell and his contemporaries in Europe pioneered the development of diverse breeds of Beef cattle, Dairy cattle, Sheep. Colour, conformation, geographical origin and some production characteristics were the main factors that differentiated these breeds. Wide geographical redistribution of animal populations was also an important factor in the formation of new breeds, as invading armies, migrating people and traders transported livestock to new lands. Animal breeding as a modern science belongs to the 20th century. Although numerous geneticists and biometricians have made significant contributions to the development of this science, J.L. Lush of Iowa State University is considered the father of the modern science of animal breeding. Lush and his students developed major scientific procedures applicable to the genetic improvement of farm animals.

SYSTEMS OF BREEDING

The ultimate aim of the breeder is to evolve outstanding and improved type of animals which can render better service to man. Selection and system of breeding constitute the only tools available to the breeder for improvement of animals. Since new genes can not be created though they can be recombined into more desirable groupings.

Systems of breeding has been broadly divided as under:

1. In breeding -breeding of the related animals.
2. Out breeding -breeding of the unrelated animals.

Inbreeding

Inbreeding is a mating system in which individuals mated are more closely related than the average of the population from which they come. It means the mating of males and females which are related. Animals deemed to be related only when they have one or more ancestors in common on the first 4-6 generations of their pedegree. The intensity of inbreeding depends upon the degree of relationship. Close inbreeding denotes mating of closely related individuals like dam to son (mother x son) or sire to daughter (father x daughter) or full brothers to full sisters.

In breeding makes more pairs of genes in the population homozygous. Wherever there is inbreeding. there will be one or more common ancestors from which, part of the gene samples (gametes) have arisen.

Close breeding

This means the mating of full sister to full brother or sire to his daughter or dam to her son. These types of matings should be only when both parents are outstanding individuals, and then only at the increased risk of bringing undesirable recessive genes into homozygous form in the progeny.

Advantages:

1. Undesirable recessive genes may be discovered and eliminated by further testing in this line.
2. The progeny are more uniform than out-bred progeny.

Disadvantage:

1. The undesirable characteristics are intensified in the progeny if unfavourable gene segregation occurs.

2. It has been observed that the progeny become more susceptible to diseases.
3. Breeding problems and reproductive failure generally increase..
4. It is difficult to find out the stage of breeding at which it should be discontinued in order to avoid the bad effects of the system.

Line breeding

This means the mating of animals of wider degrees of relationship than those selected for close breeding. It promotes uniformity in the character. Homozygosity is not reached so quickly as in close breeding. Neither desirable nor harmful characters are developed so quickly. It is a slowed method for the fixation of hereditary outstanding bull or cow and the progeny is mentioned as being line-bred to certain ancestors.

Advantages:

1. Increased uniformity.
2. The dangers involved in close breeding can be reduced.

Disadvantages:

1. The chief danger in live breeding is that the breeder will select the animal for pedegree giving no consideration to real individual merit. This may in some cases result in a few generations which receive no benefits from selection.

Consequences of in-breeding

The effects of both close breeding and line breeding are similar. The only difference is in the degree of their intensity. They are more intense in close breeding and less so in line breeding.

These may be described as follows:

1. It increases homozygosity (like alleles) and decreases heterozygosity (dissimilar genes) and hence favours the development of genetic uniformity amongst the animals.
2. It is the best method of getting true strains from unknown stocks as it sorts out the characters in the homozygous condition and thus help in the selection of undesirable individuals and culling of the undesirable individuals. Strains which breed true are not obtained in animals. however, and in plants are obtained only when close inbreeding is accompanied by intense selection.
3. The outward effects of in-breeding may include the following

Effect on growth rate: With the exception of a few laboratory animals, some moderate decrease in growth rate and mature weight has been noted. The studies on rats on the other hand have shown that they withstand inbreeding (up to 100 generations over 40 years) without any apparent bad effect on growth rate. The reproductive rate of rats is very high, however, relative to farm stock.

Effect on reproductive performance: In almost all the cases a reduction in the reproductive efficiency has been noted. It may delay testicular development and puberty, reduce gametogenesis or increase the embryonic death rate.

Effect on vigour: It has been noted that the death rates among the in-bred groups of animals are higher than those in out-bred ones. They are also adversely affected by environmental conditions. They have higher percentage of Runs (animals which are undesirable due to various reasons).

Effect on production: Production traits usually show moderate decrease with increased inbreeding.

Appearance of abnormalities: Hereditary abnormalities or lethal factors are likely to appear more often in the inbred animals than in the out bred ones.

Of the two components of inbreeding *i.e.* close breeding and live breeding, it is safer to go for the later one. A few recommendations have also been made With regard to them.

Inbreeding is useful if we want to evolve a more uniform line in any animal species. A line is often desirable for scientific purpose when animals or plants for a number of lines can be mated to get a breed with many desirable qualities. However, that is the job of a scientist or precisely a geneticist. A fine line of animals may not be commercially profitable at all. A farmer is, therefore, advised not to conduct this experiment inbreeding unless he knows genetics.

Out breeding: Out breeding is the breeding of unrelated animals and this involves the following types of breeding.

1. Out crossing
2. Cross-breeding
3. Hybridization
4. Grading up

The genetic effects of out-breeding are opposite to those of inbreeding. Whereas inbreeding increases homozygosity, out breeding tends to make more pairs of genes heterozygous. Therefore, out-bred animals are less likely to breed true than are in-bred animals.

Out-crossing: Mating of unselected animals within the same pure breed is termed out- crossing. The animals mated have no common ancestors on either side of their pedegree up to 4 -6 generations and the off-spring of such a mating is known as the outcross. Traits which are largely additively genetic and which have a high heritability, respond quickly to out crossing. In a period of many years, the various breeds in many western countries registered progress mainly though a system of out-crossing in which the best available, but unrelated sires were continuously selected for use on the females in a herd or flock.

Advantages:

1. This methods is highly effective for characters that are largely under the control of genes with additive effects *i.e.* milk production, growth rate in beef cattle, etc.

2. It is an effective system for genetic improvement if carefully combined with selection.
3. It is the best method for most herds.

Species hybridization

Crossing of individuals from two species is called hybridization. This is the widest possible out breeding system. It is possible only between related species which might have descended from common parent stock somewhere back in the evolutionary process. In the course of time, evolutionary forces might have set them off into distinct species. A well known example for species hybridization is the mule which is a cross between the jack ass and the mare. The mule is valued for its ability to work hard under the most adverse conditions. The mules are generally sterile.

Similarly, European cattle and American bison when crossed produce sterile males and fertile females. By back-crossing the females to bison and cattle attempts are being made to form a new breed of cattle-the cattalo. Species crosses are possible only when two species are related closely enough. These crosses usually result in much heterosis (a phenomenon in which the crosses of unrelated individuals often result in progeny with increased vigour (much above their parents). However, the progeny of species crosses have very low or nil fertility.

Grading is a process by which a few purebred sires can rather quickly transform a non descriptive population into a group of pure breeds. The grading process does not create anything new but it may transfer the good qualities of an improved breed. For a grading up programme, it is advisable to use a breed that I has thrived well under local condition. Otherwise graded animals may not adapt themselves to the local environment.

Advantages:

1. Pure breeds can be obtained just after a few generation (after 7th to 8th generations).
2. The start can be made with a little money in comparison to the purchase of an entire herd of purebreds.
3. It helps to prove the potentialities of the sire and adds to its market value.
4. It is a good start for new breeders who can slowly change over to pure breed systems.

Limitations of grading:

1. Pure breeds are not always better than grade or country animals for the use to be made of them.
2. Pure breed stocks which give good results in one set of environmental conditions do not always give favourable results in some different environmental set up. The pure breed dairy cattle from temperate

zones often degenerate when used in tropical areas. Moreover, their off springs fail to show vigour and constitution for high reproduction. To make grading successful the pure breed must have the ability to perfonn under the environmental set up where their offspring is going to perform.

IMPROVEMENT OF DAIRY CATTLE IN INDIA

The dairy fanning business in many tropical and subtropical countries are characterized by large number of cattle and low yield of milk. For example, India possesses over 230 million heads of cattle and buffaloes, which comes to more than one-fifth of the total world population of cattle and buffaloes. The average annual production of an Indian cow is only 173 kg as against an average yield per milking cow of 3,710 kg in Denmark, 3,250 kg in Switzerland and 3280 kg in the U.S.A.

Twenty six breeds of cattle and seven breeds of buffaloes have been recognised at present in India. However, only a small proportion of the cattle and buffaloes belong to these purebred breeds. The vast majority (more than 75 per cent) are intermixed and do not belong to any specific breed. They are classed as nondescript.

Cattle remain the draught animals in many of these regions. As a result selection over the past many centuries has been to meet the requirements of draught and agriculture. Selection pressure was applied for better capacity and fitness. This has resulted in producing excellent draught animals. Unfortunately, milk production and draught capacity are traits which are genetically negatively correlated. When milk yield capacity goes up, draught capacity goes down and vice versa. As a result attempts to develop and improve some of the dual purpose breeds which are useful for both milk and work, could not progress very far. We could have only a compromise in dual purpose breeds with medium milk production and medium work capacity. If further genetic improvement of either of the two traits is attempted, a set back with respect to the other should be expected.

In India, at present 94.3 per cent of the milch cows yield less than one kg. milk per day. Only 0.5 per cent yield more than 2 kg per day. Buffaloes are better milk producers. About 18 -20 per cent of the milch buffaloes yield more than 2 kg per day and only 19 per cent yield less than one kg per day. The reasons for such a low level of production are listed below.

1. Acute shortage of feed and fodders.
2. Excess cattle numbers.
3. Poor genetic potential for milk production.
4. The smallness of land holdings and consequently small dairy units make them economically unviable. It also acts against introduction of advanced techniques.

5. Adverse climatic conditions of the tropics.
6. Poor grazing and environmental factors.
7. Inadequate marketing facilities for the products.

The diversity of the breeding stock and the variation available in economic traits of cattle and buffaloes in the country offer greater challenge and scope for their improvement for the animal breeder. At the same time, the task of improving the genetic make up of a large number of extremely diverse, non-descript low producers is a colossal one. Anyone system of breeding can not be applied uniformly to all the animals in all the areas.

COW BREEDING: DIFFERENT BREEDS FOR DIFFERENT NEEDS

Dairy cattle are breeds that have been developed to produce milk whereas beef cattle are bred and selected primarily for the production of meat. Dual-purpose breeds do exist and they have been selected for both meat and milk production. In the last 20 years the genetic make up of the UK dairy herd has changed and is still changing. Cows are larger, earlier maturing, are producing far more milk and are more efficient at converting feed into milk.

As it is the birth of the calf that causes milk production, it is critical for the farmer that the cow becomes pregnant again during lactation and produces the next calf one year after the previous one. The average length of productive life of a dairy cow is between four and five lactations, although this varies from one or two to over ten. Since the fifties artificial insemination (AI) has increasingly replaced natural service. Embryo transplants (ET) are also being developed. A surrogate mother may have difficulty delivering, since the foreign calf is often larger than nature intended. When this happens, a rope or chain is attached to the calf's forelegs, and, with a winch-like device the newborn calf is pulled from the womb. Unnecessary suffering for both cow and calf.

Regular Calving

A high annual milk production per cow is dependent on regular calving at about 12- month intervals. Poor reproductive efficiency leads to longer calving indices and a reduction in milk sales per cow per annum. It also leads to an increase in the number of cows culled for failing to conceive. On average, in the UK, only about 85 cows calve per annum for every 100 cows available for breeding. Over thirty-five per cent of all cows culled are disposed of because of reproductive disorders.

Obtaining Semen

Photographer Henry Heap visited the Milk Marketing Board stud bull operation in Mid Wales. He provided Turning Point magazine with a complete set of photographs and commentary of his visit. "A large, handsome, tan bull stood stock still, tethered by the ring tethered through his nose to what looked like a giant circular clothes line or some grotesque merry-go-round. He was kept so motionless by a rubber hood that covered his eyes, which had a similar effect to that of hooding a falcon. A worker commented that the bull would stay still all day if they left the blindfold on. However, this bull was not the source of the semen, but instead what is known as a 'teaser'. The stud bulls are led out by the nose one at a time to mount the motionless hooded bull. After they have mounted the teaser their penis is diverted into a leather vagina which is used to collect the semen. As soon as the process is complete, another bull is led out. The hood keeps the teaser bull stationery for each successive beast to mount. These stud bulls never even see females.

During embryo transfer an embryo can be flushed from the uterus of a genetically superior cow and implanted in a receiver cow of so-called 'poorer quality' for foetal development. Embryos are also taken from dead cows at the slaughterhouse. Hormone treatment is used to induce multiple ovulations and the cow is then inseminated. A few days later several embryos can be flushed from her.

For both embryo collection and implanting the cow receives a local anaesthetic injection in the tail. Embryos of the desired breed can then be used. ET maximises the number of offspring from selected breeding females and maximises profit. Do-it-yourself embryo transfer is now taking place in the UK. Courses are run to train in direct embryo transfer techniques using the quick-thaw method for frozen embryos. Although vets are not required to be present at the time of transfer, they must check each animal is suitable for implantation. Farmers genetically improve the herd by using techniques such as MOET (multiple ovulation embryo transfer) which involves inducing a genetically 'superior' cow to produce up to 20 embryos rather than one.

For ease of management, farmers prefer cows to calf at the same time. They, therefore, need to detect when cows come into oestrous. The oestrous detection rate (ODR) is the number of cows observed in oestrus during a 3 week period, in relation to the number of cows expected to be observed. The average ODR on farms is about 60 per cent. One alternative is to synchronise the oestrous period of several animals by injections of prostaglandin or administration of progesterone as an impregnated intraveginal coil implanted for 9-12 days.

GENETIC INNOVATIONS IN ANIMAL BREEDING

While herdsmen have practiced the art of animal breeding since the beginning of domestication of livestock, the science of animal breeding is just now entering its second century. From its roots in Mendelian inheritance to the developments in quantitative genetics our field has now grown to include modern molecular sciences. The practical applications of the field have been to genetically improve the animals that produce meat, milk and other products of economic value. Animal breeding as commerce has always "protected" its endeavors commercially by developing breed societies and using pedigrees to protect the intellectual property developed by the master breeders. The advent of molecular biology, sequencing of genomes and the development of cloning, have brought with them large additional investments. These developments have come with the expectation of increased rewards by investing companies and the reality that technical improvements must be protected in order to obtain sufficient return. The purpose of this paper is to discuss forms of intellectual property, in particular patents and patent law, and to suggest ways it might impact the field of animal breeding and genetics.

FORMS OF INTELLECTUAL PROPERTY

Intellectual property (IP) is the property resulting from the creative or "inventive" work of an individual or individuals. The protection of IP is described in a large body of law that includes copyrights, trademarks, trade secrets, and patents. Copyrights protect IP such as articles, books, web pages, computer software and music. Copyright protection applies when the work is in a publishable form. The ã symbol is not required, but recommended and enforcement of rights is made easier by federally registering the mark. In trademark law, the Ô claims common law trademark rights, while the ® indicates federal registration. Trademarks are used primarily to identify the source of the owners' goods, products or services. Examples include "Le Label Rouge", "American Berkshire Gold" and the gene marker *HAL*-1843 ä that is associated with the gene test for porcine stress syndrome. A trade secret represents IP that is not divulged by the owner and confers some competitive advantage. Trade secrets can last as long as the information is kept secret. An example of a trade secret might be specific information to make specialized synthetic lines or calculate breeding values. Trade secret law (in the US called the Uniform Trade Secret Acts) helps protect a technology from theft but the degree of protection is dependent on the type of technology and the means used to keep the information secret.

Patents represent the largest form of IP and their development is rooted in a large body of law both in the US and abroad. Such laws and their interpretation are subject to change, especially when new technology is introduced. The descriptions that follow are meant to be general in nature. Patents were developed to "promote the progress of science and useful arts, by securing for limited times, for authors and inventors, the exclusive rights to their respective writings and discoveries" (Article 1, Section 8, US Constitution) in return for making details of inventions available to the public. Patenting was developed in order to support the development of science and innovation. A patent, once granted, represents a legal monopoly granted by the respective country's government to an inventor (specific to that country), permitting the patent owner to prohibit anyone else from making, using, or selling this invention for a specific period of time (now generally 20 years). Infringement of a patent is a civil wrong and the owner may sue for economic damages or injunction to stop infringement. Patents are intended to protect but not withhold technical information. Patents are very different from published papers. Inventors, unlike coauthors, must take part in the inventive process, not just participate in the research. Patents are not peer-reviewed but instead are assessed by patent examiners to determine the application's patentability.

A patent application must disclose sufficient detail to enable "one of reasonable skill in the art" to duplicate the invention. The application must be unobvious, *i.e.*, not a simple, obvious extension of existing technology. Novelty is the next criteria and refers to something created that is new. Lastly, the invention must be useful. The scope of patent rights is determined by the claims. Some patents, like gene marker patents, may enjoy very wide coverage across species while others may be limited to single polymorphisms in one breed. Patents may be related to a process, a product produced by a process or dependent on another patent. Patents are open to interpretation by both examiners and lawyers. In the US, patents can be applied for up to one year after publication or public disclosure but this is not allowed in other countries. Patents in and of themselves do not ensure income. They must be promoted and protected and can be viewed as tradable assets and licensed or assigned to third parties.

Confusion often exists relative to "international" patents (see http://www.cambiaip.org), which actually do not exist. Under the Paris Convention Treaty, a group of countries agreed to work with each other to provide a unified examination process and to make filing in its member countries easier by forming the World Intellectual Property Office (WIPO). In unison they adopted the one-year period in which to file an application in one of the other countries without losing the benefit of the original filing date. Using this procedure avoids the citation against the applicant of any "art" that became known after the original filing date in the country of origin, but before the filing date in another country.

Under this Patent Cooperation Treaty (PCT), an applicant is allowed one year to file at the WIPO office and, by designating member countries, holds the legal rights and original filing date in those designated countries without having to file in each country or pay the national expense up front. Although there are fees associated with the PCT application, they are less than one would pay for the foreign national filings at a later date. To obtain a patent in these countries, the application does eventually need to be filed in the national patent offices, fees paid, translations done and the laws and regulations of each country office followed. The applicant may have additional time from the national filing date to request examination, depending on particular country requirements. Delaying examination may provide the time desired to continue development and/or commercialization, however, during this period, fees called "annuities" must generally be paid. Due to the expense involved, applications are generally filed only in countries where protection is really needed.

A BRIEF HISTORY OF PATENTS

The early protection of intellectual property rights can be traced back to Venice in the Middle Ages when master craftsmen in guilds prohibited competition from former apprentices for a period of 20 years (Waltersheid, 1994). Such laws had considerably different economic effects for the master craftsman, the apprentice and the general public. For livestock, early breed societies were developed and monitored pedigrees to protect the IP of the master breeders.

The first modern patent act is often thought of as The US Patent Act of 1790. There was similar legislation in France in 1791. Patents related to living matter are relatively new. One of the earliest patents for living matter was granted to L. Pasteur for a yeast strain but this was done under the belief that it was an inanimate object and not living (Lesser, 2002). The first specialized patent law applied to living organisms was that of the Plant Protection Act of 1930 in the US and provided what is commonly referred to as Plant Breeders Rights (PBR) to propagate new varieties by asexual methods. In 1961, a similar law was passed in France called the UPOV (International Union for the Protection of New Varieties of Plants - Union Internationale pour la Protection des Obtentions Végétales; http://www.upov.int). Protection in the US was expanded in 1970 with the Plant Variety Protection Act to include sexually reproduced plants. The UPOV was revised in Europe in 1991. Under these laws two principles, "breeder's rights" which allows breeders to use protected varieties without permission of the owner and "farmer's privilege" which allows farmers to collect seeds from their crops and use them, were developed (Lesser, 2002). For years many seed companies have attempted to halt this latter practice by asking farmers to sign contracts prohibiting it. Recent technology, like the "terminator" technology (Kaiser, 2000), biologically prohibits the practice. Santaniello *et al.* (2000) provides an excellent review on all aspects of patent protection for plants.

The dawning and blossoming of the molecular age presented real problems for the protection of IP related to living organisms other than plants. In 1975, a French company failed to patent a "dwarf, egg-laying chicken hen produced by a process" involving a sex-linked recessive gene (Bent, 1989). In 1980, the United States Supreme Court in a 5-4 decision (*Diamond v. Chakrabarty*) declared that "anything under the sun that is made by man" is patentable. This case concerned the patenting of genetically engineered bacteria that destroyed oil sludge. In 1987, the US Patent trade office issued a pronouncement of the patentability, in principle, of non-human multicellular organisms that were not naturally occurring (Bent, 1989). This was quickly followed in 1988 by the landmark patent on the so-called "Harvard mouse" which was engineered to be susceptible to cancer. In Europe, similar laws were passed allowing patenting of animals and animal-related inventions. From 1995 to 2001, a total of 45 animal patents were granted in the US (Lesser, 2002). While these changes in patent law had large economic implications, the social and moral ramifications were also enormous.

ANIMAL BREEDING AND GENETICS

The broad area of biotechnology encompasses many of the patent applications in our field and the US Supreme Court has established guidelines that apply to this technology (Nebel *et al.,* 2002). The Court made it clear in *Brenner v. Manson* that patent utility implied usefulness and not just "any invention not positively harmful to society." The Court expressed reservations regarding a monopoly on compounds with unknown functions, and that utility must extend beyond proving that the product is a result of scientific research (Nebel *et al.,* 2002).

Technologies in the field of animal breeding that may require (or qualify for) IP protection include, but are not limited to: statistical methods for genetic improvement, DNA markers for genetic improvement, transgenic and cloned animals and methods to produce them, new methods to measure traits, methods to identify animals, computer software and other written materials.

Patent coverage in not just confined to genetic markers. Lines of pigs or chickens have been patented and can be viewed as a specialized extension of early breed development or trademarking for protection of this IP. Other patents exist for methods involving cellular and animal manipulation and involve processes like stem cell development, transgenic production (*i.e*., US 6,271,436) and cloning (*i.e*., US 6,215,041 or US 6,258,998). Several advances related to mechanical or electronic devices have been made and include new A.I. or embryo transfer tools, advanced ultra-sound equipment, formulas and methods to measure backfat and other traits in livestock (*i.e*., see early patent US 4,359,055 and more recent US 5,717,142). The increasing need for traceability of animals and animal products has spawned a number of inventions including electronic ID tags and retinal scanning methods and devices.

Considerable discussion has ensued recently from a patent entitled "Method of Bovine Herd Management" granted in the U.S. in 1994 and later in Canada (Schaeffer, 2002). The invention is for the "test-day model" and includes the gathering, mathematical treatment, and the use of the modified data by dairy producers. The novelty and unobviousness of the patent has been seriously questioned. It was pointed out that the practices of gathering, manipulating and using data by dairy producers have existed for nearly 100 years. The patent therefore claims rights to a practice that has been public knowledge for a long time. The novel idea within the patent was the specific mathematical model and procedures that Everett and co-workers developed for the analysis of test day yields.

Everett was also not the first researcher to apply a model to test day records and as has been demonstrated, the model as described in the patent, is not necessarily the best model that could be applied (Schaeffer, 2002). This patent generated considerable discussion. It has been argued "what would the field of animal breeding be if the selection index or Henderson's BLUP had been patented?" Yet while quantitative geneticists see the thought of such protection as sacrilege, molecular scientists accept (but may not like) that in a similar way the foundation patent for PCR exists and royalties must be paid for its use.

OTHER LEGAL CONSIDERATIONS

In the US and some other countries, there is no clear exception to infringement for research use of patented inventions, even in non-profit or educational institutions. Researchers are required to obtain a license or permission from the patent owner to use a protected invention for research purposes. In other countries, there is either an exemption for educational or non-commercial research or research on improving the invention. Research using the invention, however, is not allowed. Researchers should also be concerned with possible claims for "contributory infringement," *i.e.*, assisting someone else to infringe on a patent. For example, due to a claim of possible contributory infringement, the US Genome Coordinators were asked to halt the practice of supplying primers for microsatellite analysis to other researchers. To resolve the issue the Coordinators began buying the primers from a company with a research license.

ETHICAL, SOCIAL AND ECONOMIC ISSUES

Many ethical and social issues have been raised related to patenting of animals and genes (Bent, 1989; Brody, 1989; Dresser, 1988; Evans, 2002). These include: 1) patenting of animals or genes will be destructive to nature and allows man to play "God"; 2) patenting will devalue animal life and hence human life; 3) patenting will increase animal suffering; 4) patenting will lead to a decline in genetic diversity of animals and threaten species; 5) patenting speeds the trend towards commercialization of academic research and 6) patenting will undermine conventional farming and lead to increased industrial farming systems. Early humans domesticated animals and master breeders and geneticists have transformed them into productive species. Was this playing "God" or interfering with nature? The use of transgenics for making specialized animal lines for disease research is certainly adapting nature but does it devalue life and is it unethical or immoral? These are value judgements that most in society have decided are acceptable (Brody, 1989). Certainly some lines have been (and need to be) drawn to delineate what is acceptable and unacceptable. For example, most people and governments have concluded that while cloning of animals is acceptable (at least for research purposes), cloning of humans, for whatever reason, is immoral, unethical and to be avoided.

Animal welfare and animal rights issues continue to be at the forefront of livestock production and biomedical research (Evans, 2002). Individuals who believe that animals have "rights" will likely be opposed to patenting any invention derived from animal research. The most commonly cited examples are those relating to transgenic animals (*i.e.*, early transgenic pigs) in which some animals had health problems. Production and patenting of specialized lines of rodents for biomedical research that have a tendency to develop specific diseases is also considered unethical by animal rights activists (Brody, 1989). If, however, individuals believe that animal rights are subordinate to those of humans, but that they deserve proper care and welfare then the issue of patenting is much less of a concern.

It has been suggested that through the use of gene markers and highly selective breeding, or through the use of transgenics and cloning, that genetic diversity will be greatly reduced. Certainly these methods have the potential to remove within-line variation but they are likely to increase between-line variability. However, it may be argued that the patent system in fact encourages diversity as it promotes and helps establish, via patent-related deposits of biological materials, a broader genetic diversity (Bent, 1989). Related to this issue is the issue of patenting of products from animals or plants from developing countries. Should companies be allowed to sample and alter wild stocks and to subsequently earn great sums of money by then selling them to both developed countries and back to the countries from where they were obtained? Some individuals say this so-called biopiracy is encouraged by patenting and some countries are moving to address these concerns (Evans, 2002). An extension of this is the issue of species integrity. The development of transgenic pigs for xenotransplantation and other species with human genes for human protein production threatens the genetic lines that determine species. While consideration of xenotransplantation has slowed due to fears of retroviruses and diseases like mad cow disease and AIDS, as Evans (2002) points out this dispute over species integrity "is not empirical" but it is tied up in the individual and societal views on nature and "the notion of awe and wonder." It is likely, no matter how repugnant this process is to some, there will be development of transgenic lines of animals for biomedical research (not food production) and applications that do encompass genes from other species.

In a perfect world, public support of research would be 100 per cent of required funding and all IP would be publicly available in the country of origin. Many US public institutions are now funded at about 40 per cent of total budgets and pressures to obtain outside funding are growing. Companies supplying funding expect to "own", through licensing agreements or otherwise, the IP that results. Protection of IP can provide research partners a greater basis for trust and exchange of ideas and help insure that public institutions focus research on areas of high relevance and payoff. Commercial relationships also aid in technology transfer, employment opportunities for students and may allow for research opportunities not available in the public sector. Fretz and MacKenzie (2002) have suggested that: 1) such activities serve the public good; 2) short term, low quality research should not be favoured in order to obtain funding; 3) managing of IP must be a balance of serving the institution, funding agency and public; 4) independence of the researcher and institution must be maintained; 5) the mission of the public institution is not altered and 6) conflicts of interest should be avoided. Institutions must be proactive in balancing funding, IP protection and commercialization.

Does patenting increase the likelihood of "industrial farming?" Economic and governmental issues certainly play a role in the size of the average farm and the level of commercialization of farming. It can be argued that if large companies have exclusive licensing arrangements for genetic tools then small breeders will be disadvantaged. Market pressures related to size of operation and efficiency of production are much larger influences on the industrialization of farming and livestock production than patenting.

Other issues regarding patenting that have an economic basis can be discussed. Langinier and Moschini (2002) provide an excellent review of the economics of patenting and the benefits and costs that are derived. While inventors would prefer broader claims, limited claims encourage competition and further innovation. Complex products often require building on other patents. Unfortunately, development does become blocked by other patents (especially at the leading edge) and this has negative effects on both the developer as well as the public. Licensing of patents exclusively to one company may benefit that company and segments of the public but might also limit further innovation and not be in all the public's best interest. Langinier and Moschini (2002) concluded "that continued efforts are required to improve the workings of the patent system" to improve the economic performance of the system.

HORSE BREEDING

Horse breeding refers to reproduction in horses, and particularly the human-directed process of selective breeding of animals, particularly purebred horses of a given breed. While feral and wild horses breed successfully without human assistance, planned matings can be used to produce specifically desired characteristics in domesticated horses. Furthermore, modern breeding management and technologies can increase the rate of conception, a healthy pregnancy, and successful foaling.

The male parent of a horse, a stallion, is commonly known as the sire and the female parent, the mare, is called the dam. Both are genetically important, as each parent provides half of the genetic makeup of the ensuing offspring, called a foal. (Contrary to popular misuse, the word "colt" refers to a young male horse only; "filly" is a young female.) Though many amateur horse owners may simply breed a family mare to a local stallion in order to produce a companion animal, most professional breeders use selective breeding to produce individuals of a given phenotype, or breed. Alternatively, a breeder could, using individuals of differing phenotypes, create a new breed with specific characteristics.

Some breeders consider the quality of the sire to be more important than the quality of the dam. However, other breeders maintain that the mare is the most important parent. Because stallions can produce far more offspring than mares, a single stallion can have a greater overall impact on a breed. However, the mare may have a greater influence on an individual foal because its physical characteristics influence the developing foal in the womb and the foal also learns habits from its dam when young. Foals may also learn the "language of intimidation and submission" from their dam, and this imprinting may affect the foal's status and rank within the herd. Many times, a mature horse will achieve status in a herd similar to that of its dam; the offspring of dominant mares become dominant themselves.

Estrus Cycle of the Mare

The estrous cycle controls when a mare is sexually receptive towards a stallion, and helps to physically prepare the mare for conception. It generally occurs during the spring and summer months, although some mares may be sexually receptive into the late fall, and is controlled by the photoperiod (length of the day), the cycle first triggered when the days begin to lengthen. The estrus cycle lasts about 19-22 days, with the average being 21 days. As the days shorten, the mare returns to a period when she is not sexually receptive, known as anestrus. Anestrus prevents the mare from conceiving in the winter months, as that would result in her foaling during the harshest part of the year, a time when it would be most difficult for the foal to survive.

This cycle contains 2 phases:

1. *Estrus, or Follicular, phase*: 5-7 days in length, when the mare is sexually recepetive to a stallion. Estrogen is secreted by the follicle. Ovulation occurs in the final 24-48 hours of estrus.
2. *Diestrus, or Luteal, phase*: 14-15 days in length, the mare is not sexually recepetive to the stallion. The corpus luteum secretes progesterone.

Although 16 per cent of mares have double ovulations (allowing them to twin), this does not affect the length of time of estrus or diestrus.

Effects on the Reproductive System During the Estrous Cycle

Changes in hormone levels can have great affects on the physical characteristics of the reproductive organs of the mare, thereby preparing, or preventing, her from conceiving.

- *Uterus*: Increased levels of estrogen cause edema within the uterus, making it feel heavier, and the uterus loses its tone. This edema decreases following ovulation, and the muscular tone increases. High levels of progesterone do not cause edema within the uterus. The uterus becomes flaccid during anestrus.
- *Cervix*: The cervix relaxes right before estrus occurs, and the secretions of the cervix increase. High progesterone levels (during diestrus) cause the cervix to close.
- *Vagina*: The portion of the vagina near the cervix becomes engorged with blood right before estrus. The vagina becomes relaxed and secretions increase.
- *Vulva*: Relaxes right before estrus begins. Becomes dry, and closes tightly, during diestrus.

Hormones Involved in the Estrus Cycle, During Foaling, and after Birth

The cycle is controlled by several hormones which regulate the estrous cycle, the mare's behaviour, and the reproductive system of the mare. The cycle begins when the increased day length causes the pineal gland to reduce the levels of melatonin, thereby allowing the hypothalamus to secrete GnRH.

- *GnRH (Gonadotropin releasing hormone):* Secreted by the hypothalamus, causes the pituitary to release of 2 gonadotrophins: LH and FSH.
- *LH (Luteinizing hormone):* Levels are highest 2 days following ovulation, then slowly decrease over 4-5 days, dipping to their lowest levels 5-16 days after ovulation. Stimulates the follicle to develop, which in turn causes an increase in estrogen. Unlike most mammals, the mare does not have an increase of LH right before ovulation.
- *FSH (Follicle-stimulating hormone):* Secreted by the pituitary, causes the ovarian follicle to develop. Levels of FSH rise slightly at the end of estrus, but have their highest peak about 10 days before the next ovulation. FSH is inhibited by inhibin, at the same time LH and estrogen levels rise, which prevents immature follicles from continuing their growth.
- *Estrogen:* Secreted by the developing follicle, it causes the pituitary gland to secrete more LH (therefore, these 2 hormones are in a positive feedback loop). Additionally, it causes behavioural changes in the mare, making her more recpetive towards the stallion, and causes physical changes in the cervix, uterus, and vagina to prepare

the mare for conception. Estrogen peaks 1-2 days before ovulation, and decreases within 2 days following ovulation.

- *Inhibin:* Secreted by the developed follicle right before ovulation, "turns off" FSH, which is no longer needed now that the follicle is larger.
- Progesterone: Prevents conception and decreases sexual receptibility of the mare to the stallion. Progesterone is therefore lowest during the estrus phase, and increases during diestrus. It decreases about 15 days after ovulation, when the corpus luteum begns to decrease in size.
- Prostaglandin: Secreted by the endrometrium 13-15 days following ovulation, causes luteolysis and prevents the corpus luteum from secreting progesterone
- eCG (equine chorionic gonadotropin - formerly called PMS/PMSG pregnant mare serum and pregnant mare serum gonadotropin): Chorionic gonadotropins secreted if the mare conceives. First secreted by the endometrial cups around the 36th day of gestation, peaking around day 60, and decreasing after about 120 days of gestation. Also help to stimulate the growth of the fetal gonads.
- *Prolactin*: Stimulates lactation
- *Oxytocin*: Stimulates the uterus to contract

While horses in the wild mate and foal in mid to late spring, many horses domestically bred for competitive purposes, especially horse racing and various futurities, are usually born as close to January first as possible, so as to be at an advantage in size and maturity when competing against other horses in the same age group. When an early foal is desired, barn managers will put the mare "under lights" by keeping the barn lights on in the winter to simulate a longer day, thus bringing the mare into estrus sooner than she would in nature. Mares signal estrus and ovulation by urination in the presence of a stallion, raising the tail and revealing the vulva. A stallion, approaching with a high head, will usually nicker, nip and nudge the mare, as well as sniff her urine to determine her readiness for mating.

Once fertilized, the oocyte (egg) remains in the oviduct for approximately 5.5 more days, and then descends into the uterus. The initial single cell combination is already dividing and by the time of entry into the uterus, the egg might have already reached the blastocyst stage. The gestation period lasts for about eleven months, or about 342 days (normal average range 320-370 days). During the early days of pregnancy, the embryo (up to about 35 days), or fetus (after 35-40 days), is mobile, moving about in the uterus until about day 16 when "fixation" occurs.

True implantation - invasion into the endometrium of any sort - does not occur until about day 35 of pregnancy with the formation of the endometrial cups, and true placentation (formation of the placenta) is not initiated until about day 40-45 and not completed until about 140 days of pregnancy. A heartbeat can be detected on day 21, and the fetus gender can be determined by day 70 of the gestation. Halfway through gestation the fetus is the size of between a rabbit and a beagle. The most dramatic fetus development occurs in the last 3 months of pregnancy when 60 per cent of fetal growth occurs.

Care of the Pregnant Mare

Domestic mares receive specific care and nutrition to ensure that they and their foals are healthy. Mares are given vaccinations against diseases such as the Rhinopneumonitis (EHV-1) virus (which can cause abortions) as well as vaccines for other conditions that may occur in a given region of the world. Mares can be used for riding or driving during most of their pregnancy, and it's healthy for them to have exercise.

But only moderate exercise, especially when they become heavy in foal. Exercise in excessively high temperatures has been suggested as being detrimental to pregnancy maintenance during the embryonic period - it should however be noted that ambient temperatures encountered during the research were in the region of 100 degrees F and the same results may not be encountered in regions with lower ambient temperatures.

Mares are generally fed more than other horses and are provided supplemental vitamins and minerals because their bodies require extra nutrition to form (and later to nurse) a foal. Many feeds designed for pregnant and lactating mares also have extra fat and protein. Overfeeding the pregnant mare should also be avoided, as excess weight may contribute to difficulties foaling or fetal/ foal related problems. A special foaling stall or shed that is large and clutter free provides the mare with a safe place to give birth. Most mares foal at night or early in the morning, and prefer to give birth alone when possible. Labour is rapid, often no more than 30 minutes, and from the time the feet of the foal appear to full delivery is often only about 15 to 20 minutes. Once the foal is born, the mare will lick the newborn foal to clean it and help blood circulation. In a very short time, the foal will attempt to stand and get milk from its mother. A foal should stand and nurse within the first hour of life.

To create a bond with her foal, the mare licks and nuzzles the foal, enabling her to distinguish hers from others. Some mares are aggressive when protecting their foals, and may attack other horses or unfamiliar humans that come near their newborns. After birth, a foal's navel is dipped in iodine to prevent infection, it is sometimes given an enema to help clear the meconium from its digestive tract, and the newborn is monitored to ensure that it stands and nurses without difficulty. While most horse births happen without complications, many owners have first aid supplies prepared and a veterinarian on call in case of a birthing emergency.

People who supervise foaling should also watch the mare to be sure that she passes the placenta in a timely fashion, and that it is complete with no fragments remaining in the uterus, where they could cause a serious infection. If the placenta is not removed from the stall after it is passed, a mare will often eat it, an instinct from the wild, where blood would attract predators. Foals develop rapidly, and within a few hours a wild foal can travel with the herd. In domestic breeding, the foal and dam are usually kept separate for a while, but within a few weeks are typically pastured with the other horses. A foal will begin to eat hay, grass and grain alongside the mare at about 4 weeks old; by 10-12 weeks the foal requires more nutrition than the mare's milk can supply. Foals are typically weaned at 4-8 months of age, although in the wild a foal may nurse for a year.

How Breeds Develop

Beyond the appearance and conformation of a specific type of horse, breeders aspire to improve physical performance abilities. This concept, known as matching "form to function," has led to the development of not only different breeds, but also families or bloodlines within breeds that are specialists for excelling at specific tasks. For example, the Arabian horse of the desert naturally developed speed and endurance to travel long distances and survive in a harsh environment, and domestication by humans added a trainable disposition to the animal's natural abilities. In the meantime, in northern Europe, the locally adapted heavy horse with a thick, warm coat was domesticated and put to work as a farm animal that could pull a plow or wagon.

This animal was later adapted through selective breeding to create a strong but ridable animal suitable for the heavily-armored knight in warfare. Then, centuries later, when people in Europe wanted faster horses than could be produced from local horses through simple selective breeding, they imported Arabians and other oriental horses to breed as an outcross to the heavier, local animals. This led to the development of breeds such as the Thoroughbred, a horse taller than the Arabian and faster over the distances of a few miles required of a European race horse or light cavalry horse. Another cross between oriental and European horses produced the Andalusian, a horse developed in Spain that was powerfully built, but extremely nimble and capable of the quick bursts of speed over short distances necessary for certain types of combat as well as for tasks such as bullfighting.

Later, the people who settled the Americas needed a hardy horse that was capable of working with cattle. Thus, Arabians and Thoroughbreds were crossed on Spanish horses, both domesticated animals descended from those brought over by the Conquistadors, and feral horses such as the Mustangs, descended from the Spanish horse, but adapted by natural selection to the ecology and climate of the west. These crosses ultimately produced new breeds such as the American quarter horse and the Criollo of Argentina. In modern times, these breeds themselves have since been selectively bred to further specialize at certain tasks.

One example of this is the American quarter horse. Once a general-purpose working ranch horse, different bloodlines now specialize in different events. For example, larger, heavier animals with a very steady attitude are bred to give competitors an advantage in events such as team roping, where a horse has to start and stop quickly, but also must calmly hold a full-grown steer at the end of a rope. On the other hand, for an event known as cutting, where the horse must separate a cow from a herd and prevent it from rejoining the group, the best horses are smaller, quick, alert, athletic and highly trainable. They must learn quickly, have conformation that allows quick stops and fast, low turns, and the best competitors have a certain amount of independent mental ability to anticipate and counter the movement of a cow, popularly known as "cow sense." Another example is the Thoroughbred. While most representatives of this breed are bred for horse racing, there are also specialized bloodlines suitable as show hunters or show jumpers.

The hunter must have a tall, smooth build that allows it to trot and canter smoothly and efficiently. Instead of speed, value is placed on appearance and upon giving the equestrian a comfortable ride, with natural jumping ability that shows bascule and good form. A show jumper, however, is bred less for overall form and more for power over tall fences, along with speed, scope, and agility. This favours a horse with a good galloping stride, powerful hindquarters that can change speed or direction easily, plus a good shoulder angle and length of neck. A jumper has a more powerful build than either the hunter or the racehorse. The history of horse breeding goes back millennia. Though the precise date is in dispute, humans could have domesticated the horse as far back as approximately 4500 BCE. However, evidence of planned breeding has a more blurry history.

One of the earliest people known to document the breedings of their horses were the Bedouin of the Middle East, the breeders of the Arabian horse. While it is difficult to determine how far back the Bedouin passed on pedigree information via an oral tradition, there were written pedigrees of Arabian horses by A.D. 1330. The Akhal-Teke of West-Central Asia is another breed with roots in ancient times that was also bred specifically for war and racing. The nomads of the Mongolian steppes bred horses for several thousand years as well. The types of horses bred varied with culture and with the times. The uses to which a horse was put also determined its qualities, including smooth amblers for riding, fast horses for carrying messengers, heavy horses for plowing and pulling heavy wagons, ponies for hauling cars of ore from mines, packhorses, carriage horses and many others. Medieval Europe bred large horses specifically for war, called destriers. These horses were the ancestors of the great heavy horses of today, and their size was preferred not simply because of the weight of the armor, but also because a large horse provided more power for the knight's lance. Weighing almost twice as much as a normal riding horse, the destrier was a powerful weapon in battle. On the other hand, during this same time, lighter horses were bred in northern Africa and the Middle East by Muslim warriors, who preferred a faster, more agile horse. The lighter horse suited the raids and battles of the Bedouins, allowing them to outmaneuver rather than overpower the enemy. When Muslim warriors and European knights collided in warfare, the heavy knights were frequently outmaneuvered. The Europeans, however, soon made up for the lack of speed of their native breeds by incorporating genetic traits from captured oriental horses such as the Arabian, Barb to their stables. This cross-breeding led both to a nimbler war horse, such as today's Percheron, but also to created a type of horse known as a Courser, a predecessor to the Thoroughbred, which was used as a message horse.

During the Renaissance, horses were bred not only for war, but for haute ecole riding, derived from the most athletic movements required of a war horse, and popular among the elite nobility of the time. Breeds such as the Lipizzan were developed from Spanish-bred horses for this purpose, and also became the preferred mounts of cavalry officers, who were derived mostly from the ranks of the nobility. It was during this time that gunpowder was developed, and so the light cavalry horse, a faster and quicker war horse, was bred for a "shoot and run" tactic rather than the close hand-to-hand fighting seen in the Middle Ages. After Charles II retook the British throne in 1660, horse racing, which had been banned by Cromwell, was revived. The Thoroughbred was developed 40 years later, bred to be the ultimate racehorse, through the lines of 3 foundation Arabian stallions.

The 17th and 18th centuries saw more of a need for fine carriage horses in Europe, bringing in the dawn of the warmblood. The warmblood breeds have been exceptionally good at adapting to changing times, and from their carriage horse beginnings they easily transitioned during the 1900s into a sport horse type. Today's warmblood breeds, although still used for competitive driving, are more often seen competing in the show jumping or dressage arenas. The Thoroughbred continues to dominate the horseracing world, although its lines have been more recently used to improve warmblood breeds and to develop sport horses. The predecessor of the American Quarter Horse was developed in the 1700s, mainly for quarter racing (racing ¼ of a mile). The breed was later adapted for work in the west, and "cow sense" was particularly bred for as their use for herding cattle increased.

However, because there was also a need for animals suitable for sprint racing, the modern Quarter Horse has two distinct types: the sleeker racing type and the stock horse type. The racing type most resembles the finer-boned ancestors of the first racing Quarter Horses, and the type is still used for ¼-

mile races. The stock horse type, used in western events, is bred for a shorter stride, docile temperament, and cow sense. The need for horses for heavy draft and carriage work continued until the industrial revolution and the advent of the automobile and the tractor. After this time, draft and carriage horse numbers dropped significantly, though light riding horses remained popular for recreational pursuits. Draft horses today are used on a few small farms, but today are seen mainly for pulling and plowing competitions rather than farm work. Heavy harness horses are now used as an outcross with lighter breeds, such as the Thoroughbred, to produce the modern warmblood breeds popular in Olympic and sport horse disciplines. Breeding a horse is an endeavor where the owner, particularly of the mare, will usually need to invest considerable time and money.

For this reason, a horse owner needs to consider several factors, including:

- Does the proposed breeding animal have valuable genetic qualities to pass on?
- Is the proposed breeding animal in good physical health, fertile, and able to withstand the vigours of reproduction?
- For what purpose will the foal be used?
- Is there a market for the foal in the event that the owner does not wish to keep the foal for its entire life?
- What is the anticipated economic benefit, if any, to the owner of the ensuing foal?
- What is the anticipated economic benefit, if any, to the owner(s) of the sire and dam or the foal?
- Does the owner of the mare have the expertise to properly manage the mare through gestation and parturition?
- Does the owner of the potential foal have the expertise to properly manage and train a young animal once it is born?

There are value judgements involved in considering whether an animal is suitable breeding stock, hotly debated by breeders. Additional personal beliefs may come into play when considering a suitable level of care for the mare and ensuing foal, the potential market or use for the foal, and other tangible and intangible benefits to the owner. If the breeding endeavor is intended to make a profit, there are additional market factors to consider, which may vary considerably from year to year, from breed to breed, and by region of the world. In many cases, the low end of the market is saturated with horses, and the law of supply and demand thus allows little or no profit to be made from breeding unregistered animals or animals of poor quality, even if registered.

The minimum cost of breeding for a mare owner includes the stud fee, and the cost of proper nutrition, management and veterinary care of the mare throughout gestation, parturition, and care of both mare and foal up to the time of weaning. Veterinary expenses may be higher if specialized reproductive technologies are used or health complications occur. Making a profit in horse breeding is often difficult. While some owners of only a few horses may keep a foal for purely personal enjoyment, many individuals breed horses in hopes of making some money in the process.

A general rule of thumb is that a foal intended for sale should be worth three times the cost of the stud fee if it were sold at the moment of birth. From birth forward, the costs of care and training are added to the value of the foal, with a sale price going up accordingly. If the foal wins awards in some form of competition, that may also enhance the price.On the other hand, without careful thought, foals bred without a potential market for them may wind up being sold at a loss, and in a worst-case scenario, sold for "salvage" value — a euphemism for sale to slaughter as horsemeat.

Therefore, a mare owner must consider their reasons for breeding, asking hard questions of themselves as to whether their motivations are based on either emotion or profit and how realistic those motivations may be.The stallion should be chosen to complement the mare, with the goal of producing a foal that has the best qualities of both animals, yet avoids having the weaker qualities of either parent. Generally, the stallion should have proven himself in the discipline or sport the mare owner wishes for the "career" of the ensuing foal. Mares should also have a competition record showing that they also have suitable traits, though this does not happen as often. A purebred horse is usually worth more than a horse of mixed breeding, though this matters more in some disciplines than others. The breed of the horse is sometimes secondary when breeding for a sport horse, but some disciplines may prefer a certain breed or a specific phenotype of horse. Sometimes, purebred bloodlines are an absolute requirement: For example most Racehorses in the United States must be recorded with a breed registry in order to race. Bloodlines are often considered, as some bloodlines are known to cross well with others. If the parents have not yet proven themselves by competition or by producing quality offspring, the bloodlines of the horse are often a good indicator of quality and possible strengths and weaknesses.

Some bloodlines are known not only for their athletic ability, but could also carry a conformational or genetic defect, poor temperament, or for a medical problem. Some bloodlines are also fashionable or otherwise marketable, which is an important consideration should the mare owner wish to sell the foal. Horse breeders also consider conformation, size and temperament. All of these traits are heritable, and will determine if the foal will be a success in its chosen discipline. The offspring, or "get," of a stallion are often excellent indicators of his ability to pass on his characteristics, and the particular traits he actually passes on. Some stallions are fantastic performers but never produce offspring of comparable quality. Others sire fillies of great abilities but not colts. At times, a horse of mediocre ability sires foals of outstanding quality. Mare owners also look into the question of if the stallion is fertile and has successfully "settled" (*i.e.* impregnated) mares. A stallion may not be able to breed naturally, or old age may decrease his performance. Mare care boarding fees and semen collection fees can be a major cost. Breeding a horse can be an expensive endeavor, whether breeding a backyard competition horse or the next Olympic medalist.

Costs may include:

- The stud and booking fee
- Fees for collecting, handling, and transporting semen (if AI is used and semen is shipped)
- *Mare exams*: to determine if she is healthy enough to breed, to determine when she ovulates, and (if AI is used) to inseminate her
- Mare transport, care, and board if the mare is bred live cover at the stallion's residence
- Veterinary bills to keep the pregnant mare healthy while in foal
- Possible veterinary bills during pregnancy or foaling should something go wrong
- Veterinary bills for the foal for its first exam a few days following foaling

Stud fees are determined by the quality of the stallion, his performance record, the performance record of his get (offspring), as well as the sport and general market that the animal is standing for. The highest stud fees are generally for racing Thoroughbreds, which may charge from two to three thousand dollars for a breeding to a new or unproven stallion, to several hundred thousand dollars for a breeding to a stakes winner. Sport horse stallions generally range from $1000 to $3000, although the top stallions may reach $4000 for one breeding.

The lowest stud fees may only be $100-$200, but there are trade-offs: the horse will probably be unproven, and probably much less athletic than a horse with a stud fee that is a bit higher. As a stallion's career, either performance or breeding, improves, his stud fee tends to increase in proportion. If one or two offspring are especially successful, winning several stakes races or an Olympic medal, the stud fee will generally greatly increase. Younger, unproven stallions will generally have a lower stud fee earlier on in their careers.

To help decrease the risk of financial loss should the mare die or abort the foal while pregnant, many studs have a live foal guarantee (LFG), allowing the owner to have a free breeding to their stallion the next year. However, this is not offered for every breeding.

DOG BREEDING

The dog is a domesticated subspecies of the wolf, a mammal of the Canidae family of the order Carnivora. The term encompasses both feral and pet varieties and is also sometimes used to describe wild canids of other subspecies or species. The domestic dog has been one of the most widely-kept working and companion animals in human history, as well as being a food source in some cultures. There are estimated to be 400,000,000 dogs in the world. The dog has developed into hundreds of varied breeds. Height measured to the withers ranges from a few inches in the Chihuahua to a few feet in the Irish Wolfhound; colour varies from white through grays to black, and browns from light to dark in a wide variation of patterns; and, coats can be very short to many centimeters long, from coarse hair to something akin to wool, straight or curly, or smooth.

In breeding circles, a male canine is referred to as a *dog*, while a female canine is called a *bitch*. The father of a litter is called the *sire*, and the mother of a litter is called the *dam*. Offspring are generally called *pups* or *puppies* until they are about a year old. A group of offspring is a *litter*. The process of birth is *whelping*. Many terms are used for dogs that are not purebred.

Origin and Evolution

Based on DNA evidence, the wolf ancestors of modern dogs diverged from other wolves about 100,000 years ago, and dogs were domesticated from those wolf ancestors about 15,000 years ago. This date would make dogs the first species to be domesticated by humans. Evidence suggests that dogs were first domesticated in East Asia, possibly China, and some of the peoples who entered North America took dogs with them from Asia. As humans migrated around the planet a variety of dog forms migrated with them.

The agricultural revolution and subsequent urban revolution led to an increase in the dog population and a demand for specialization. These circumstances would provide the opportunity for selective breeding to create specialized working dogs and pets. This ancient mosaic, likely Roman, shows a large dog with a collar hunting a lion. Molecular systematics indicate that the domestic dog descends from one or more populations of wild wolves. As reflected in the nomenclature, dogs are descended from the wolf and are able to interbreed with wolves. The relationship between human and canine has deep roots. Converging archaeological and genetic evidence indicate a time of domestication in the late Upper Paleolithic close to the Pleistocene/Holocene boundary, between 17,000 and 14,000 years ago. Fossil bone morphologies and genetic analysis of current and ancient dog and wolf populations have not yet been able to conclusively determine whether all dogs descend from a single domestication event, or whether dogs were domesticated independently in more than one location.

Domesticated dogs may have interbred with local populations of wild wolves on several occasions (a process known in genetics as introgression). The earliest dog fossils, two crania from Russia and a mandible from Germany, date from 13,000 to 17,000 years ago. Their likely ancestor is the large northern Holarctic wolf, *Canis lupus lupus*. Remains of smaller dogs from Mesolithic cave deposits in the Middle East, dated to around 12,000 years ago, have been interpreted as descendants of a lighter Southwest Asian wolf, *Canis lupus arabs*.

Rock art and skeletal remains indicate that by 14,000 years ago, dogs were present from North Africa across Eurasia to North America. Dog burials at the Mesolithic cemetery of Svaerdborg in Denmark suggest that in ancient Europe dogs were valued companions. Genetic analyses have so far yielded divergent results. The ancestors of dogs split off from other wolves between 75,000 and 135,000 years ago, while a subsequent analysis indicated a "common origin from a single gene pool for all dog populations" between 40,000 and 15,000 years ago in East Asia. Both sets of dates must be reevaluated in light of recent findings showing that poorly calibrated molecular clocks have systematically overestimated the age of geologically recent events. On balance, and in agreement with the archaeological evidence, 15,000 years ago is the most likely time for the wolf-dog divergence.

Development of Dog Breeds

Dogs have been bred into a variety of shapes, colours and sizes. Variation can be wide even within a breed, as with these Cavalier King Charles Spaniels. There are numerous dog breeds, with over 800 being recognized by various kennel clubs worldwide. Many dogs, especially outside the United States of America and Western Europe, belong to no recognized breed. A few basic breed types have evolved gradually during the domesticated dog's relationship with humans over the last 10,000 or more years, but all modern breeds are of relatively recent derivation. Many of these are the product of a deliberate process of artificial selection. Because of this, some breeds are highly specialized, and there is extraordinary morphological diversity across different breeds. Despite these differences, dogs are able to distinguish dogs from other kinds of animal.

The definition of a dog breed is a matter of some controversy. Depending on the size of the original founding population, closed gene pool breeds can have problems with inbreeding, specifically due to the founder effect. Dog breeders are increasingly aware of the importance of population genetics and of maintaining diverse gene pools. Health testing and new DNA tests can help avoid problems, by providing a replacement for natural selection. Without selection, inbreeding and closed gene pools can increase the risk of severe health or behavioural problems. Some organizations define a breed more loosely, such that an individual may be considered of one breed as long as 75 per cent of its parentage is of that breed. These considerations affect both pets and the show dogs entered in dog shows.

Even prize-winning purebred dogs sometimes possess crippling genetic defects due to founder effect or inbreeding. These problems are not limited to purebred dogs and can affect cross-breed populations. The behaviour and appearance of a dog of a particular breed can be predicted to a degree, while mixed-breed dogs show a broader range of innovative appearance and behaviour. Mixed-breed dogs or Mongrels (also called "mutts") are dogs that do not belong to specific breeds, being mixtures more than two in variant percentages. Mixed breed dogs and purebred dogs are both suitable as companions, pets, working dogs, or competitors in dog sports. Sometimes different breed dogs are deliberately bred, to create cross-breeds such as the Cockapoo, a mixture of Cocker Spaniel and Miniature Poodle. Such deliberate crosses may display some degree of hybrid vigour and other desirable traits, but may or may not inherit any of the desired traits of their parents, such as temperament or a particular colour or coat. Without genetic testing of the parents, the crosses can end up inheriting genetic defects that occur in both parental breeds. A breed is a group of animals that possesses a set of inherited characteristics that distinguishes it from other animals within the same species. Deliberately crossing two or more breeds is also a manner of establishing new breeds, but it is only a breed when offspring will reliably demonstrate that particular set of characteristics and qualities.

Breed Popularity

Breed popularity varies widely over time and in different parts of the world and different segments of the population. Counting by American Kennel Club (AKC) registration (not by licensing registration or by United Kennel Club (UKC) registration, which could present different statistics), the Labrador Retriever has been the United States's most commonly registered breed of dog since 1991. However, even within parts of the United States, popularity varies; for example, in 2005 the most-registered breed in New York City was the Poodle while the Yorkshire Terrier was the second-most-registered breed in Houston.

However, animal shelters in many parts of the United States report that the most-commonly available dog for adoption is the American Pit Bull Terrier or pit bull-type mixes, making up as much as 20 per cent of dogs available for adoption, none of which would be registered with the AKC. Two decades ago, in 1983, the AKC's top two registered breeds were the American Cocker Spaniel and the Poodle.

In the United Kingdom, The Kennel Club reports that the most-registered breed from at least 1999 to 2005 was the Labrador Retriever. It rounds out the top three for 1999 to 2005 with the German Shepherd Dog, also popular in the States, and the English Cocker Spaniel, which is no longer even in the top ten in the States. In the United Kingdom a national dog adoption and rescue service indicates that the most common breed appearing in shelters is the Greyhound followed by the Staffordshire Bull Terrier.

Physical Characteristics

Modern dog breeds show more variation in size, appearance, and behaviour than any other domestic animal. Within the range of extremes, dogs generally share attributes with their wild ancestors, the wolves. Dogs are predators and scavengers, possessing sharp teeth and strong jaws for attacking, holding, and tearing their food. Although selective breeding has changed the appearance of many breeds, all dogs retain basic traits from their distant ancestors. Like many other predatory mammals, the dog has powerful muscles, fused wristbones, a cardiovascular system that supports both sprinting and endurance, and teeth for catching and tearing. Compared to the bone structure of the human foot, technically, dogs walk on their toes.

Differences from Other Canids

Compared to equally sized wolves, dogs tend to have 20 per cent smaller skulls and 10 per cent smaller brains, as well as proportionately smaller teeth than other canid species. Dogs require less calories to function than wolves. Their diet of human refuse in antiquity made the large brains and jaw muscles needed for hunting unnecessary. It is thought by certain experts that the dog's limp ears are a result of atrophy of the jaw muscles. Dogs tend to be poorer than wolves and coyotes at observational learning, being more responsive to instrumental conditioning.

Dogs have thicker skins than similairly sized wolves. Inuits tend to favour dog skin over wolf skin due to the formers resistance to tear. Unlike wolves, but like coyotes, domestic dogs have sweat glands on their pawpads. The paws of a dog are half the size of those of a wolf, and their tails tend to curl upwards, another trait not found in wolves. Feral dogs show little of the complex social structure or dominance heirarchy present in wolf packs. For dogs, other members of their kind are of no help in locating food items, and are more like competitors. Dogs differ from wolves and most other large canid species by the fact that they do not regurgitate food for their young, nor the young of other dogs in the same territory.

Many dogs, such as this American Water Spaniel, have had their natural hunting instincts suppressed or altered to suit human needs. Like most mammals, dogs are dichromats and have colour vision equivalent to red-green colour blindness in humans. Different breeds of dogs have different eye shapes and dimensions, and they also have different retina configurations. Dogs with long noses have a *"visual streak"* which runs across the width of the retina and gives them a very wide field of excellent vision, while those with short noses have an *"area centralis"*—a central patch with up to three times the density of nerve endings as the *visual streak*—giving them detailed sight much more like a human's. Some breeds, particularly the sighthounds, have a field of vision up to 270° (compared to 180° for humans), although broad-headed breeds with short noses have a much narrower field of vision, as low as 180°. Dogs detect sounds as low as the 16 to 20 Hz frequency range (compared to 20 to 70 Hz for humans) and above 45 kHz (compared to 13 to 20 kHz for humans), and in addition have a degree of ear mobility that helps them to rapidly pinpoint the exact location of a sound. Eighteen or more muscles can tilt, rotate and raise or lower a dog's ear. Additionally, a dog can identify a sound's location much faster than a human can, as well as hear sounds up to four times the distance that humans are able to. Those with more natural ear shapes, like those of wild canids like the fox, generally hear better than those with the floppier ears of many domesticated species.

Dogs have nearly 220 million smell-sensitive cells over an area about the size of a pocket handkerchief (compared to 5 million over an area the size of a postage stamp for humans). Some breeds have been selectively bred for excellence in detecting scents, even compared to their canine brethren. What information a dog actually detects when he is scenting is not perfectly understood; although once a matter of debate, it now seems to be well established that dogs can distinguish two different types of scents when trailing, an air scent from some person or thing that has recently passed by, as well as a ground scent that remains detectable for a much longer period. The characteristics and behaviour of these two types of scent trail would seem, after some thought, to be quite different, the air scent being intermittent but perhaps less obscured by competing scents, whereas the ground scent would be relatively permanent with respect to careful and repetitive search by the dog, but would seem to be much more contaminated with other scents.

In any event, it is established by those who train tracking dogs that it is impossible to teach the dog how to track any better than it does naturally; the object instead is to motivate it properly, and teach it to maintain focus on a single track and ignore any others that might otherwise seem of greater interest to an untrained dog. An intensive search for a scent, for instance searching a ship for contraband, can actually be very fatiguing for a dog, and the dog must be motivated to continue this hard work for a long period of time. Domestic dogs often display the remnants of counter-shading, a common natural camouflage pattern. The general theory of countershading is that an animal that is lit from above will appear lighter on its upper half and darker on its lower half where it will usually be in its own shade. This is a pattern that predators can learn to watch for. A countershaded animal will have dark colouring on its upper surfaces and light colouring below.

This reduces the general visibility of the animal. One reminder of this pattern is that many breeds will have the occasional "blaze", stripe, or "star" of white fur on their chest or undersides. Dogs can generate large amounts of energy for a short period of time. A dog's heart and lungs are oversized relative to its body and its normal everyday needs. A dog also has relatively more red blood cells than a human. Most of the time the dog will keep the extra red blood cells stored in its spleen. When the animal enters into a situation where its full metabolism is required, such as play, catching game, or fighting other dogs, the extra cells are released into the bloodstream. The "oversized" heart and lungs will now be running at full capacity, and the animal will have an enhanced ability to engage in aerobic activity. This activity will produce internal heating. Dogs, being covered in fur, are limited in their ability to cool down. After a short time the animal must either cease its athletic activity or risk harming itself from overheating. One can easily observe this pattern of intense activity followed by rest periods in puppies. During the rest phase the spleen collects red blood cells and the animal may pant to cool down.

Dogs are very social animals, but their personality and behaviour vary with breed as well as how they are treated by their owners and others who come in contact with them. It is not uncommon for dogs to attack humans and other animals; however, this is usually because of lack of care or improper upbringing by its owner. Many dogs can be trained to skillfully perform tasks not natural to canines, such as in this dog agility competition. Dogs are valued for their intelligence. This intelligence is expressed differently with different breeds and individuals, however. For example, Border Collies are noted for their ability to learn commands, while other breeds may not be so motivated towards obedience, but instead show their cleverness in devising ways to steal food or escape from a yard. Being highly adaptable animals themselves, dogs have learned to do many jobs as required by humans over the generations.

Dogs are employed in various roles across the globe, proving invaluable assets in areas such as search-and-rescue; law enforcement (including attack dogs, sniffer dogs and tracking dogs); guards for livestock, people or property; herding; Arctic exploration sled-pullers; guiding the blind and acting as a pair of ears for the deaf; assisting with hunting, and a great many other roles which they may be trained to assume. Most dogs rarely have to deal with complex tasks and are unlikely to learn relatively complicated activities (such as opening doors) unaided. Some dogs (such as guide dogs for the visually impaired) are specially trained to recognize and avoid dangerous situations. The meaning of "intelligence" in general, not only in reference to dogs, is hard to define. Some tests measure problem-solving abilities and others test the ability to learn in comparison to others of the same age. Defining it for dogs is just as difficult. It is likely that dogs do not have the ability to premeditate an action to solve a problem.

For example, the ability to learn quickly could be a sign of intelligence. Conversely it could be interpreted as a sign of a desire to please. In contrast, some dogs who do not learn very quickly may have other talents. An example is breeds that are not particularly interested in pleasing their owners, such as Siberian Huskies. Huskies are often fascinated with the myriad of possibilities for escaping from yards, catching small animals, and often figuring out on their own numerous inventive ways of doing both. Assistance dogs are also required to be obedient at all times. This means they must learn a tremendous number of commands, understand how to act in a large variety of situations, and recognize threats to their human companion, some of which they might never before have encountered. Many owners of livestock guardian breeds believe that breeds like the Great Pyrenees or the Kuvasz are not easily trained because their stubborn nature prevents them from seeing the point of such commands as "sit" or "down".

Hounds may also suffer from this type of ranking. These dogs are bred to have more of a "pack" mentality with other dogs and less reliance on a master's direct commands. While they may not have the same kind of intelligence as a Border Collie, they were not bred to learn and obey commands quickly, but to think for themselves while trailing game. Dogs are highly social animals. This can account for their trainability, playfulness, and ability to fit into human households and social situations. These attributes have earned dogs a unique position in the realm of interspecies relationships despite being one of the most effective, voracious, and potentially dangerous predators.

Dogs and humans at times co-operate in some of the most effective hunting in the animal world; in that context, dogs are superpredators. The loyalty and devotion that dogs demonstrate as part of their natural instincts as pack animals closely mimics the human idea of love and friendship, leading many dog owners to view their pets as full-fledged family members. Conversely, dogs seem to view their human companions as members of their pack, and make few, if any, distinctions between their owners and fellow dogs.

Dogs fill a variety of roles in human society and are often trained as working dogs. For dogs that do not have traditional jobs, a wide range of dog sports provide the opportunity to exhibit their natural skills. In many countries, the most common and perhaps most important role of dogs is as companions. Dogs have lived and worked with humans in so many roles that their loyalty has earned them the unique sobriquet "man's best friend". However, some cultures consider dogs to be unclean. In some parts of the world, dogs are raised as livestock to produce dog meat for human consumption. In many places, consumption of dog meat is discouraged by social convention or cultural taboo.

Laughter might not be confined or unique to humans, despite Aristotle's observation that "only the human animal laughs". The differences between chimpanzee and human laughter may be the result of the adaptations that enable human speech. However, some behavioural psychologists argue that self-awareness of one's situation, or the ability to identify with somebody else's predicament, are prerequisites for laughter, so animals are not really laughing in the same way that humans do. The dog-laugh sounds similar to a normal pant. However by analyzing the pant using a spectrograph, this pant varies with bursts of frequencies, resulting in a laugh. When this recorded dog-laugh vocalization is played to dogs in a shelter setting, it can initiate play, promote pro-social behaviour, and decrease stress levels. In a study by Simonet, Versteeg, and Storie, one hundred and twenty subject dogs residing in a mid-size county animal shelter were observed. Dogs ranging from 4 months to 10 years of age were compared with and without exposure to a dog-laugh recording. The stress behaviours measured included panting, growling, salivating, pacing, barking, cowering, lunging, play-bows, sitting, orienting and lying down. The study resulted in positive findings. Exposure to the dog laughing recording resulted in the following: significantly reduced stress behaviours, increased tail wagging and the display of a play-face when playing was initiated, and more frequent pro-social behaviour such as approaching and lip licking. This research suggests exposure to dog-laugh vocalizations can calm and possibly increase shelter adoptions.

SELECTIVE BREEDING IN DOMESTICATED ANIMALS

Selective breeding in domesticated animals is the process of a breeder developing a cultivated breed over time, and selecting qualities within individuals of the breed that will be best to pass on to the next generation. Breeding techniques such as inbreeding, linebreeding and outcrossing are utilized by breeders in the maintenance and improvement of their chosen breeds.

Breeding Stock

"Breeding stock" is a term used to describe a group of animals used for purpose of planned breeding. When an individual is looking to breed animals, he or she is looking for certain valuable traits in purebred stock for a certain purpose, or may intend to use some type of crossbreeding to produce a new type of stock with different, and presumably superior abilities in a given area of endeavor.

To breed chickens, a typical breeder intends to receive eggs, meat, and new, young birds for further reproduction. Thus the breeder has to study different breeds and types of chickens and analyse what can be expected from a certain set of characteristics before he or she starts breeding them. Accordingly, when purchasing initial breeding stock, the breeder seeks a group of birds that will most closely fit the purpose intended.

Purebred Breeding

The idea of *breed purity* may strike an unpleasant chord with some individuals because it is reminiscent of nineteenth-century eugenics notions of a "superior strain" in humans, supposedly exemplified by members of an aristocratic social class or of certain races. The application of theories of eugenics had far-reaching consequences for human beings, particularly in cases such as The Holocaust. However, the idea of a superior strain in animals, such as that incorporated in the breeding of Thoroughbred race horses, was less controversial. By "breeding the best to the best," employing a certain degree of careful inbreeding, considerable culling, and selection for "superior" qualities, one could develop a bloodline or "breed" superior in certain respects to the original base stock or landrace which had been produced primarily by natural selection.

This process forms the basis for a "true-breeding" line of animals, which today is generally called a "purebred" line. Once created, a purified line is "closed" and preserved from dilution and possible "debasement" by unrelated stock. However, this theory is not without some negative consequences. When the ideal of purified lineage or purely aesthetic type is seen as an end in itself, without consideration of usefulness or vigour, a breed can suffer over time. Also, if a breed registry is closed with too small a gene pool, over time, degradation due to inbreeding depression may occur. For many purebred breeders, the need for a periodic outcross is recognized and incorporated into an overall plan. The observable phenomenon of hybrid vigour stands in contrast to the notion of breed purity. However, on the other hand, indiscriminate breeding of crossbred or hybrid animals may also result in degradation of quality, particularly when animals are mass-produced without concern for overall health or longevity.

Backyard Breeding

The term backyard breeder is a general term, sometimes considered derisive, used to describe people who are "home breeders", and breed only a few animals for personal use or occasional sale. However, while some backyard breeders pay strict attention to quality, the term usually describes those who allow animals, particularly dogs or horses, to procreate regardless of physical, genetic, and/or emotional health. It is also used to describe unreputable commercial breeders. In the process of careless breeding, many backyard breeders produce genetically weak animals that can be predisposed to genetic disorders or debilitating physical deformities, particularly those that result from inbreeding.

In dog breeding, these individuals focus on only a limited aspect of the dog, while ignoring the original function and temperament of the breed. When such breeding is carried out on a large scale, the venue is called a *puppy mill* or *puppy farm*. Backyard breeding is popularly blamed for the proliferation of aggressive dogs for the sports of dog baiting and dog fighting. Dog fanciers generally believe that such ill-bred dogs are the reason for the bad public reputation of certain dog breeds, and the increasingly frequent enactment of breed-specific legislation. Because of the time and expense of feeding and caring for horses, which produce one foal per year, there are fewer horse breeders who produce animals *en masse,* though some individuals do engage in animal hoarding and breed far more animals than they can support.

In the horse world, over breeding of grade animals that cannot be sold except for very low prices raises concerns that such animals will be slaughtered for horsemeat. Some, though not all, colour breeders make errors similar to puppy mill breeders by focusing on only one characteristic to the detriment of health, soundness, and quality conformation. In addition, industrial breeders, such as those who produce *pregnant mare urine* for the manufacture of premarin, allow many mares to be bred indiscriminately, simply so that they are able to produce the urine laced with reproductive hormones that forms the base for the drug. Their foals are generally considered an unwanted byproduct, particularly males (colts), and sold off, frequently to the slaughter market.

IN-BREEDING AND OTHER BREEDING METHODS

One of the most hotly talked about topics with regard to pure-bred dogs is the use of in-breeding. This is a term that is often misused and is extremely misunderstood. Part of the misunderstandings comes from differences in the way the terms are used within the scientific/medical field, and how it is commonly used by breeders. These are the most commonly accepted definitions used by serious dog breeders and will be the definitions used within this article.

- *In-breeding:* This is the breeding of closely related animals. Brother-Sister, Parent-Offspring.
- *Line-breeding:* This is the breeding of animals that share common ancestors but are not closely related. The dogs may share a common great-grandparent.
- *Out-cross:* This is generally considered the breeding of animals with no common ancestors within the first 4 or 5 generations.

Common Misconceptions

- *In-breeding causes genetic diseases:* Breeding closely related animals increases the possibility that any bad genes in a line will show up. It does not 'cause' genetic disease.
- *Out-crossed dogs are healthier:* This is only partly true. There is a known phenomenon called Hybrid Vigour. Two animals of unrelated strains breed and the offspring is often bigger and grows faster than it's purebred cousins. This method is often used by farmers in order to get their animals to market sooner. But one of the biggest misconceptions of hybrid vigour is that it applies to all animals of mixed heritage. Hybrid Vigour only applies to the animals that are the direct offspring of the crossing of the unrelated strains. In other words if you continue to breed animals of different strains there

generally will not be any additional increase in hybrid vigour. If the unrelated strains share common genes for genetic disorders, hybrid vigour will not over ride the risk of the disorder showing up. Out-crossing can also cause problems if widely divergent physical types are mixed due to differences in growth rates and bone and muscle sizes.

BENEFITS OF EACH TYPE OF BREEDING

By definition, purebred dogs have a smaller gene pool to draw on than mixed breed dogs. That smaller gene pool gives the breed its individual characteristics, such as physical appearance and temperament. It is what makes a poodle a Poodle and a Golden Retriever a Golden Retriever. But there is considerable controversy with regard to whether the gene pools of the modern pure-bred have become too small.

Inbreeding

In-breeding is more likely to help "set" or "fix" a particular trait within a breed or a line by narrowing the gene pool to favour those traits. So if a breeder is looking to set a particular desirable feature of their line then in-breeding and choosing the offspring most strongly possessing that trait can be beneficial. In the first pedigree Sydney is the maternal son of Annie, thus doubling on Annie's genes. Or another way to look at it is Annie is both Sydney's mother and Grand-mother. In the second example we have a full brother/sister in-breeding so we are doubling up on the genes from both Kiley and Annie.

In-breeding can also help identify those bad genes that exist within a line. Dogs possessing the bad genes can be eliminated from a breeding programme and carriers also identified. Intermittent in-breeding within a line or breed is not damaging to the long term health of the animals. However, in-breeding over successive generations can lead to reduced fitness and fertility problems among the offspring, resulting in phenomena known as In-breeding Depression. It can take many generations to show up depending on the traits involved. To use this method responsibly a breeder would not want to in-breed on animals with known genetic disorders, temperaments not in keeping with its given breed, or known serious structural faults, or to in-breed frequently even on healthy-superior specimens.

Line-Breeding

Line-breeding is another way to help "set" or "fix" desirable traits. With line-breeding we breed animals that are related, but we are also routinely introducing genes from other lines into the genetic mix. It takes longer to fix the desirable traits this way, but doing so lowers the risk of those problems associated with repeated in-breeding.

Loose line-breeding over successive generations will result in more variations of physical appearance than would in-breeding or tight line-breeding, but will keep the physical look and structure within the same general size and shape, it also carries fewer long term risks.

According to geneticists: Line-breeding can be carried on for many many generations without deleterious effects on the line or breed as long as the individuals involved have few hidden genetic disorders.

Out-Crossing

Out-crossing in terms of pure-bred dogs is the breeding of unrelated dogs. On a pedigree no names will be repeated within the first 5 generations. This type of breeding has both advantages and disadvantages. Which as it turns out are flip sides of the same argument. The greatest genetic diversity are maintained by out -ccrossing, but this also leads to the least consistency in terms of physical appearance and other traits.

Out-crossing does not guarantee that the animals won't develop genetic disorders, but it does tend to reduce the numbers of affected offspring. Best chance of getting an animal that is less prone to developing a genetic disorder comes more from finding a conscientious breeder that screens their animals for hereditary disorders and breeds for the betterment of the breed. All three methods of breeding have their place in a long term, well thought out breeding programme. Talk to the breeder, ask questions as to what their goal is in doing a particular breeding. Ask about the risks and what problems are known to that line. And all lines have some because the perfect dog and the perfect lines are still goals of the future for all breeders.

GENETIC IMPROVEMENT OF DAIRY CATTLE

Dairymen are constantly concerned with improvement of their herds. Dairy cattle improvement is expressed most frequently as increased milk production and less often in terms of changes in body conformation or "type." Improvement in milk production can be considered as being environmental or genetic. Both environmental and genetic factors contribute to variation in milk production. There are environmental and genetic limitations of production. More importantly, management of environment and genetic factors can result in improved performance, *i.e.,* greater milk production.

Genetic Principles of Repeatability

Understanding the concept and knowing the repeatability of economically important traits help considerably in deciding which cows to cull. Repeatability also is important in sire proofs. All repeatability estimates represent some type of correlation. The trait one is considering, and the times and places it was measured, influence the estimate. The correlation between 305-day mature equivalent milk records of first-calf heifers, and their records as second-calf heifers, is about 0.5.

So milk records are said to have a repeatability of 0.5. Conception rate from year-to-year with the same cow has a repeatability of about zero, indicating that there is little or no relationship (correlation) between the number of services it takes for a cow to become pregnant each time. These estimates are useful in making culling and management decisions, since they help us make the best possible guess at how well the cow will do next year if we know how well she did this year. For milk yield, 50 per cent of the apparent superiority (or inferiority) of the cow for this year will appear next year on the average.

If a cow gave 1000 pounds more than her herdmates this year, the best prediction of her yield next year will be 500 pounds more. Naturally, what she actually does is another story; she could be much better or much worse than our guess. The same process is used for below average cows; only one-half of their apparent inferiority will appear the following year on the average. Repeatability is included in the calculation of estimated relative producing ability in DHI. Only a fraction of the cow's performance is permanent (repeatable). Superior cows may do even better next year, but on the average they do worse; inferior cows on the average do better.

Genetic Principle of Heritability

Heritability is a measure of the degree to which a trait such as milk yield is genetically determined. Although not all of the genetic contribution is included, essentially all which is available to a breeder who is trying to select within a breed is included. Obviously heritability is an important factor among the several factors determining how much genetic improvement can be made in any characteristic. One of the disadvantages with the dairy cow, in attempts to make genetic progress, is that most of the economically important traits can be measured only in the female. Another disadvantage is that heritabilities are low or close to zero in several cases. A zero heritability would mean that no genetic change could be expected from usual selection techniques.

A practical idea of the meaning of heritability can be obtained easily. A genetically average group of males is mated to a large group of high producing first calf heifers. These heifers average 1000 pounds of milk more than their contemporaries. Assume also that no extraordinary changes in environment have occurred by the time that daughters from these matings freshen and complete their first record.

How much milk will the daughters produce compared to their herdmates? Answer: about 100 to 150 pounds more. Reason: milk yield has a heritability of about 20 to 30 per cent, and since the sires used were genetically average, and sires and dams contribute about equally to the genotype of their offspring, only 10 to 15 per cent of the superiority of the dams will be seen in the offspring.

A good practical definition of heritability is, therefore, the fraction of the merit (or lack of merit) observed in the parents which can be seen in the offspring, with merit being defined as a deviation from the population average. Note that several reproductive traits have very low heritabilities. Milk yield would be said to be moderately heritable; fat and protein percentages are highly heritable.

One possibility exists for genetic improvement of important traits with zero or very low heritability. This revolves around what geneticists call non-additive genetic variance. If there is appreciable non-additive genetic variance, it can be taken advantage of by crossbreeding. Heterosis (hybrid vigour) is a reflection of this type of genetic variability. Unfortunately, although there is some heterosis in reproductive performance, longevity, livability and milk yield, it is not a large amount.

Crossbreeding has several disadvantages for dairy cattle, although this discussion cannot be covered here. Few people are ready to advise taking advantage of non-additive genetic variance in dairy cattle at the present time, and few cattleman are attempting to do so. Reproductive efficiency can be improved considerably by improving management practices while selection is being practiced for more highly heritable traits.

CORRELATED RESPONSE AND SELECTION

Traits which are heritable presumably can be changed by selection, but the degree of heritability is only one of three factors which determine the progress which can be made. The other two are the culling level (intensity of selection) and the variability of the trait. These are logical since the more severely we can cull, the more progress we can make. Also, the more variable a trait is, the higher the population average will be for the selected individuals for any given culling percentage.

It turns out that the amount of progress which can be made in one generation is the product of intensity of selection, heritability for that trait, and the observed variability. Note that if any of the three components is zero, selection progress will be zero also. Good estimates are available for how much progress can be made from selection for most of the economically important characteristics of dairy cattle. Of course, there are several traits which are important, so there is a need to look further than anticipated results from selection for a single trait.

Knowing what the genetic relationships are among the important traits, plus a few other things, permits us to predict what will happen to a trait such as fat percentage if we ignore it completely and select for milk yield, or vice versa. Though these results pertain to Holsteins, they are similar to results for other breeds. The changes expected in the trait being selected for; values off the diagonal are changes expected to happen in other traits. Heritabilities and variabilities used in these calculations agreed with those generally accepted by dairy cattle geneticists. Intensity of selection in the estimations was fairly conservative. If more severe culling were practiced, greater progress would be made. If milk yield alone is selected for at a reasonable level of intensity, an increase is expected in genetic merit in the next generation of 607 pounds. Since milk yield, fat yield and protein yield are genetically fairly closely related (that is, their genetic correlations are high and positive), an increase is expected in fat and protein yields, even though these were ignored in the selection programme.

Because of the negative genetic correlations between milk yield and fat percentage, and milk yield and protein percentage, these two percentages are expected to go down. Selecting for milk yield and ignoring the other four traits would be expected to result in an increase per generation of 607 pounds of milk, 23.2 pounds of fat and 13.7 pounds of protein, but decreases of 0.036 per cent fat and 0.018 per cent protein. If one attempted to increase fat percentage genetically in his herd by selecting for it and ignoring other traits, he could increase fat percentage. A reasonable increase per generation to be expected is 0.190 per cent. Along with this, however, there would be an actual decrease in milk yield of 287 pounds, although fat yield would increase 24.1 pounds.

Present pricing systems for milk in the United States are based mostly on its weight and fat percentage, although additional factors such as protein percentage may be included also. So the dairyman needs to decide what combination of milk yield and fat percentage and other changes is most profitable for him. He also needs to produce milk that passes minimum legal standards; this is an important factor in herds which already are near the minimum. Beyond this, there are a number of other important traits which must be considered. Dairymen, of course, mentally put all of these factors together in their selection programmes and, depending on their present herd status and present milk pricing, give each trait a certain amount of emphasis. In other words, they weight each characteristic they think is important. Geneticists do the same thing and attempt to establish numerical values for the weights; a numerical index, called a selection index, then can be formulated for each economic situation.

Milk Yield and Composition

Just about every index formulated for the United States gives yield more emphasis than any other trait, with enough emphasis placed on fat percentage to meet legal standards. Protein has become more important in pricing systems in recent years.

Reproductive Performance

Efficient reproduction is very important economically because a cow must first reproduce to produce. However, most measures of heritability are zero or close to zero, suggesting that little or no improvement can be made by selection.

There already is considerable selection pressure being exerted on improving reproductive performance without any conscious effort being made by the breeder. This occurs because sterile males and females leave no offspring, and those with low efficiency leave a lot fewer than those of higher reproductive performance. So within a given herd culling for reproductive failure would have little effect on the genetic merit for reproduction of the herd. Now that large numbers of records are being obtained on reproduction of daughters of AI sires, however, it may turn out that this information can be taken advantage of in selecting which proven AI sires to use in the herd. Even though heritability of most single measures of reproduction are close to zero, estimates of average genetic merit of a large group of daughters for reproductive efficiency are much more reliable.

Desirable Body Type

Desirable type has appealed to dairymen for centuries, and major changes in appearance in dairy cattle have occurred over the years, accompanied by increases in milk yield. This fact does not itself mean that the two traits have (or ever have had) an appreciable, positive genetic correlation. Regardless of what has gone before, the breeder must concern himself with the present facts of life in his selection programme. The genetic relationship between type and production seems close to zero and may even be negative. There have been a number of changes in the breed association type classification programmes and policies in recent years, and evaluation of all such changes cannot yet be made. A number of scoring systems also have been developed by others such as universities and AI studs. Yet the final value of desirable type must be in its relationship with production and longevity of the cow.

Except for the breeder who derives an important fraction of his income from sale of breeding stock, dairymen should realise (according to research) that only 2 to 6 per cent of their income is related to variability in body confirmation. USDA researchers estimated several years ago that giving equal emphasis to type and production would reduce the possible increase in milk yield by 29 per cent. Though perhaps the final judgement on emphasis to be placed on desirable type cannot yet be made, it appears that if major attempts are made to improve type, they will be made at the sacrifice of possible genetic increases in milk yield.

Longevity

Longevity may be defined as the ability to avoid death or culling, the former being mostly for disease or accident and the latter for reproductive failure, low milk yield, mastitis, etc. Few cows die simply from old age. Of the various possible reasons for loss, some have a genetic basis but few are highly heritable. In any event, high producing cows remain in the herd longer than low producers.

Natural selection pressures also work in favour of the ability to resist death or culling. Selection for high milk yield, plus natural selection, seem to be exerting sufficient selection pressure for increased longevity such that the breeder need not give it major emphasis in his programme.

Body Size and Weight

Within each dairy breed, large cows give more milk, on the average, than small cows, but they also require more feed. Research suggests that the additional costs involved with larger cows about equal the additional income derived from them. In the case of two cows of equal production but different size, one actually would expect the smaller cow to be more profitable. Most measures of size (weight, height, etc.) show appreciable heritability and variability, so larger cows could doubtless be developed if desired.

However, most research now suggests that milk yield in dairy cows can be increased considerably without there being any increase in size. Selection for increased body size could be successful, but the correlated response in milk yield though probably positive would be close to zero. Though North Carolina researchers indicated that it might still be wise to check cows considerably above or below the breed average in body weight for productivity and longevity, they concluded that "undue emphasis on larger cows in the short run could lead to the development of less profitable animals."

Other Traits

Many others are economically important and are heritable to some degree. Many qualitative (simply inherited) traits, ranging from lethals on one extreme to aesthetic traits on the other, can be important under some circumstances. Artificial insemination, particularly frozen semen, has caused concern because of possible widespread dissemination of undesirable genes. The genetics of the situation have been studied, however, and with present knowledge potential dangers apparently can be avoided. Traits other than those mentioned also are important. Mastitis resistance, is of extreme economic value and has measurable heritability. A certain amount of direct pressure for resistance will be exerted from culling of mastitic animals, although clinical cases may remain in a dairy herd for years under proper treatment.

Present knowledge of genetics of dairy cattle suggests that major selection emphasis should be placed on milk yield, with enough selection pressure on fat percentage to maintain composition of milk at legal standards. Other components of milk may become important if pricing systems change to emphasize them. Care should be taken to include as few traits as practical in a selection programme, since progress in milk yield may be sacrificed to a degree such that total profit per cow may decrease. This is especially true with traits of low heritability and only slight genetic relationship to milk yield. Reproductive efficiency and resistance to mastitis may appear as economically important traits in future sire proving systems.

The Animal Model

The Animal Model does many things notably different from previous evaluation systems. These include simultaneous evaluation of all sires and cows. This means all information on all relatives is used to increase the accuracy of any one animal's evaluation. This also means that cow families are directly included in the evaluation of all sires and cows, effectively eliminating the basis for previous objections by some breeders. All female ancestors and descendants are used in the evaluation of both sires and cows, weighted by how closely they are related. Since all known relatives of both registered and grade cows are used, computing the evaluations for the Holstein breed requires simultaneous solution of about 50 million equations. Other changes include correction for genetic merit of mates, use of a new genetic base, new ways of using records on cows missing first records and cows changing herds, and definition of different management groups for registered and grade cows.

There is some new terminology with the Animal Model emphasizing that new methods are truly different. Predicted Difference (PD) changes to Predicted Transmitting Ability (PTA), or one-half the breeding value. This is the prediction of what the sire transmits to his offspring. Cow Index changes to Predicted Transmitting Ability (PTA).

Repeatability (RPT) changes to reliability. Repeatability always has been a misnomer as used by the USDA in this context anyway. Since reliability includes contributions from additional relatives, reliability figures should be higher than the old repeatabilities. The biggest difference is for sires of cows with few daughters compared to those with many. A sire with no daughters would have a reliability of 37 per cent. per centile is sire's ranking based on the active AI bull population; for cows it will be based on cows with recent lactations. The rankings are based on an index including milk, fat, and protein evaluations. Previously they were based on an index including only milk and fat. As good as the new Animal Model system is, it cannot correct for preferential treatment of daughters of certain sires. No genetic evaluation system can. Random AI sampling of young sires and wide daughter distribution will continue to be important in establishing accurate first evaluations. This also is true for sires with relatively few daughters.

Economic Evaluation of Sires and Their Semen

Predicted transmitting ability for dollar value (PTA PRO$) is based on milk, fat, and protein evaluations. Current national prices are used. For the July 1989 run these were $11.20 per hundredweight of milk containing 3.5 per cent milk fat and 3.2 per cent protein. Hence the equation for estimating PTA PRO$ is

PTA PRO$ = $.0242 (PTA milk) + $ 1.54 (PTA fat) + $ 1.09 (PTA protein). Note that the dollar value is computed from pounds fat and protein produced, not percentages. per centile rankings are based on this index.

In Florida, the milk price is higher than almost everywhere else in the United States, and the dollar values of superior sires should be proportionally higher. The ranking of bulls by evaluations would not change much, but the net value over their semen price would. Taking advantage of considerable previous research, Marsha Wilcox and coworkers developed techniques to aid dairymen in the selection of semen to purchase for herd use. The concept is that of net present value (discounted net income) of semen. Although the method is computationally elaborate, availability of microcomputers to dairymen will permit each to develop a system of evaluating semen based on a set of selection criteria tailored to individual needs. Information necessary for semen selection include sire proof for milk, fat, protein, type (PDT), interest rate, and such management parameters for the individual herd as conception rate, calving interval and female mortality rate. Outcome of this linear net merit index for semen is Present Value Dollars (PV$). These researchers then evaluated a number of possible economic and management situations to determine the effects on sire selection; considerable shifting of rank occurred with changes in criteria. For one sample selection programme (all selection pressure on milk, none on desirable type), rank correlations of PV$ and PDM, and PV$ and PD$ + semen price, were.56 and.52. Such procedures should prove useful and profitable to dairymen who are able to characterize their own management parameters and to define their selection goals. Thus, the concepts of proving sires are not difficult. Efforts to obtain better estimates, thus increasing rates of genetic improvement, continue to make the exact calculations and formulas very complex.

FACTORS AFFECTING BREEDING EFFICIENCY

Reproduction is an important consideration in the economics of cattle production. In the absence of regular breeding and calving at the appropriate time cattle rearing will not be profitable. A healthy calf each year is the usual goal. This is possible only by increasing the reproductive efficiency of the animals. Successful reproduction encompasses the ability to mate, the capacity to conceive and to nourish the embryo and deliver the viable young ones at the end of a normal gestation period. In fact, interruption in this chain of events leads to failure of the cow either to conceive or the embryo to die or to have a premature delivery of the foetus.

The reproductive efficiency is a complex phenomenon controlled by both genetic and non-genetic factors, the non- genetic factors being climate, nutrition, and level of management. The reproductive efficiency varies not only between species and breeds but also among the animals within the Same breed. Even the best feeding and management can not coax performance beyond the genetic limit of an inferior animal. Improving the genetic merits of livestock populations is important at all levels of management. A sound breeding programme is a necessary part of the total animal production system. It is absolutely imperative to improve the productive capacity and physical appearance of the animal population.

The factors which influence the breeding efficiency of cattle are as follows:

Number of ova

The first limitation on the breeding efficiency of fertility of an animal is the number of functional ova released during each cycle of ovulation. Ovulation is the process of shedding of ovum from the Graffian follicle. In the case of cow, usually a single ovum is capable of undergoing fertilization only for a period of 5-10 hours. Therefore, the time of mating in relation to ovulation is important for effective fertilization.

percentage of fertilization

The second limitation is fertilization of ova. Failure to be fertilized may result from several causes. The spermatozoa may be few or low in vitality. The service may be either too early or too late. so that the sperms and eggs do not meet at the right moment, to result in fertilization.

Embryonic death

From the time of fertilization till birth, embryonic mortality may occur due to a variety of reasons. Hormone deficiency or imbalance may cause failure of implantation of fertilized ova which die subsequently. Death may occur as a result of lethal genes for which the embryos are homozygous. Other causes may be accidents in development, over-crowding in the uterus, insufficient nutrition or infections in tile uterus.

Age of first pregnancy

Breeding efficiency may be lowered seriously by increasing the age of first breeding. Females bred at a lower age are likely to appear stunted during the first lactation, but their mature size is affected little by their having been bred early.

Frequency of pregnancy

The breeding efficiency can be greatly enhanced by lowering the interval between successive pregnancies. The wise general policy is to breed for the first time at an early age and to rebreed at almost the earliest opportunity after each pregnancy. In this way the lifetime efficiency is increased. Cows can be rebred in 9-12 weeks after parturition.

Longevity

The length of life of the parent is an important part of breeding efficiency, because the return over feed cost is greater in increased length of life. Also, it affects the possibility of improving the breed. The longer the life of the parents, the smaller the percentage of cows needed for replacement every year.

MANAGEMENT PRACTICES TO IMPROVE BREEDING EFFICIENCY

Some of the management suggestions which will tend to improve breeding efficiency of cattle are listed below.

1. Keep accurate breeding records of dates of heat, service and parturition. Use records in predicting the dates of heat and observe the females carefully for heat.
2. Breed cows near the end of heat period.
3. Have females with abnormal discharges examined and treated by veterinarian.
4. Call a veterinarian to examine females not settled after three services.
5. Get the females checked for pregnancy at the proper time after breeding.
6. Buy replacements only from healthy herds and test them before putting them in your herd.
7. Have the females give birth in isolation, preferably in a parturition room and clean up and sterilize the area once parturition is over.
8. Follow a programme of disease prevention, test and vaccination for diseases affecting reproduction and vaccinate the animals against such diseases.
9. Practice a general sanitation programme.
10. Supply adequate nutrition.
11. Employ the correct technique.
12. Provide suitable shelter management.
13. Detect silent or mild heat, by using a teaser bull.

SELECTION AND CULLING

Selection and culling are the two sides of the same coin. Selection is the process in which certain individuals in a population are included for becoming the parents of the next generation. Automatically some are excluded for the purpose which are culled. Natural selection has been going on since ages where animals which were stronger, which had better survivability and which were in more unison with the environment around them, found a better chance to reproduce. Thus certain genes for certain characters got more chance to be selected to form individuals in the subsequent generations. Since domestication of cattle, man has been looking for superior phenotypes in traits useful to him and selecting such animals to form the parental generation. This is man made artificial selection. Now man has progressed one step further in making estimates of genotypes from the study of phenotypes and making use of that information (in artificial) selection.

Selection methods

There is only one way to select and that is to "keep the best and cull the poorest. The various selection methods are techniques for identifying or estimating the genetic values of individual candidates for selection. The procedure discussed here apply to selection for quantitative trails.

Performance testing

Performance test is a measure of the phenotypic value of the individual candidates for selection. Since the phenotypic value is determined by both genetic and environmental influences, the performance test is an estimate, not a measure of the genetic value. The occurrence of this estimate depends upon the heritability of the trait *i.e.* on the degree to which the genetic value is modified by the environmental influences.

Advantages:

a. Among simple procedures, the performance test is the most accurate.
b. Environmental influences can be minimised by testing candidates for selection in the same pen or in similar environmental conditions.
c. The measure is direct, not on a relative basis.
d. All candidates for selection can be tested in contrast to progeny testing where only a parent can be tested.
e. Generation intervals are usually short.
f. Testing can usually be done on the farm under normal management conditions.

Disadvantages:

a. Accuracy become low when heretability is low.
b. Phenotypes are not available for one sex or in sex limited traits such as milk yield.
c. Traits which are not expressed until maturity may become expensive or difficult to manage by performance tests since most selection decisions must be made before maturity.

Performance tests should be the backbone of most selection programmes. Although much publicity has been given to other selection methods, it remains a fact that most of the progress in livestock improvement to date has been due to selection on the individual's Own phenotype *i.e.* performance test.

Pedegree selection

A pedegree is a record of an individual's ancestors including its parents. This information is valuable because each individual possesses a sample half of the genes from each parent. If we can precisely know an individual's phenotype, little is gained by considering pedegree in selection. Pedegree considerations are useful when we do not have sufficient accurate records of production of the individual.

Also, it is useful in the early selection when the traits in question might not have expressed themselves. It is also useful for selection of males when the traits selected for are expressed only by the female (sex-limited) such as milk production in dairy cattle.

Advantages:

a. It provides information when performance tests are not available for the candidates.
b. It provides information to supplement performance test information.
c. It allows selection to be completed at a young age. Pedegree records may be used to select animals for performance or progeny testing in multi-stage selection scheme.
d. It allows selection of box sexes for sex-limited traits *i.e.* bulls can be selected on the milk records of their female relatives.

Disadvantages:

a. Accuracy, relative to alternative selection procedures is usually low.
b. Too much emphasis on relatives, especially remote relatives, greatly reduces genetic progress.
c. Progeny of favoured parents are often environmentally favoured.
d. Relatives often make records under quite different environments, thus introducing non random bases into the selection system.

Progeny testing

In this method we evaluate the breeding value by a study of the expression of the trait in its off springs. Individuality tells us what an animal seems to be, his pedegree tells us what he ought to be, but the performance of his progeny tells us what he is.

Progency testing is, of course, a two-stage selection system because some preliminary selection determines which animals first produce progeny followed by further culling of these which produce poor progeny.

Advantages:

a. High accuracy when many progeny are obtained.

Disadvantages:

a. Long generation interval.
b. Requires high reproductive rate.
c. Low selection intensity.

Show ring selection

Selection on the basis of show ring performance has had considerable value in the past. Essentially this selection has been directed towards bringing the conformation of the animal to some ideal conformation. This improvement has been based on two goals: (i) improvement conformation, and (ii) correlated response. Improvement of conformation has economic value because a part of the sale price is determined by the conformation of the individual. The ideal type was chosen so that, in the opinion of the judges, the animal possessing this conformation was most likely to be a profitable producer. In other words, the judges were attempting to stress traits of conformation which are corrected with productive ability.

With the advent of record keeping it was found that direct selection for performance traits resulted in much faster progress than selection through correlated conformation traits. Also, when subjected to intensive study, many of the correlations between performance and show ring were found to be of non-genetic origin.

If the correlations are of genetic origin, direct selection for performance should improve conformation as well as the reverse situation. The show ring has been a good forum for discussion of what constitutes ideal type and good management and has produced dramatic changes in the conformation of some species.

This has resulted primarily from education of the breeders, however, for most animals which are presented in the ring are good and selection differential among these animals is usually so small as to produce little change.

Advantages:

1. It enables breeders to exchange ideas and experience.
2. It allows comparisons among superior animals both within and between breeds.
3. It allows new breeders to make contact with established breeders.

Disadvantages:

1. Emphasis is usually placed on traits of little economic importance.
2. Clever fitting and showmanship can mask defects of various kinds.
3. Differences between exhibited animals are usually small.
4. Conformation and production traits usually have low genetic correlations.

Choosing Traits for selection

Many factors enter into the choice of traits to be selected for. The following ones are the most important.

1. The goal of the selection programme
2. The habitability of the traits
3. The economic value of improvement in each trait.
4. The range in variation of each trait.
5. Correlation among the traits.
6. The cost of the selection programme.

Selection goals

Often the goal of the selection programme makes the choice of traits quite obvious. The breeder of the race horses must select for speed if he is to be successful and his choice of traits are limited to alternative ways to measure speed.

Similarly, the breeder of dairy cattle generally sets out to breed cows with superior milk production cllaracteristics. Thus, his choice of traits is specified by his selection goals.

Heritability

Heritability is defined to be the fraction of the superiority of parents which is, on the average, transmitted to their off-springs. To explain habitability in simpler words: Heritability tells us how much of the observable differences in the animal is caused by genes and how much by environment. Heritability for the same characteristics may vary from one population to another and also may vary from one characteristic to another even ink the same population.

The ability to recognise the breeding values or transmitting abilities of animals is closely associated with heritability. If the heritability is high for a trait, we can proceed straight way to adopt a system of mass selection of superior animals, with little attention to pedegree information, collateral relatives, progeny test or inbreeding and genetic improvement in that trait is low, genetic progress may be disappointing with mass selection and greater attention should be paid to pedegree records, family information and use of progeny tests.

Variability of the trait

Selection operates on the variability in expression of the trait uniform for a trait. there will be little selection response because any selected groups of parents will not be much better than those not selected. Some traits are much more valuable than others. thus the innate variation of the traits should be carefully considered in choosing traits for selection. Variation can be increased by improving exotic types and sometimes this can result in new combination of genes which are superior to either parent type.

Correlated traits

Sometimes traits tend to be inherited together. These correlations may arise in several ways.

1. The traits may be of different measures of some underlying trait. For example. weight and height are both measures of body size. thus taller animals are usually heavier and these two traits are said to be correlated.
2. If the same genes produce response in several traits. those traits will be correlated. This condition is referred to as pleiotrophy.

Correlated responses are common. Selection for increased milk yield produces a correlated decrease in the per cent of fat in the milk of dairy cows. Thus. both direct and correlated responses result from selection and some correlated responses are positive while others negative. Correlated response may be advantageously used in selection programme. For example feed efficiency is expensive to measure because it requires both weight gain and feed intake on each individual, whereas weight gain requires neither feed weight nor individual feeding.

In summary. definite goals are essential for a successful selection programme. The success in achieving these goals depends on the existence of genetic differences. the degree to which phenotype differences are heritable and the correlated responses in other traits. In comparing the selection programme, the breeder must evaluate the value of the expected response and the cost of the programme relative to the costs and responses of alternative selection programmes.

ARTIFICIAL INSEMINATION

Artificial insemination (AI) now accounts for around 75 per cent of British cattle births. The cow is tied up and one hand of the inseminator manipulates the cervix through the rectum wall whilst the other discharges semen into the vagina and cervix using an inseminating gun. A regional manager with AI firm Genus the problems associated with the procedure. "Placing a quarter of a cc of diluted semen into an area as small as a penny coin, when the target area is inside 500kg of unco-operative cow, requires a high degree of expertise. The more the cow is able to move about, the harder it becomes to hit the target.

Poor handling facilities cause stress, the cow, the inseminator and both our bank balances." There are many invasive procedures carried out on farm animals that the general public rarely witness. One of many includes making sure the cow is in calf. This is carried out using an ultrasound device much like a small TV set with a long thick wire attachment. A vet will insert their hand and arms length into the vagina of the cow.

The animal is held in a holding device to keep her still. The screen shows the small calf in the middle of a dark shaped uterus. Herds usually calve at the same time (due to human intervention) so all the cows on the farm will usually be checked at the same time.

Jersey Cattle

A small, honey-brown breed of dairy cattle, the Jersey is renowned for the high butterfat content of its milk, as well as a genial disposition. The Jersey cow is quite small, ranging from only 800 to 1200 pounds (360 to 540 kg).

The main factor contributing to the popularity of the breed has been their greater economy of production, due to:

- The ability to carry a larger number of effective milking cows per unit area due to lower body weight, hence lower maintenance requirements
- High butterfat conditions, and to thrive on locally produced food.

Bulls are also small, ranging from 1200 to 1800 pounds (540 to 820 kg), and are notoriously very aggressive. Castrated males can be trained into fine oxen which, due to their small size and gentle nature make them popular with young teamsters. Jersey oxen are not as strong as larger breeds however and are generally out of favour among competitive teamsters. Due to the small size, curious docile character and attractive features of the Jersey cow, small herds were imported into England by aristocratic landowners as props for aesthetic landscapes. As its name implies, the Jersey was bred on the British Channel Island of Jersey.

It apparently descended from cattle stock brought over from the nearby Norman mainland, and was first recorded as a separate breed around 1700. Frank Falle has speculated (on the basis of DNA evidence) that the oldest settlers in Jersey were Danish vikings who had been to Nantes with Hatain, where it is recorded that tribute was given by the King of France to Hatain of 500 cattle to leave that area, whereupon they settled in the Normandy area. The marked resemblance between Jersey cattle and the Nantaise cow would seem to bear this out. While the breed is isolated from outside influence today, this was not always the case. Before 1789 cows would be given as dowry for inter-island marriages between Jersey and Guernsey. This was, however, not widespread. Since 1789, imports of foreign cattle into Jersey have been forbidden by law to maintain the purity of the breed, although exports of cattle and semen have been an important economic resource for the island.

The restriction on the import of cattle was initially introduced in 1789 to prevent a collapse in the export price. The United Kingdom levied no import duty on cattle imported from Jersey. Cattle were being shipped from France to Jersey and then being shipped onward to England to circumvent the tariff on French cattle. The increase in the supply of cattle, sometimes of inferior quality, was bringing the price down and damaging the reputation of Jersey cattle. The import ban stabilised the price and enabled a more scientifically controlled programme of breeding to be undertaken.

Sheep Breeds

The domestic sheep is a multi-purpose animal, and the more than 200 breeds now in existence were created to serve these diverse purposes. Some sources give a count of a thousand or more breeds, but concrete documentation has not been produced for such a number. Almost all sheep are classified as being best suited to furnishing a certain product: wool, meat, milk, hides or a combination thereof (*i.e.* dual-purpose breeds). Other features that are important when classifying sheep include face colour (generally white or black), the length of the tail, the presence or lack of horns, and the topography for which the breed has been developed.

This last point is especially stressed in the United Kingdom, where breeds are described as either upland (hill or mountain) or lowland breeds. A sheep may also be of a fat-tailed breed, which is a type of dual-purpose sheep common in Africa and Asia with larger deposits of fat within its tail, or the meat from such a breed. Ovine breeds are also grouped based on how well they are suited to producing a certain type of breeding stock. Generally, sheep are thought to be either "ewe breeds" or "ram breeds". Ewe breeds are those that are hardy, and have good reproductive and mothering capabilities—they are for replacing breeding ewes in standing flocks. Ram breeds are selected for rate of weight gain and carcass quality, and are mated with ewe breeds to produce meat lambs. Many breeds, especially rare or primitive ones, fall into neither category strictly. Lowland and upland breeds are also crossed in this fashion, with the hardy hill ewes crossed with larger, fast-growing lowland rams to produce a type of ewe called a mule, which can then be crossed with meat-type rams to produce prime market lambs.

Sheep breeds are categorized by the type of wool they exhibit, irrespective of whether they are favoured for meat or wool production. Fine wool breeds are those that have wool of great crimp and density, and are consequently preferred by the textile industry. Most of these breeds were derived from Merino sheep, and Merinos continue to dominate the world sheep industry. The record for the most valuable sheep ever belongs to an Australian Merino ram which was sold for $16,000 AUD. Medium wool breeds have wool between the extremes, and are typically fast-growing meat and ram breeds with dark faces. Some major medium wool breeds, such as the Corriedale, are dual-purpose crosses of long and fine-wooled breeds created for high-production commercial flocks. Long wool breeds are the largest of sheep, with lengthy wool and a slow rate of maturation. Long wool sheep are most valued by modern agriculture for crossbreeding to improve the attributes of other sheep types. The American Columbia breed was developed through crossing Lincoln rams—a long wool breed—with fine-wooled Rambouillet ewes.

Coarse or carpet wool sheep are those with a medium to long length wool of characteristic coarseness. Breeds traditionally used for carpet wool show more variability in general traits than other types, the chief requirement being a wool that will not break down under heavy use (as would that of the finer breeds). As the demand for carpet-quality wool declines, some breeders of this type of sheep are attempting to reposition a few of the traditional breeds to alternative purposes. Others have always been primarily meat-class sheep. A last group of sheep breeds is that of fur or hair sheep, which do not grow wool at all. Hair sheep are similar to the very early type of domesticated sheep kept before wooly breeds were developed, and are primarily raised for meat and pelts. Some modern breeds of hair sheep, such as the Dorper, are the result of crosses between wooled and hair breeds. For producers uninterested in selling wool, hair sheep are highly advantageous, as those who keep hair sheep need not go to the expense of shearing. Hair sheep are also resistant to parasites and hot weather, and are consequently increasing in popularity among a subset of producers.

With the modern rise of corporate agrobusiness and the decline of localized family farms, many breeds of sheep are in danger of extinction. Preference for only those breeds with uniform characteristics and fast growth rates has pushed heritage (or heirloom) breeds to the margins of the sheep industry. Those that remain are maintained through the efforts of conservation organizations, breed registries, and individual farmers dedicated to their preservation.

SELECTING SIRES FOR A SINGLE HERD

Most of the factors involved in sire selection have been covered in earlier sections, but one major question remains. Will the sire which has been proven to be superior in other herds also be superior in my herd? Older breeders will remember the term "nicking," which is not too often heard anymore. If a sire did well in one herd and poorly in a second, he was said to nick well in the first, poorly in the second. Today, researchers look at the importance of nicking by studying the importance of sire by farm interactions. Such interactions arise if there are real changes in the relative performance of sires (usually measured by the milk yield of their daughters) as they are used on various farms. Should there be an appreciable number of sires which rank high in some herds or farms, and low in others, and vice versa for other sires, this would show up as a sire-by-herd interaction.

The importance of this question is apparent; if there were such interactions, a dairymen would be unable to rely on USDA sire proofs, since he could not depend on sires with superior proofs to be superior in his herd. This was an area that obviously needed research, and during the past several years considerable research has been published on the subject. A classic study by North Carolina scientists involved sire-by-region interactions. The two regions involved were North Central and Southeastern United States and the question essentially was, "Do AI sires used in these two areas rank about the same, or are there some who are outstanding in one region but poor in the other, and vice versa?" Much of the semen used by Florida dairymen is from sires bred, house and proven in the Eastern and North Central United States, so this question was of immediate concern.

Results showed that this particular type of interaction was not an important factor, and indeed was almost non-existent. The authors stated, "provided adequate unbiased progeny information is utilized, a bull may be safely selected and used for artificial insemination in one region, even though he was evaluated in the other region." Studies on the importance of sire by herd interactions have been in very close agreement, with one being conducted by Florida researchers on records of Florida DHIA Guernsey, Holstein and Jersey first-calf heifers. Everyone recognizes that even the best bull may have a certain number of poor daughters and that a dairyman who has only a few daughters of a particular sire may have been unlucky, by chance, and gotten one or more of them. The opposite can happen also with poor sires.

The problem is to determine how important the interactions are overall, realizing that a certain amount of shifting in rankings of sires can occur by chance, particularly if a sire is represented in one or more herds by only a small number of daughters. The research found that under Florida DHIA herd conditions; these interactions were of little or no practical or theoretical consequence for milk and fat yields. There was some evidence among Guernsey's and Jerseys of an interaction for fat percentage; this has been suggested in other research studies also, but many of the scientists who have studied interactions have not looked at fat percentage. In summary, it appears that sires proven superior for milk and fat yields in other herds can be used with confidence in Florida dairy herds, without concern for possible sire by herd interactions. The possibility exists for the presence of interactions with fat percentage.

FINDING SUPERIOR FEMALES

Measurements for most of the economically important traits can be made directly on females. It frequently takes a long time to get the measurements of interest, however. Mature body size, would have to wait until the cow is 6 to 8 years old. An estimate of milk yield must wait until the first lactation is completed (at about 3 years of age), unless one accepts an estimate of the lactation based on the first several months.

These points are mentioned because genetic progress is influenced by the time at which measurements are made, and their reliability. Take milk yield as an example. An estimate of the milk producing ability of a cow comes with her first milk record. Certainly, however, one knows more about her when there are two records rather than one, and even more with three records. Generally, the more records available, the more reliable the estimate of her real ability; but it takes longer to get more records. An estimate of the producing ability of a cow was called "Most Probable Producing Ability" years ago by Dr. J.L. Lush, well known animal geneticist. This measure takes into consideration the idea of repeatability, and shows how the number of records a cow has should influence the estimate of her producing ability. If a heifer produces more than her herdmates during her first lactation, we will certainly guess that she is a better than average animal.

If the average of her first two records is above herdmates, one would be even more confident that she is truly above average, and her good performance did not just happen from some favourable accident which probably will not happen again. The calculations are simple and can be applied to any trait of interest. Although the concept is rarely mentioned today, since availablity of computers makes more sophisticated procedures feasible, MPPA serves as the basis for modern methods. State the milk record, or any other measurement, as a deviation from herdmates. If the cow has two measurements, state each as a deviation, and take the average. MPPA then can be calculated as:

$$\text{MPPA} = (\text{Average deviation}) \times [(nr)/(1 + nr - r)]$$

where n = the number of measurements and r = the repeatability. How much better two measurements or records are than one, as far as reliability, is determined by a similar equation.

$$\text{Gain in accuracy} = 1 - [(1 + nr - r) / n]$$

with n and r as in MPPA. The more records one has on a cow, the more one can accept her records at face value. If she has six records averaging 2000 pounds better than her herdmates, the best guess of her MPPA is 1714 pounds better.

Which cow should be selected as a dam for a future herd sire, a cow with two records averaging 3000 pounds above herdmates, or one with four records averaging 2400 pounds above? The two estimates of MPPA are close, being 2000 pounds for the cow with two records and 1920 pounds for the older cow. One probably would look at other characteristics of the two animals, particularly fat per cent but possibly others, before making a decision. The more sophisticated and useful application of this concept, of course, is that it is incorporated into calculations of ERPA.

GENETIC CHANGE IN MILK YIELD

The theoretical upper limits of genetic change in milk yield have been known for quite some time. Knowing the various repeatabilities, heritabilities, generation intervals and selection pressures (intensities), one can calculate what would be expected to happen. This tells nothing about what is actually going on, of course, but it does tell what one can expect under the best conditions.

Potential

If milk yield is given maximum emphasis in the selection programme, with little or no pressure being exerted on other traits, and efficient use is made of an AI sire proving programme, the maximum attainable genetic change in milk yield is slightly over 2.0 per cent per year. Most of this, over 90 per cent, theoretically would come directly or indirectly from sire selection. In a Holstein population averaging 15,000 pounds per cow, this would amount to 300 pounds per year. Though this estimate has appeared disappointingly small to some breeders, it is more than the actual total change which has been attained in the United States.

Observed

It is only in recent years that it has been possible to obtain reliable estimates of the actual genetic change which has occurred in milk production. A major problem in the actual measurement of such progress is that environmental and genetic changes are confounded in such a manner over time as to make their separation extremely difficult. With the availability of high-speed large-capacity electronic computers, many of the problems can now be handled. Use of the animal model has improved the accuracy of the estimates. As far as progress arising from selection of males and females, theoretical estimates has been right on target.

About 90 to 92 per cent of all genetic change in milk yield has occurred from sire selection; this represents the selection of males and females to produce herd sires and the selection of males to produce herd dams. The remaining 8 to 10 per cent of genetic progress arises from selection of females to produce herd dams. The fact that we must raise nearly all female calves born alive in a population to maintain population size, and the difficulties of evaluating the true genetic merit of dams, precludes much selection pressure from selection of dams.

Average annual genetic change in milk yield of tested dairy cattle has been variable but seems to be about 0.7 per cent of their average levels. This would amount to about 105 pounds per cow in a 15,000 pound herd, or 140 pounds in a 20,000 pound herd. In individual herds, one would expect to find genetic changes which were considerably higher or lower. If 2.0 per cent annual genetic change is possible, as theory suggests, dairymen are progressing genetically only about one-third as fast as possible.

6

Drug Resistance and Medicinal Fungi

Drug resistance is the reduction in effectiveness of a drug such as an antimicrobial or an antineoplastic in curing a disease or condition. When the drug is not intended to kill or inhibit a pathogen, then the term is equivalent to *dosage failure* or drug tolerance. More commonly, the term is used in the context of resistance that pathogens have "acquired", that is, resistance has evolved. When an organism is resistant to more than one drug, it is said to be multidrug-resistant. In a broad sense the immune system of an organism is a drug delivery system, albeit autonomous, and faces the same arms race problems as external drug delivery.

The development of antibiotic resistance in particular stems from the drugs targeting only specific bacterial proteins. Because the drug *is* so specific, any mutation in these proteins will interfere with or negate its destructive effect, resulting in antibiotic resistance.

Bacteria are capable of not only altering the enzyme targeted by antibiotics, but also by the use of enzymes to modify the antibiotic itself and thus neutralise it. Examples of target-altering pathogens are *Staphylococcus aureus*, vancomycin-resistant enterococci and macrolide-resistant *Streptococcus*, while examples of antibiotic-modifying microbes are *Pseudomonas aeruginosa* and aminoglycoside-resistant *Acinetobacter baumannii*.

In short, the lack of concerted effort by governments and the pharmaceutical industry, together with the innate capacity of microbes to develop resistance at a rate that outpaces development of new drugs, suggests that existing strategies for developing viable, long-term anti-microbial therapies are ultimately doomed to failure. Without alternative strategies, the acquisition of drug resistance by pathogenic microorganisms looms as possibly one of the most significant public health threats facing humanity in the 21st century. Resistance to chemicals is only one aspect of the problem, another being resistance to physical factors such as temperature, pressure, sound, radiation and magnetism, and not discussed in this article, but found at Physical factors affecting microbial life.

INTRODUCTION

Drug or toxin or chemical resistance is a consequence of evolution and is a response to pressures imposed on any living organism. Individual organisms vary in their sensitivity to the drug used and some with greater fitness may be capable of surviving drug treatment. Drug-resistant traits are accordingly inherited by subsequent offspring, resulting in a population that is more drug-resistant.

Unless the drug used makes sexual reproduction or cell-division or horizontal gene transfer impossible in the entire target population, resistance to the drug will inevitably follow. This can be seen in cancerous tumors where some cells may develop resistance to the drugs used in chemotherapy. Chemotherapy causes fibroblasts near tumors to produce large amounts of the protein WNT16B. This protein stimulates the growth of cancer cells which are drug-resistant. Malaria in 2012 has become a resurgent threat in South East Asia and sub-Saharan Africa, and drug-resistant strains of *Plasmodium falciparum* are posing massive problems for health authorities. Leprosy has shown an increasing resistance to dapsone.

A rapid process of sharing resistance exists among single-celled organisms, and is termed horizontal gene transfer in which there is a direct exchange of genes, particularly in the biofilm state. A similar asexual method is used by fungi and is called "parasexuality". Examples of drug-resistant strains are to be found in microorganisms such as bacteria and viruses, parasites both endo- and ecto-, plants, fungi, arthropods, mammals, birds, reptiles, fish, and amphibians.

In the domestic environment, drug-resistant strains of organism may arise from seemingly safe activities such as the use of bleach, tooth-brushing and mouthwashing, the use of antibiotics, disinfectants and detergents, shampoos, and soaps, particularly antibacterial soaps, hand-washing, surface sprays, application of deodorants, sunblocks and any cosmetic or health-care product, insecticides, and dips.

The chemicals contained in these preparations, besides harming beneficial organisms, may intentionally or inadvertently target organisms that have the potential to develop resistance.

- "Drug resistance develops naturally, but careless practices in drug supply and use are hastening it unnecessarily." - Center for Global Development
- "The overuse of antibacterial cleaning products in the home may be producing strains of multi-antibiotic-resistant bacteria." - Better Health Channel - Australian Government
- "The use and misuse of antimicrobials in human medicine and animal husbandry over the past 70 years has led to a relentless rise in the number and types of microorganisms resistant to these medicines -

leading to death, increased suffering and disability, and higher health care costs." - World Health Organisation 2010

- "Deaths from acute respiratory infections, diarrhoeal diseases, measles, AIDS, malaria, and tuberculosis account for more than 85 per cent of the mortality from infection worldwide. Resistance to first-line drugs in most of the pathogens causing these diseases ranges from zero to almost 100 per cent. In some instances resistance to second- and thirdline agents is seriously compromising treatment outcome. Added to this is the significant global burden of resistant, hospital-acquired infections, the emerging problems of antiviral resistance and the increasing problems of drug resistance in the neglected parasitic diseases of poor and marginalised populations." - WHO Global Strategy for Containment of Antimicrobial Resistance 2010

MECHANISMS OF DRUGS

The four main mechanisms by which microorganisms exhibit resistance to antimicrobials are:

1. *Drug inactivation or modification*: *e.g.*, enzymatic deactivation of *Penicillin* G in some penicillin-resistant bacteria through the production of â-lactamases.
2. *Alteration of target site*: *e.g.*, alteration of PBP — the binding target site of penicillins — in MRSA and other penicillin-resistant bacteria.
3. *Alteration of metabolic pathway*: *e.g.*, some sulfonamide-resistant bacteria do not require para-aminobenzoic acid (PABA), an important precursor for the synthesis of folic acid and nucleic acids in bacteria inhibited by sulfonamides. Instead, like mammalian cells, they turn to utilising preformed folic acid.
4. *Reduced drug accumulation*: by decreasing drug permeability and/or increasing active efflux (pumping out) of the drugs across the cell surface.

METABOLIC PRICE

Biological cost or *metabolic price* is a measure of the increased energy metabolism required to achieve a function. Drug resistance has a high metabolic price in pathogens for which this concept is relevant (bacteria, endoparasites, and tumor cells.) In viruses, an equivalent "cost" is genomic complexity.

TREATMENT

The chances of drug resistance can sometimes be minimised by using multiple drugs simultaneously. This works because individual mutations can be independent and may tackle only one drug at a time; if the individuals are still killed by the other drugs, then the mutations cannot persist. This was used successfully in tuberculosis. However, cross resistance where mutations confer resistance to two or more treatments can be problematic.

For antibiotic resistance, which represents a widespread problem nowadays, destroying the resistant bacteria can be achieved by phage therapy, in which specific bacteriophage (virus that kill bacteria) are being used.

MEDICINAL MYCOLOGY

For centuries, certain mushrooms have been documented as a folk medicine in China, Japan, and Russia. Although the use of mushrooms in folk medicine is largely centered on the Asian continent, people in other parts of the world like the Middle East, Poland and Belarus have been documented using mushrooms for medicinal purposes. Certain mushrooms, especially polypores like Reishi were thought to be able to benefit a wide variety of health ailments. Medicinal mushroom research in the United States is currently active, with studies taking place at City of Hope National Medical Center, as well as the Memorial Sloan–Kettering Cancer Center.

Current research focuses on mushrooms that may have hypoglycemic activity, anti-cancer activity, anti-pathogenic activity, and immune system enhancing activity. Recent research has found that the oyster mushroom naturally contains the cholesterol-lowering drug lovastatin, mushrooms produce large amounts of vitamin D when exposed to UV light, and that certain fungi may be a future source of taxol. To date, penicillin, lovastatin, ciclosporin, griseofulvin, cephalosporin, ergometrine, and statins are the most famous pharmaceuticals which have been isolated from the fungi kingdom.

MEDICINAL MUSHROOMS

Medicinal mushrooms are mushrooms used or studied as possible treatments for disease. Research indicates mushrooms have potential antiviral, antimicrobial, anticancer, antihyperglycemic, hypocholesterolemic, and anti-inflammatory activities. The pleuromutilin antibiotics are derivatives of the mushroom isolate pleuromutilin. The mushroom isolates lentinan, PSK, and PSP, are licensed pharmaceuticals in certain countries. Fungi that do not produce mushrooms, are currently being used to manufacture paclitaxel, artemisinin, and were the original source of the first beta-lactam antibiotics, statins, immunosuppressants like ciclosporin, and antifungals like griseofulvin.

HISTORY

Mushrooms, fermentation molds, mycelia, and lichens, have a history of medicinal use spanning millennia. The mushroom with the longest record of medicinal use *Ganoderma lucidum*, is known in Chinese as líng zhî ("spirit plant"), and in Japanese as mannentake ("10,000 year mushroom"). In ancient

Japan, *Grifola frondosa* was worth its weight in silver. *Inonotus obliquus* was used in Russia as early as the 16th century, and its medicinal properties were described by Nobel laureate Alexandr Solzhenitsyn. Ancient Egyptians considered mushrooms food for royalty. Ötzi the Iceman was found carrying *Fomes fomentarius* and *Piptoporus betulinus*. A Hadith states, "Truffles are manna which Allah sent to the people of Israel through Moses, and its juice is a medicine for the eyes."

DEFENSIVE ACTIVITIES

Mushrooms have antimicrobial, antiviral, and nematicidal, activities *in vitro*. Antibiotics retapamulin, tiamulin, and valnemulin, are derivatives of the mushroom isolate pleuromutilin. Plectasin is a mushroom derived defensin antibiotic. The immunosuppressants ciclosporin, mycophenolic acid, mizoribine, the antibiotics penicillin, cephalosporins, fusafungine, usnic acid, fusidic acid, fumagillin, brefeldin A, verrucarin A, alamethicin, and the antifungals griseofulvin, echinocandins, strobilurin, azoxystrobin, caspofungin, micafungin, were discovered in fungi that do not produce mushrooms. Codinaeopsin is an antimalarial fungal isolate, and Sanofi uses fungi to manufacture artemisinin.

CHOLESTEROL BIOSYNTHESIS INHIBITORS

Pleurotus creates the statin lovastatin. A study noted an unidentified statin in *Agaricus bisporus*. Zaragozic acids were discovered in fungi, and the pravastatin precursor mevastatin, was discovered in *Penicillium*. The red yeast rice fungus *Monascus purpureus* creates lovastatin, mevastatin, and the simvastatin precursor, monacolin J. Mushroom isolate eritadenine and its 3-deaza derivatives have hypocholesterolemic activity *in vivo*.

ANTICANCER ACTIVITIES

Lentinan, PSK, and PSP, are registered anticancer immunologic adjuvants/BRMs. Irofulven and acylfulvene are anticancer derivatives of illudin S. *Clavariadelphus truncatus* creates the FTI clavaric acid. *Inonotus obliquus* creates the betulinic acid precursor betulin. *Flammulina velutipes* creates asparaginase. *Agaricus* creates conjugated linoleic acid. BMS manufactures paclitaxel using *Penicillium* and plant cell fermentation (PCF). Endophytic fungi can synthesize podophyllotoxin and camptothecin, precursors to etoposide, teniposide, topotecan, and irinotecan.

VITAMIN D AND ERGOTHIONEINE

Mushrooms exposed to UV-light create significant amounts of vitamin D_1, vitamin D_2, and vitamin D_4. This is notable, as non-animal sources of vitamin D are practically nonexistent. Although ergothioneine is only synthesized by mushrooms, other forms of fungi, and certain bacteria, human gene SLC22A4 ensures accumulation.

ANTIHYPERGLYCEMIC ACTIVITIES

Mushrooms have antihyperglycemic activity *in vivo*. *Inonotus obliquus* creates DPP-4 inhibitors. *Grifola frondosa* creates α-glucosidase inhibitors. *Trametes versicolour* creates α-amylase inhibitors, and the antihyperglycemic cyclic heptapeptide ternatin.

PSYCHOTROPIC ACTIVITIES

Over 200 species of mushrooms synthesize psilocybin, and its degradation product psilocin. Due to a chemical resemblance to serotonin, psilocin is a non-selective, partial serotonin agonist. *Claviceps* isolates and/or derivatives, were used to develop the headache medications cafergot, dihydroergotamine, methysergide, methylergometrine, the dementia medications hydergine, nicergoline, the Parkinson's disease medications lisuride, bromocriptine, cabergoline, pergolide, and the controlled substance lysergic acid diethylamide.

The mycorrhizal mushroom *Polyozellus multiplex* synthesizes prolyl endopeptidase inhibitors polyozellin, thelephoric acid, and kynapcins –12, –13, –24, –28. The mycorrhizal mushroom *Boletus badius* synthesizes theanine. *Hericium erinaceus* isolates promote nerve growth factor synthesis *in vitro*.

OTHER ACTIVITIES

ACE inhibitor peptides were isolated from *Pleurotus*, *Agaricus*, *Ganoderma*, and *Flammulina*. Cordycepin, inotilone, quercinol, antcin A, and benzocamphorin F, are mushroom anti-inflammatory compounds. *Pholiota squarrosa* contains xanthine oxidase inhibitors.

Coprinopsis atramentaria contains coprine, which metabolises to the acetaldehyde dehydrogenase inhibitor 1-aminocyclopropanol. Mushroom extracts inhibit 5-alpha reductase and/or aromatase *in vitro*. Other researched fungal isolates include, integrasone, ergosterol peroxide, enniatins, oudenone, velleral, peptaibols, kojic acid, beauvericin, pullulan, lichenin, gliotoxin, and asperlicin.

DRUGS FROM FUNGI

Fungi make an extraordinarily important contribution to managing disease in humans and other animals. At the beginning of the 21st century, Fungi were involved in the industrial processing of more than 10 of the 20 most profitable products used in human medicine. Two anti-cholesterol statins, the antibiotic penicillin and the immunosuppressant cyclosporin A are among the top 10. Each of these has a turn over in excess of $1 billion annually. Drug discovery continues. The following have recently been approved for human use: Micafungin is an antifungal agent; mycophenolate is used to prevent tissue rejection; rosuvastatin is usd to reduce cholesterol; and cefditoren as an antibiotic.

Fungi are extremely useful organisms in biotechnology. Fungi construct unique complex molecules using established metabolic pathways. Different taxa produce sets of related molecules, each with slightly different final products. Metabolites formed along the metabolic pathway may also be biologically active. In addition, the final compounds are often released into the environment. Manipulation of the genome, and environmental conditions during formation of compounds, enable the optimisation of product formation.

On the negative side, single isolates of fungi in manufacture may lose their capacity to form or release the target molecules. Indeed, the target compound may only be expressed under specific conditions, or at a specific point in the life cycle of the fungus. It is amazing that so many biologically active compounds have been discovered and taken to the point where they are medically important. Indeed, attempts to 'discover' new and exciting molecules remains the core activity of many research groups.

The role of fungi was established early in history. Yeasts have been used in the making of bread and alcohol since the beginning of civilization. In modern times, the discovery of penicillin marked the beginning of a new approach to microbial diseases in human health. More recent approaches include the application of hydrophobins to surfaces leading to biocompatibility of implants, and to emulsion formation improving drug delivery. The established importance of fungi is being expanding way beyond their capacity to transform and protect.

ANTIBIOTICS FROM FUNGI

In 1941, penicillin from the fungus Penicillium chrysogenum was first used successfully to treat an infection caused by a bacterium. Use of penicilin revolutionised the treatment of pathogenic disease. Many formally fatal diseases caused by bacteria became treatable, and new forms of medical intervention were possible. When penicillin was first produced, the concentration of active ingredient was approximately 1 microgram per ml of broth solution. Today, improved strains and highly developed fermentation technologies produce more than 700 micrograms per ml of active ingredient.

In the early broths, several closely related molecules were present. These molecules are beta lactam rings fused to five-membered thiazolidine rings, with a side chain. The side chain can be chemically modified to provide slightly different properties to the compound. The natural penicillins have a number of disadvantages. They are destroyed in the acid stomach, and so cannot be used orally. They are sensitive to beta lactamases, which are produced by resistant bacteria, thus reducing their effectiveness. Also, they only act on gram positive bacteria.

Modifications to manufacturing conditions have resulted in the development of oral forms. However, antibiotic resistance among bacteria is becoming an extremely important aspect determining the long-term use of all antibiotics.

Cephalosporins also contain the beta lactam ring. The original fungus found to produce the compounds was a Cephalosporium, hence the name. As with penicillin, the cephalosporin antibiotics have a number of disadvantages. Industrial modification of the active ingredients has reduced these problems.

The only broadly useful antifungal agent from fungi is griseofulvin. The original source was Penicillium griseofulvin. Griseofulvin is fungistatic, rather than fungicidal. It is used for the treatment of dermatophytes, as it accumulates in the hair and skin following topical application. More recently, several new groups have been developed. Strobilurins target the ubihydroquinone oxidation centre, and in mammals, the compound from fungi is immediately excreted. Basidiomycetes, especially from tropical regions, produce an enormous diversity of these compounds. Sordarins are structurally complex molecules that show a remarkably narrow range of action against yeasts and yeast-like fungi. The compounds inhibit protein biosynthesis and so may become important agents against a number of fungal pathogens of humans. Echinocandins are cyclic peptides with a long fatty acid side chain. They target cell wall formation. Semi-synthetic members of the group of compounds include pneumocandins which are in use in humans.

IMMUNE SUPPRESSANTS

Cyclosporin A is a primary metabolite of several fungi, including Trichoderma polysporum and Cylindrocarpon lucidum. Cyclosporin A has proven to be a powerful immunosuppressant in mammals, being widely used during and after bone marrow and organ transplants in humans.

Cyclosporin A is a cyclic peptide consisting of 11 mainly hydrophobic amino acids. Its inhibition of lymphocytes was first discovered during the 1970s. Subsequently, the mode of action was elucidated.

Cyclosporin A binds to a cytosolic protein called cyclophilin. Cyclophilin is found amongst many different organisms and its form appears highly conserved. Cyclophilin is involved with folding the protein ribonuclease. However, the Cyclosporin A/cyclophilin complex also binds to calcineurin. Calcineurin dephosphorylates a transcription factor, thereby triggering transcription of numerous genes associated with T cell proliferation. When the complex binds to calcineurin, T cell proliferation is suppressed. The inhibition of T cells proliferation results in the suppression of the activation process associated with invasion by foreign bodies. As a consequence, transplant tissues, which are foreign bodies, are not rejected.

Calcineurin is also highly conserved amongst phylogenetically diverse organisms. In fungi such as the human pathogen Cryptococcus neoformans, calcineurin is necessary for recovery from cell cycle arrest, growth in hypertonic solutions and regulation of the calcium pump. Thus the interaction of the Cyclosporin A/cyclophilin complex with calcineurin in Cryptococcus will result in death of the pathogen. However, in humans, cyclosporin also suppresses the immune system. The side effect is an unacceptable risk, and Cyclosporin A is not used as a fungicide in humans at present.

Gliotoxins also have immunological and antibiotic activity. Produced by many fungi including Aspergillus fumigatus, gliotoxins belong to a class of compounds called epipolythiodioxopiperazines. The antibiotic activity is widely recognised and considered uninteresting. However, its effect on the immune system, especially macrophages, is being re-examined. A wide range of other compounds with antibiotic activity are also known. They have been rejected for use in medicine because of unwanted side effects, or instability of the active compound.

ERGOT ALKALOIDS

Claviceps purpurea is the causal agent of St Anthonies fire, a scourge of the middle ages when ergots contaminated flour. The ergots contain many alkaloids. Their effects are quite variable. They act on the sympathetic nervous system resulting in the inhibition of noradrenaline and sclerotin, causing dilation of blood vessels.

They also act directly on the smooth muscles of the uterus causing contractions, thus their early use to induce abortion. Their strongest effect is intoxication, caused by lysergic acid amides, one of which is the recreational (and illegal) drug, LSD. Ergot alkaloids have a number of medicinal uses. Perhaps the most widespread use is in the treatment of migraines. The vasodilator activity reduces tension during an attack. The drugs also reduce blood pressure, though with untoward side effects. Alkaloids are now produced in culture by strains of C. fusiformis and C. paspalii.

STATINS

Aspergillus terreus, a soil-borne fungus, produces a secondary metabolite called lovastatin and Phoma sp produces squalestatin. Statins have been used to reduce or remove low density lipoproteins from blood vessels in humans. In fact, the compounds all act via an enzyme in the liver that makes cholesterol, lovastatin inhibits HMG CoA reductase and squalestatin inhibits squalene synthase. By blocking the enzyme, the body removes cholesterol complexes from the inside of blood vessels. This has the effect of reducing or removing blockages in arteries, and thereby reducing the chance of a heart attack, strokes and diabetes.

In addition, statins have been implicated in attracting stem cells to damaged tissues. The stem cells then appear to regenerate the tissue. Some statins induce problems. One form of the drug has been associated with muscle wastage. Others appear to lack side effects and have been recommended for wide spread use to control heart disease.

MEDICINAL FUNGI

AGARICUS SUBRUFESCENS

Agaricus subrufescens (syn. Agaricus blazei, Agaricus brasiliensis or Agaricus rufotegulis) is a species of mushroom, commonly known as almond mushroom, mushroom of the sun, God's mushroom, mushroom of life, royal sun agaricus, jisongrong or himematsutake and by a number of other names. Agaricus subrufescens is a choice edible, with a somewhat sweet taste and fragrance of almonds. The fungus is also well known as a medicinal mushroom, for its purported medicinal properties, due to research which indicates it may stimulate the immune system.

Taxonomy

Agaricus subrufescens was first described by the American botanist Charles Horton Peck in 1893. During the late 19th and early 20th centuries, it was cultivated for the table in the eastern United States. It was discovered again in Brazil during the 1970s, and misidentified as *Agaricus blazei* Murrill, a species originally described from Florida. It was soon marketed for its purported medicinal properties under various names, including ABM (for *Agaricus blazei* Murill), *cogumelo do sol* (mushroom of the sun), *cogumelo de Deus* (mushroom of God), *cogumelo de vida* (mushroom of life), *himematsutake*, royal sun agaricus, *Mandelpilz*, and almond mushroom. In 2002, Didukh and Wasser correctly rejected the name *A. blazei* for this species, but unfortunately called the Brazilian fungus *A. brasiliensis*, a name that had already been used for a different species, *Agaricus brasiliensis* Fr. (1830). Richard Kerrigan undertook genetic and interfertility testing on several fungal strains, and showed that samples of the Brazilian strains called *A. blazei* and *A. brasiliensis* were genetically similar to, and interfertile with, North American populations of *Agaricus subrufescens*. These tests also found European samples called *A. rufotegulis* to be of the same species. Because *A. subrufescens* is the oldest name, it has taxonomical priority. Note that *Agaricus blazei* Murrill is a perfectly valid name, but for a completely different mushroom. *Agaricus silvaticus* Schaeff. is also a perfectly valid name for a common, north temperate, woodland mushroom. Neither is a synonym of *Agaricus subrufescens*.

Description

The cap is initially hemispherical, later becoming convex, with a diameter of 5 to 18 cm (2.0 to 7.1 in). The cap surface is covered with silk-like fibres, although in maturity it develops small scales (squamulose). The colour of the cap may range from white to grayish or dull reddish-brown; the cap margin typically splits with age. The flesh of *A. subrufescens* is white, and has the taste of "green nuts", with the odour of almonds. The gills are not attached to the stalk (free), narrow, and crowded closely together. They start out whitish in colour, then later pinkish and finally black-brown as the spores mature. Spores are ellipsoid, smooth, dark purplish-brown when viewed microscopically, with dimensions of 6–7.5 by 4–5 μm. The stipe is 6 to 15 cm (2.4 to 5.9 in) by 1 to 1.5 cm (0.4 to 0.6 in) thick, and bulbous at the base. Initially solid, the stipe becomes hollow with age; it is cottony (floccose) to scaly towards the base. The annulus is abundant and double-layered; it is bent downwards towards the stem, smooth and whitish on the upper side, and covered with cottony scales on the lower side.

Fig. The Floccose Stipe and Annulus of *A. Subrufescens*

Distribution and Habitat

Agaricus subrufescens forms fruitbodies singly or in clusters in leaf litter in rich soil, often in domestic habitats. Originally described from the northeastern United States and Canada, it has been found growing in California, Hawaii, Great Britain, the Netherlands, Taiwan, Philippines and Brazil.

Edibility

Aroma

Agaricus subrufescens is a choice edible, with a somewhat sweet taste and fragrance of almonds. The almond smell of the mushroom is mostly due to the presence of benzaldehyde, benzyl alcohol, benzonitrile, and methyl benzoate.

Commercial Use

Because *Agaricus subrufescens* contains a high level of beta glucans, compounds known for stimulating the immune system, the fungus is used in oncological therapy in Japan and Brazil. In addition to beta glucans, the mushroom's effect on the immune system is believed to be due to other polysaccharides, such as alpha glucans. In Japan, *Agaricus subrufescens* is sold under the brand names KingAgaricus 100, Sen-Sei-Ro Gold, and ABMK, and is used by an estimated 500,000 people In Japan, *Agaricus subrufescens* is also the most popular complementary and alternative medicine used by cancer patients. Although *Agaricus subrufescens* is cultivated in the United States, the largest exporters are China and Brazil. As noted in a scientific review of *A.*

subrufescens research, the range of quality in *A. subrufescens* cultivation can affect the mushroom's ability to impact cells of the immune system.

Recently, Watanabe *et al.* published a report in the *Biological and Pharmaceutical Bulletin* on a novel hybrid of *A. subrufescens* called *Basidiomycetes-X* (BDM-X) and a US patent was issued on a novel hybrid of the *A. subrufescens* edible mushroom which was crossbred (hybridised) with another medicinal mushroom, resulting in a new hybrid claimed to possess 10 to 3000 times the potency of similar but unpatented mushrooms.

Research

Many researchers have studied *Agaricus subrufescens*, as well as other medicinal mushrooms for close to 50 years, due to laboratory tests which show they may stimulate immune system cells and the production of immune system cytokines. Below is a summary of this research, which is often based on animal or cellular models.

Cancer Research

Table. Animal and Cellular Research has Shown that *Agaricus Subrufescens* Application or Consumption Appears to offer Anticancer Properties.

Experimental Model	Experimental Effect (Cell Culture)	Experimental Effect (Animal Research)	Experimental Effect (Clinical)
Colourectal cancer			Benefited hematological and immunological parameters
Fibrosarcoma	Inhibited growth via apoptosis	Inhibited growth	
Sarcoma		Inhibited angiogenesis, inhibited growth	
Gynaecological cancer			Increased NK cell activity, quality of life
Ovarian cancer	Inhibited growth and metastasis via apoptosis induction		Inhibited metastasis, growth
Lung cancer	Inhibited growth via apoptosis	Inhibited metastasis, growth	
Leukaemia	Inhibited growth via apoptosis	Inhibited growth	
Myeloma		Inhibited growth	
Hepatocarcinoma	Inhibited abnormal collagen formation	Inhibited growth	
Stomach cancer	Inhibited growth via apoptosis		
Prostate cancer	Inhibited growth via apoptosis	Inhibited growth	

Skin cancer	Inhibited growth when applied orally or topically

Immune System

Cellular and animal research has shown *A. subrufescens* may stimulate immune system cells and the production of cytokines, such as interferons and interleukins (reviewed by G. Hetland).

Direct Antiviral Properties

Agaricus subrufescens mushrooms are known to have antiviral properties in cell culture. The ability of *Agaricus subrufescens* to inhibit viruses in the human body has not been studied.

Other Possible Effects

Besides evidence *Agaricus subrufescens* may up-regulate the immune system, additional research suggests the mushroom has a beneficial effect on cholesterol, inhibiting pathogenic factors, and inhibiting angiogenesis. Limited clinical and animal research suggests *Agaricus subrufescens* consumption may lower blood glucose levels and improve insulin resistance. Dietary studies have listed *A. subrufescens* as a non-animal source of conjugated linoleic acid (CLA), which contains combined trans- and a far greater amount of cis-fat isomers, the latter which is believed to be responsible for an increase in the metabolising of fatty tissue in the body (resting metabolic rate) and an increase in the development of muscle mass (which burns more calories than fat mass), resulting in healthy weight loss and better overall health.

ANTRODIA

Antrodia is a genus of fungi in the family Fomitopsidaceae. Antrodia species have fruiting bodies that typically lie flat or spread out on the growing surface, with the hymenium exposed to the outside; the edges may be turned so as to form narrow brackets. Most species are found in temperate and boreal forests, and cause brown rot.

Description

Antrodia are effused-resupinate, that is, they lie stretched out on the growing surface with the hymenium exposed on the outer side, but turned out at the edges to form brackets. When present, these brackets are typically white or pale brown. The pores on the surface of the hymenium may be round or angular. The context is white or pale. All species cause brown-rot.

Typically, basidiospores are thin-walled, cylindrical, and narrowly ellipsoidal or fusiform in shape. Most species grow on the wood of coniferous trees, except for *A. albida*, with grows the dead wood on deciduous trees.

Phylogeny

In order to reliably identify the various species and strains of medicinal *Antrodia*, genetic markers have been developed and phylogenetic analyses performed. This analysis showed that there are three distinct phylogenetic lineages with the *Antrodia* genus.

Classification

The modern definition of the genus was described by Gilbertson and Ryvarden (1986) in their book *North American Polypores*.

Distribution

Roughly twenty-nine species are known from Europe, 21 species in North America, and 18 species in East Asia, although more new species have been reported since the time of these publications.

Species

There are approximately 50 *species in this genus*: Antrodia albida, Antrodia albidoides, Antrodia albobrunnea, Antrodia alpina, Antrodia aurantia, Antrodia bondartsevae, Antrodia calceus, Antrodia carbonica, Antrodia cinnamomea, Antrodia citrina, Antrodia conchata, Antrodia crassa, Antrodia daedaleiformis, Antrodia destructor, Antrodia eutelea, Antrodia ferox, Antrodia flava, Antrodia formosana, Antrodia gossypina, Antrodia gossypium, Antrodia heteromorpha, Antrodia infirma, Antrodia juniperina, Antrodia lalashana, Antrodia leucaena, Antrodia lindbladii, Antrodia luteola, Antrodia macra, Antrodia macrospora, Antrodia malicola, Antrodia mellita, Antrodia oleracea, Antrodia pictilis, Antrodia pini-cubensis, Antrodia plicata, Antrodia porothelioides, Antrodia primaeva, Antrodia pseudosinuosa, Antrodia pulvinascens, Antrodia ramentacea, Antrodia rupamii, Antrodia sandaliae, Antrodia saxonica, Antrodia serialiformis, Antrodia serialis, Antrodia serpens, Antrodia sinuosa, Antrodia sitchensis, Antrodia sordida, Antrodia stratosa, Antrodia subalbidoides, Antrodia submalicola, Antrodia subramentacea, Antrodia taxa, Antrodia terrei, Antrodia variiformis, Antrodia xantha.

ASTRAEUS HYGROMETRICUS

Astraeus hygrometricus, commonly known as the hygroscopic earthstar, the barometer earthstar, or the false earthstar, is a species of fungus in the Diplocystaceae family. Young specimens resemble a puffball when young and unopened. In maturity, the mushroom displays the characteristic earthstar shape that is a result of the outer layer of fruit body tissue splitting open in a star-like manner. The false earthstar is an ectomycorrhizal species that grows in association with various trees, especially in sandy soils. *A. hygrometricus* has a cosmopolitan distribution, and is common in temperate and tropical regions.

Its common names refer to the fact that it is hygroscopic (water-absorbing), and can open up its rays to expose the spore sac in response to increased humidity, and close them up again in drier conditions. The rays have an irregularly cracked surface, while the spore case is pale brown and smooth with an irregular slit or tear at the top. The gleba is white initially, but turns brown and powdery when the spores mature. The spores are reddish-brown, roughly spherical with minute warts, measuring 7.5–11 micrometers in diameter.

Despite a similar overall appearance, *A. hygrometricus* is not related to the true earthstars of genus *Geastrum*, although historically, they have been taxonomically confused. The species was first described by Christiaan Hendrik Persoon in 1801 as *Geastrum hygrometricus*. In 1885, Andrew P. Morgan proposed that differences in microscopic characteristics warranted the creation of a new genus *Astraeus* distinct from *Geastrum*; this opinion was not universally accepted by later authorities. Several Asian populations formerly thought to be *A. hygrometricus* were renamed in the 2000s once phylogenetic analyses revealed they were unique *Astraeus* species, including *A. asiaticus* and *A. odoratus*. Research has revealed the presence of several bioactive chemical compounds in the fruit bodies. North American field guides typically rate *A. hygrometricus* as inedible.

Taxonomy, Naming, and Phylogeny

Because this species resembles the earthstar fungi of *Geastrum*, it was placed in that genus by early authors, starting with Christian Hendrik Persoon in 1801 (as *Geaster*, an alternate spelling of *Geastrum*). According to the American botanist Andrew P. Morgan, however, the species differed from those of *Geastrum* in not having open chambers in the young gleba, having larger and branched capillitium threads, not having a true hymenium, and having larger spores. Accordingly, Morgan set Persoon's *Geaster hygrometricum* as the type species of his new genus *Astraeus* in 1889. Despite Morgan's publication, some authorities in the following decades continued to classify the species in *Geastrum*. The New-Zealand based mycologist Gordon Herriot Cunningham explicitly transferred the species back to the genus *Geastrum* in 1944, explaining:

The treatment of this species by certain taxonomists well illustrates the pitfalls that lie in wait for those who worship at the shrine of ontogenic classification... The only feature of those outlined in which the species differs from others of *Geastrum* is the somewhat primitive hymenium. In the developing plant the glebal cavities are separated by tramal plates so tenuous as to be overlooked by the uncritical worker. Each cavity is filled with basidia somewhat irregularly arranged in clusters (like those of *Scleroderma*) and not in the definite palisade of the species which have been studied. This difference disappears as maturity is reached, when plants resemble closely the fructification of any other member of the genus. The taxonomist is then unable to indicate any point of difference by which "*Astraeus*" may be separated from *Geastrum*, which indicates that the name should be discarded.

Cunningham's treatment was not followed by later authorities, who largely considered *Astraeus* a distinct genus. According to the taxonomical authority MycoBank, synonyms of *Astraeus hygrometricus* include *Lycoperdon stellatus* Scop. (1772); *Geastrum fibrillosum* Schwein. (1822); *Geastrum stellatum* (Scop.) Wettst. (1885); and *Astraeus stellatus* E.Fisch. (1900). *Astraeus hygrometricus* has been given a number of colloquial names that allude to its hygroscopic behaviour, including the "hygrometer earthstar", the "hygroscopic earthstar", the "barometer earthstar", and the "water-measure earthstar". The resemblance to *Geastrum* species (also known as true earthstars) accounts for the common name "false earthstar". The specific name is derived from the Greek words Qãñüò (*hygros*) "wet" and ìÝôñïí (*metron*) "measure". The German Mycological Society selected the species as their "Mushroom of the Year" in 2005.

Studies in the 2000s showed that several species from Asian collection sites labelled under the specific epithet *hygrometricus* were actually considerably variable in a number of macroscopic and microscopic characteristics. Molecular studies of the DNA sequences of the ITS region of the ribosomal DNA from a number of *Astraeus* specimens from around the world have helped to clarify phylogenetic relationships within the genus. Based on these results, two Asian "*hygrometricus*" populations have been described as new species: *A. asiaticus* and *A. odoratus* (synonymous with Petcharat's *A. thailandicus* described in 2003). Preliminary DNA analyses suggests that the European *A. hygrometricus* described by Persoon is a different species than the North American version described by Morgan, and that the European population may be divided into two distinct phylotypes, from France and from the Mediterranean. A 2010 study identified a Japanese species, previously identified as *A. hygrometricus*, as genetically distinct; it has yet to be officially named.

A form of the species found in Korea and Japan, *A. hygrometricus* var. *koreanus*, was named by V.J. Stan□k in 1958; it was later (1976) published as a distinct species—*A. koreanus*—by Hanns Kreisel. As pointed out by Fangfuk and colleagues, clarification of the proper name for this taxon must await analysis of *A. hygrometricus* var. *koreanus* specimens from the type locality in North Korea.

Description

Young specimens of *A. hygrometricus* have roughly spherical fruit bodies that typically start their development partially embedded in the substrate. A smooth whitish mycelial layer covers the fruit body, and may be partially encrusted with debris. As the fruit body matures, the mycelial layer tears away, and the outer tissue layer, the exoperidium, breaks open in a star-shaped (stellate) pattern to form 4–20 irregular "rays". This simultaneously pushes the fruit body above ground to reveal a round spore case enclosed in a thin papery endoperidium.

Fig. A Collection of Fruit Bodies

The rays open and close in response to levels of moisture in the environment, opening up in high humidity, and closing when the air is dry. This is possible because the exoperidium is made of several different layers of tissue; the innermost, fibrous layer is hygroscopic, and curls or uncurls the entire ray as it loses or gains moisture from its surroundings. This adaptation enables the fruit body to disperse spores at times of optimum moisture, and reduce evaporation during dry periods. Further, dry fruit bodies with the rays curled up may be readily blown about by the wind, allowing them to scatter spores from the pore as they roll.

- "This veritable barometer is the most theatrical of all the earthstars. A few minutes immersion in water will open up old, dried-up specimens that seem as tightly closed as clenched fists."

The fruit body is 1–8 cm (0.4–3.1 in) in diameter from tip to tip when expanded. The exoperidium is thick, and the rays are typically areolate (divided into small areas by cracks and crevices) on the upper surface, and are dark grey to black. The spore case is sessile (lacking a stalk), light gray to tan colour and 1 to 3 cm (0.4 to 1.2 in) broad with a felt-like or scurfy (coated with loose scaly crust) surface; the top of the spore case is opened by an irregular slit, tear or pore.

The interior of the spore case, the gleba, is white and solid when young, and divided into oval locules—a characteristic that helps to distinguish it from *Geastrum*.

The gleba becomes brown and powdery as the specimen matures. Small dark hairlike threads (rhizomorphs) extend from the base of the fruit body into the substrate. The rhizomorphs are fragile, and often break off after maturity.

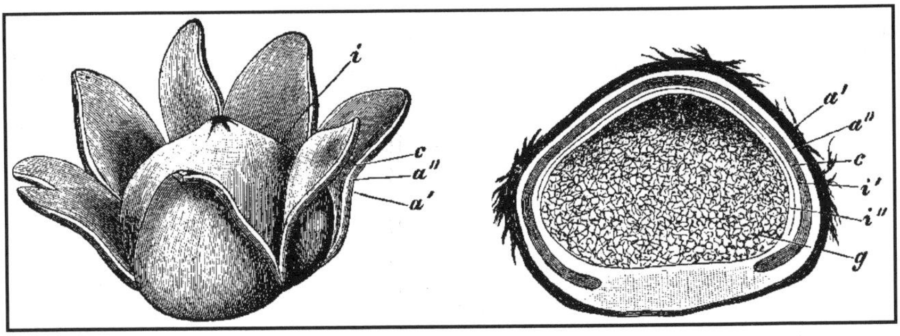

Fig. Mature Fruit Body (Left); Cross-section of Young Fruit Body (Right).

The spores are spherical or nearly so, reddish-brown, thick-walled and verrucose (covered with warts and spines). The spores' dimensions are 7–11 μm; the warts are about 1 μm long. The spores are non-amyloid, and will not stain with iodine from Melzer's reagent. The use of scanning electron microscopy has shown that the spines are 0.90–1.45 μm long, rounded at the tip, narrow, tapered, and sometime joined together at the top. The capillitia (masses of thread-like sterile fibres dispersed among the spores) are branched, 3.5–6.5 μm in diameter, and hyaline (translucent). The basidia (spore-bearing cells) are four- to eight-spored, with very short sterigmata. The basidia are arranged in long strings of clusters; individual basidia measure 11–15 by 18–24 μm. The threads of the capillitia arise from the inner surface of the peridium, and are thick-walled, long, interwoven, and branched, measuring 3–5.5 μm thick. The exoperidium (the outer layer of tissue, comprising the rays) is made of four distinct layers of tissue: the mycelial layer contains branched hyphae that are 4–6 μm in diameter; the hyphae of the fibrous layer are 6–8 μm diameter and branched; the collenchyma-type layer has branched hyphae of 3–4 μm diameter; the soft layer contains hyphae that are 3–6 μm in diameter.

Edibility

Fig. Specimen from Japan

North American sources describe *A. hygrometricus* as being of either unknown edibility, or too tough to be edible. However, they are regularly consumed in Asia, including Nepal and South Bengal, where "local people consume them as delicious food". They are collected from the wild and sold in the markets of India.

A study of a closely related southeast Asian *Astraeus* species concluded that the fungus contained an abundance of volatile eight-carbon compounds (including octanol, 1-octen-3-ol, and 1-octen-3-one) that imparted a "mushroom-like, earthy, and pungent odour that was evident as an oily and moss-like smell upon opening the caps."

The study's authors further noted that the fruit bodies after cooking have a "roasted, maillard, herbal, and oily flavour." Volatile compounds detected after cooking the mushroom samples included furfural, benzaldehyde, cyclohexenone, and furanyl compounds. The regional differences in opinions on edibility are from sources published before it was known that North American and Asian versions of *A. hygrometricus* were not always the same; in some cases Asian specimens have been identified as new species, such as *A. asiaticus* and *A. odoratus*.

Similar Species

Fig. Astraeus Pteridis is Similar in Appearance, but Larger; it is found in North America and the Canary Islands.

Although *A. hygrometricus* bears a superficial resemblance to member of the "true earthstars" *Geastrum*, it may be readily differentiated from most by the hygroscopic nature of its rays. Hygroscopic earthstars include *G. arenarium*, *G. corollinum*, *G. floriforme*, *G. recolligens*, and *G. kotlabae*. Unlike *Geastrum*, the young fruit bodies of *A. hygrometricus* do not have a columella (sterile tissue in the gleba, at the base of the spore sac). *Geastrum* tends to have its spore sac opening surrounded by a peristome or a disc, in contrast with the single lacerate slit of *A. hygrometricus*.

There are also several microscopic differences: in *A. hygrometricus*, the basidia are not arranged in parallel columns, the spores are larger, and the threads of the capillitia are branched and continuous with the hyphae of the peridium. Despite these differences, older specimens can be difficult to distinguish from *Geastrum* in the field. One species of *Geastrum*, *G. mammosum*, does have thick and brittle rays that are moderately hygroscopic, and could be confused with *A. hygrometricus*; however, its spores are smaller than *A. hygrometricus*, typically about 4 µm in diameter.

Astraeus pteridis is larger, 5 to 15 cm (2.0 to 5.9 in) or more when expanded, and often has a more pronounced areolate pattern on the inner surface of the rays. It is found in North America and the Canary Islands. *A. asiaticus* and *A. odoratus* are two similar species known from throughout Asia and Southeast Asia, respectively. *A. odoratus* is distinguished from *A. hygrometricus* by a smooth outer mycelial layer with few adhering soil particles, 3–9 broad rays, and a fresh odour similar to moist soil.

The spore ornamentation of *A. odoratus* is also distinct from *A. hygrometricus*, with longer and narrower spines that often joined together. *A. asiaticus* has an outer peridial surface covered with small granules, and a gleba that is purplish-chestnut in colour, compared to the smooth peridial surface and brownish gleba of *A. hygrometricus*.

The upper limit of the spore size of *A. asiaticus* is larger than that of its more common relative, ranging from 8.75–15.2 µm. *A. koreanus* differs from the more common form in its smaller size, paler fruit body, and greater number of rays; microscopically, it has smaller spores (between 6.8 and 9 ìm in diameter), and the spines on the spores differ in length and morphology. It is known from Korea and Japan.

Habitat, Distribution, and Ecology

Astraeus hygrometricus is an ectomycorrhizal fungus and grows in association with a broad range of tree species. The mutualistic association between tree roots and the mycelium of the fungus helps the trees extract nutrients (particularly phosphorus) from the earth; in exchange, the fungus receives carbohydrates from photosynthesis. In North America, associations with oak and pine are usual, while in India, it has been noted to grow commonly with chir pine (*Pinus roxburghii*) and sal (*Shorea robusta*). The false earthstar is found on the ground in open fields, often scattered or in groups, especially in nutrient-poor, sandy or loamy soils. It has also been reported to grow on rocks, preferring acid substrates like slate and granite, while avoiding substrates rich in lime. In Nepal, fruit bodies have been collected at elevations of 3,000 m (9,800 ft). Fruit bodies typically appear in autumn, although the dry fruit bodies are persistent and may last up to several years. *Gelatinipulvinella astraeicola* is a leotiaceous fungus with minute, gelatinous, pulvinate (cushion-shaped) apothecia, known to grow only on the inner surface of the rays of dead *Astraeus* species, including *A. hygrometricus*.

The species has a cosmopolitan distribution except for arctic, alpine and cold temperate regions; it is common in temperate and tropical regions of the world. It has been collected in Africa, Asia, Australia, Europe, North America, and South America.

Bioactive Compounds

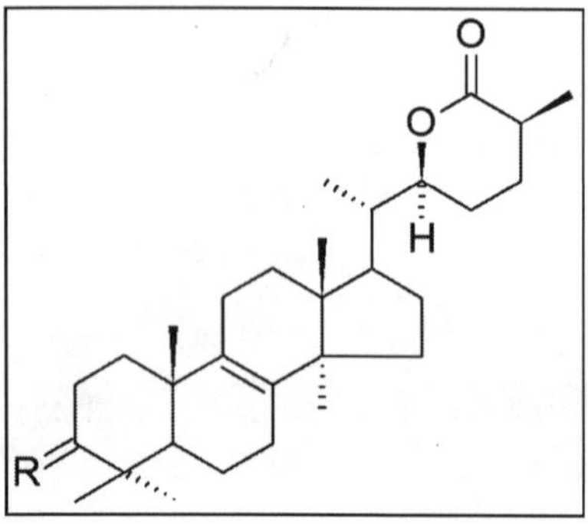

Fig. Backbone Chemical Structure upon which the Unique Triterpenes Astrahygrol, 3-*epi*-Astrahygrol, and Astrahygrone are Formed.

Mushroom polysaccharides from a number of species have attracted research interest for their immunomodulatory and antitumor properties. Extracts from *A. hygrometricus* containing the polysaccharide named AE2 were found to inhibit the growth of several tumor cell lines in laboratory tests, and stimulated the growth of splenocytes, thymocytes, and bone marrow cells from mice.

The extract also stimulated mouse cells associated with the immune system; specifically, it enhanced the activity of mouse natural killer cells, stimulated macrophages to produce nitric oxide, and enhanced production of cytokines. The activation of macrophages by AE2 might be mediated by a mitogen-activated protein kinase pathway of signal transduction. AE2 is made of the simple sugars mannose, glucose, and fucose in a 1:2:1 ratio.

In addition to the previously known steroid compounds ergosta-7,22-diene-3-ol acetate and ergosta-4,6,8-(14),22-tetraene-3-one, three unique triterpenes—derivatives of 3-hydroxy-lanostane—have been isolated from fruit bodies of *A. hygrometricus*. The compounds, named astrahygrol, 3-*epi*-astrahygrol, and astrahygrone (3-oxo-25*S*-lanost-8-eno-26,22-lactone), have δ-lactone (a six-membered ring) in the side chain—a chemical feature previously unknown in the Basidiomycetes. A previously unknown steryl ester (3β, 5α-dihydroxy-(22*E*, 24*R*)-ergosta-7,22-dien-6α-yl palmitate) has been isolated from mycelia grown in liquid culture. The compound has a polyhydroxylated ergostane-type nucleus.

Ethanol extracts of the fruit body are high in antioxidant activity, and have been shown in laboratory tests to have anti-inflammatory activity comparable to the drug diclofenac. Studies with mouse models have also demonstrated hepatoprotective (liver-protecting) ability, possibly by restoring diminished levels of the antioxidant enzymes superoxide dismutase and catalase caused by experimental exposure to the liver-damaging chemical carbon tetrachloride.

Traditional Beliefs

This earthstar has been used in traditional Chinese medicine as a hemostatic agent; the spore dust is applied externally to stop wound bleeding and reduce chilblains. Two Indian forest tribes, the Baiga and the Bharia of Madhya Pradesh, have been reported to use the fruit bodies medicinally. The spore mass is blended with mustard seed oil, and used as a salve against burns. The Blackfoot of North America called the fungus "fallen stars", considering them to be stars fallen to the earth during supernatural events.

AURICULARIA AURICULA-JUDAE

Auricularia auricula-judae, known as the Jew's ear, jelly ear or by a number of other common names, is a species of edible Auriculariales fungus found worldwide. The fruiting body is distinguished by its noticeably ear-like shape and brown colouration; it grows upon wood, especially elder. Its specific epithet is derived from the belief that Judas Iscariot hanged himself from an elder tree; the common name "Judas's ear" eventually became "Jew's ear", while today "jelly ear" and other names are sometimes used. The fungus can be found throughout the year in temperate regions worldwide, where it grows upon both dead and living wood.

In the west, *A. auricula-judae* was used in folk medicine as recently as the 19th century for complaints including sore throats, sore eyes and jaundice, and as an astringent. Although it is not widely consumed in the west, it has long been popular in China, to the extent that Australia exported large volumes to China in the early twentieth century. Today, the fungus is a popular ingredient in many Chinese dishes, such as hot and sour soup, and also used in Chinese medicine. It is also used in Ghana, as a blood tonic. Modern research into possible medical applications have variously concluded that *A. auricula-judae* has antitumour, hypoglycemic, anticoagulant and cholesterol-lowering properties.

Taxonomy and Naming

The species was first mentioned in the scientific literature as *Tremella auricula* by Carl Linnaeus in his 1753 *Species Plantarum*, and later (1789) described by Jean Baptiste François Pierre Bulliard as *Tremella auricula-judae*. However, the genus *Tremella* is now reserved for fungal species that live as parasites on other fungi. *Tremella auricula-judae* is now considered a basionym. In 1791, Bulliard transferred the species to the genus *Peziza*. In 1822, Elias Magnus Fries transferred the species to *Exidia*, and, in so doing, sanctioned the name. In 1860, Miles Joseph Berkeley described the species as a member of *Hirneola*, a genus described by Fries in 1848, now considered synonymous with *Auricularia*. The species was given the name *Auricularia auricula-judae* in 1888 by Joseph Schröter. The specific name of *A. auricula-judae* comprises *auricula*, the Latin word meaning *ear*, and *Judae*, meaning *of Judas*. Under binomial nomenclature, a species name can comprise only two words; but the taxonomists responsible for this naming hyphenated the specific name to "bend the rules" and keep the name "within the letter of the law". The name was criticised by mycologist Curtis Gates Lloyd, who said "Auricularia auricula-Judae is cumbersome and in addition is a slander on the Jews". Though critical of Lucien Marcus Underwood, saying he "would probably not have known the Jew's ear from the calves' liver", he followed him in using *Auricularia auricula*, which was in turn used by Bernard Lowy in an article on the entire genus. Despite this, *Auricularia auricula-judae* is the currently recognised name for the species by many sources, though *Auricularia auricula* is still occasionally used. As well as the obligate synonyms from Bulliard, Fries and Berkeley, there are numerous other synonymous names. Mycologist George Willard Martin, writing in 1943, noted that the species was known by at least 12 binomials, of which none appeared to be valid, and noted that "the citations given for the various names are extremely erratic". Mycologist Mary F. Barrett attributes "such multiplication of names" to "the wide distribution of the Judas' ear, its ability to grow upon many different kinds of decaying wood, and to its great variation in size, colour and shape".

The fungus is associated with Judas Iscariot because of the belief that he hanged himself on an elder tree after his betrayal of Jesus Christ. Folklore suggests that the ears are Judas's returned spirit, and are all that are left to remind us of his suicide. The common name of the fungus was originally Judas's ear, but this was later shortened to Judas ear and, in the late 19th century, shortened again to Jew's ear. Common names for the fungus which refer to Judas can be traced back to at least the end of the 16th century; for instance, in the 17th century, Thomas Browne wrote of the species:

In *Jews' ears* something is conceived extraordinary from the name, which is in propriety but *fungus sambucinus*, or an excrescence about the roots of elder, and concerneth not the nation of the Jews, but Judas Iscariot, upon a conceit he hanged on this tree; and is become a famous medicine in quinsies, sore throats, and strangulations, ever since. While the term Jew's meat was a deprecatory term used for all fungi in the Middle Ages, the term is unrelated to the name Jew's ear. A further change of name to jelly ear was recommended in the *List of Recommended Names for Fungi*. The idea was rejected by mycologist Patrick Harding who considered it "to be the result of political correctness where it is not necessary", and who "will continue to call [the species] Jew's ear", explaining that, while anti-Semitism was commonplace in Britain, the name Jew's ear is in reference to Judas, who was a Jew. However, the name has been adopted in some recent field guides. Unrelated common names include the ear fungus, common ear fungus, the Chinese Fungus, the pig's ear, the wood ear, the black wood ear, the tree ear and Kikurage. The species was known as "fungus sambuca" among herbalists, in reference to *Sambuca*, the generic name for elder.

Description

The fruit body of *A. auricula-judae* is normally 3 to 8 centimetres (1.2 to 3.1 in) across, but can be as much as 12 centimetres (4.7 in). It is distinctively shaped, typically being reminiscent of a floppy ear, though the fruit bodies can also be cup-shaped. It is normally attached to the substrate laterally and sometimes by a very short stalk. The species has a tough, gelatinous, elastic texture when fresh, but it dries hard and brittle. The outer surface is a bright reddish-tan-brown with a purplish hint, often covered in tiny, downy hairs of a grey colour. It can be smooth, as is typical of younger specimens, or undulating with folds and wrinkles. The colour becomes darker with age. The inner surface is a lighter grey-brown in colour and smooth. It is sometimes wrinkled, again with folds and wrinkles, and may have "veins", making it appear even more ear-like.

Microscopic Features

The spores of *A. auricula-judae* are long and sausage shaped, ranging in size from 16 to 18 micrometres (µm) long by 6 to 8 ìm thick. The spores themselves are white, cream or yellowish, and are hyaline. The spores can sometimes be seen in a whitish mass on the underside of the fruit body. The species has elongated cylindrical basidia with three transverse septa (internal cross-walls dividing the hyphae). Basidia 60-72 × 4-7.5µm; sterigmata lateral, well developed, 3-4.5µm long. Spores smooth, hyaline, reniform to allantoid, 14-18 × 6-8µm, guttulate. Hairs on the fruit body are from 85 to 100 µm in length, and 5 to 6 µm in diameter. They are hyaline, lack a central strand and have rounded tips. They do not grow in dense tufts.

Similar Species

Auricularia auricula-judae is similar to *A. fuscosuccinea* in colour and texture, and "may be confused with it if only external features are considered". The spore and basidia sizes of the two species are slightly different, but this is not a reliable way to tell them apart. *A. cornea* is another similar species in the same genus, but has distinct internal differences, is normally more pilose (more covered in soft hair) and tends to fruit in larger numbers.

Habitat, Ecology and Distribution

Auricularia auricula-judae grows upon the wood of deciduous trees and shrubs, favouring elder. In up to 90 per cent of cases, the mushroom is found on elder, but it is often incorrectly assumed to grow exclusively on elder. It has also been recorded on *Acer pseudoplatanus* (known in the United Kingdom as sycamore), beech, ash, spindle, and in one particular case, the sycamore draining board of an old sink in Hatton Garden. Recently, *A. auricula-judae* has been recorded from semi-evergreen to evergreen and wet evergreen shola forests in the Western Ghats, India.

Fig. Auricularia Auricula-judae Fruit Bodies can be Found in Large Numbers on Old Wood.

This species occurs scattered and in clusters on dead or dying branches of trees, on main trunk, decaying logs, etc. This species occurs during the monsoon period in large imbricate clusters and under high humid conditions produces exceptionally large sized basidiomes. *A. auricula-judae* growing in wetevergreen and shola forests shows remarkable variation in size, shape and colour. In Australia, it is found in *Eucalyptus* woodland and rainforests; in the rainforests, it can grow in very large colonies on fallen logs. It favours older branches, where it feeds as a saprophyte (on dead wood) or a weak parasite (on living wood), and it causes white rot.

Commonly growing solitarily, it can also be gregarious (in a group) or caespitose (in a tuft). Spores are ejected from the underside of the fruit bodies with as many as several hundred thousand an hour, and the high rate continues when the bodies have been significantly dried. Even when they have lost some 90 per cent of their weight through dehydration, the bodies continue to release a small number of spores. It is found all year, but is most common in autumn. It is widespread throughout temperate and sub-tropical zones worldwide, and can be found across Europe, North America, Asia, Australia, South America and Africa. There has been some debate about the appearance of the species in the tropics; while it has been frequently reported there, Bernard Lowy, in an article on *Auricularia*, said that "of the specimens I have examined, none could be assigned here".

Uses

Food

Auricularia auricula-judae has a soft, jelly-like texture. Though edible, it was not held in high culinary regard in the west for many years. It has been likened to "eating an Indian rubber with bones in it", while in 19th century

Britain, it was said that "it has never been regarded here as an edible fungus". It has a mild flavour, and is useful for mixed mushroom recipes, but is still considered bland in the west. It can be dried and rehydrated, sometimes swelling to a very large size. Young specimens are best, but the species is not edible when raw, needing to be cooked thoroughly. The whole fruit body can be eaten, but should be thoroughly washed before cooking. Cooking can sometimes take a comparatively long time.

The nutritional content of 100 g (3.5 oz) of dried fungus includes 370 kcal, 10.6 g of protein, 0.2 g of fat, 65 g of carbohydrate, 5.8 g ash, and 0.03 per cent mg of carotene. Fresh mushrooms contain about 90 per cent moisture. Dried specimens may be ground up into a powder and used to absorb excess liquid in soups and stews, as it rehydrates into tiny fragments. Both *A. auricula-judae* and the similar *A. polytricha* are popular in China, where the medicinal use of food is common; a soup containing the species is used medicinally for dealing with colds and fevers by reducing the heat of the body.

There is evidence that the species were being cultivated in China as early as the Tang dynasty (618–907). Li Shih-chen, in his *Pen Tsao Kang Mu*, quotes Tang Ying-chuan from that period as saying "... put the steamed bran on logs, cover with straw, Wood Ear will grow". In the early 20th century, large volumes of *A. auricula-judae* were exported from Australia to China, and it is today still sold in Asian food shops for culinary use. It is also popular in Vietnam, although the climate is there more suited to *A. polytricha*. A report on small-scale fungi cultivation concluded that *A. auricula-judae* would be suitable for cultivation only in cooler climates. According to a 2010 publication, the annual production of *Auricularia* species worldwide is the fourth highest among all industrially cultivated culinary and medicinal mushrooms, and in China, the estimated output was roughly 1.655 million tonnes (based on 2003 data), most of which are *A. polytricha* however (which has by and large replaced *A. auricula-judae* in international trade).

Auricularia auricula-judae is also in cultivation elsewhere in the world, for instance, in Ghana. In the Brong-Ahafo and Ashanti regions, it is grown with what is referred to as the "plastic bag method". Sawdust is packed into polypropylene bags and then sterilised by steam for several hours. Once the sawdust has cooled, *Sorghum* grain spawn is added, and the bags are kept in moderately dark conditions. Once the sawdust is exposed to a humid environment, *A. auricula-judae* fruit bodies begin to grow. Elsewhere in the world, a study on the use of mushrooms by the Bini people inhabiting a remote village in southern Nigeria found that the local inhabitants collected and ate *A. auricula-judae*, but that it was not one of the mushrooms they used medicinally.

Collection of the mushroom for culinary use has also been documented in Nepal. However, the Nepalese do not consider it a choice mushroom for eating; of the three grades given to edible mushrooms, it was given the worst. Again, unlike other mushrooms, no medicinal use was reported. The mushroom has also been the only non-morel species exported from Nepal for culinary use.

Other places where *A. auricula-judae* has been recorded as commonly consumed include Poland, Mozambique and Indonesia, while, in Bolivia, Goeldi's monkeys have been recorded as commonly eating the sporocarps.

Folk Medicine

Auricularia auricula-judae has been used as a medicinal mushroom by many herbalists. It was used as a poultice to treat inflammations of the eye, as well as a palliative for throat problems. The 16th-century herbalist John Gerard, writing in 1597, recommended *A. auricula-judae* for a very specific use; other fungi were used more generally.

He recommends the preparation of a liquid extract by boiling the fruit bodies in milk, or else leaving them steeped in beer, which would then be sipped slowly in order to cure a sore throat.

The resultant broth was probably not dissimilar to the Chinese soups that use *A. polytricha*. Carolus Clusius, writing in 1601, also said that the species could be gargled to cure a sore throat, and John Parkinson, writing in 1640, reported that boiling in milk or steeping in vinegar was "the onely use the are put unto that I know".

Writing in 1694, herbalist John Pechey described *A. auricula-judae* by saying "It grows to the Trunk of the Elder-Tree. Being dried it will keep a good year. Boyl'd in Milk, or infus'd in Vinegarm 'tis good to gargle the Mouth or Throat in Quinsies, and other inflammations of the Mouth and Throat. And being infus'd in some proper Water, it is good in Diseases of the Eyes." The species also saw use as an astringent due to its ability to absorb water.

There are recorded medicinal usages from Scotland, where it was again used as a gargle for sore throats, and from Ireland, where, in an attempt to cure jaundice, it was boiled in milk.

The medicinal use of *A. auricula-judae* continued until at least 1860, when it was still sold at Covent Garden; at the time, it was not considered edible in the United Kingdom.

Medicinal use in Indonesia was also recorded in the 1930s, and was more recently reported in modern-day Ghana. A report for the 2005 Commonwealth Forestry Conference examining the possible effects of deforestation in southern Ghana on medicinal and edible fungi found that *A. auricula-judae* was in use as a blood tonic.

Pharmacology

Auricularia auricula-judae has been the subject of research into possible medicinal applications. Experiments in the 1980s concluded that two glucans isolated from the species showed potent antitumour properties when used on mice artificially implanted with Sarcoma 180 tumours. This was despite the conclusion of earlier research indicating that, while aqueous extracts from several other fungal species had antitumour effects, extracts from *A. auricula-judae* did not.

Further, research on genetically diabetic mice showed that a polysaccharide extracted from *A. auricula-judae* had a hypoglycemic effect; mice fed with food including the polysaccharide showed reduced plasma glucose, insulin, urinary glucose and food intake.

Another chemical extracted from the species was an acidic polysaccharide (made up of mostly mannose, glucose, glucuronic acid and xylose) which showed anticoagulant properties.

The article concluded that "the polysaccharides from these mushrooms may constitute a new source of compounds with action on coagulation, platelet aggregation and, perhaps, on thrombosis". Another study reported that the species may be effective in stopping platelet binding *in vitro*, with possible uses regarding hypercholesterolemia. Research has shown that *A. auricula-judae* can be used to lower cholesterol levels generally, and, in particular, is one of two fungi shown to reduce the level of bad cholesterol.

BETA-GLUCAN

β-Glucans (beta-glucans) are polysaccharides of D-glucose monomers linked by β-glycosidic bonds. β-glucans are a diverse group of molecules that can vary with respect to molecular mass, solubility, viscosity, and three-dimensional configuration.

They occur most commonly as cellulose in plants, the bran of cereal grains, the cell wall of baker's yeast, certain fungi, mushrooms and bacteria. Some forms of beta glucans are useful in human nutrition as texturing agents and as soluble fibre supplements, but can be problematic in the process of brewing. Oat is a rich source of the water-soluble fibre (1,3/1,4) β-glucan, and its effects on health have been extensively studied the last 30 years. Oat β-glucans are the only dietary fibre being recognised by the European Food Safety Authority (EFSA) to be able to reduce a disease risk. Oat β-glucans can be highly concentrated in different types of oat brans. "Barley has more beta glucan fibre than any other grain" claims a report on DiabetesHealth web site; 11 sources are listed.

Yeast and medicinal mushroom derived β-glucans are notable for their ability to modulate the immune system. One study has shown that insoluble (1,3/1,6) β-glucan, has greater biological activity than that of its soluble (1,3/1,4) β-glucan counterparts. The differences between β-glucan linkages and chemical structure are significant in regards to solubility, mode of action, and overall biological activity.

Overview

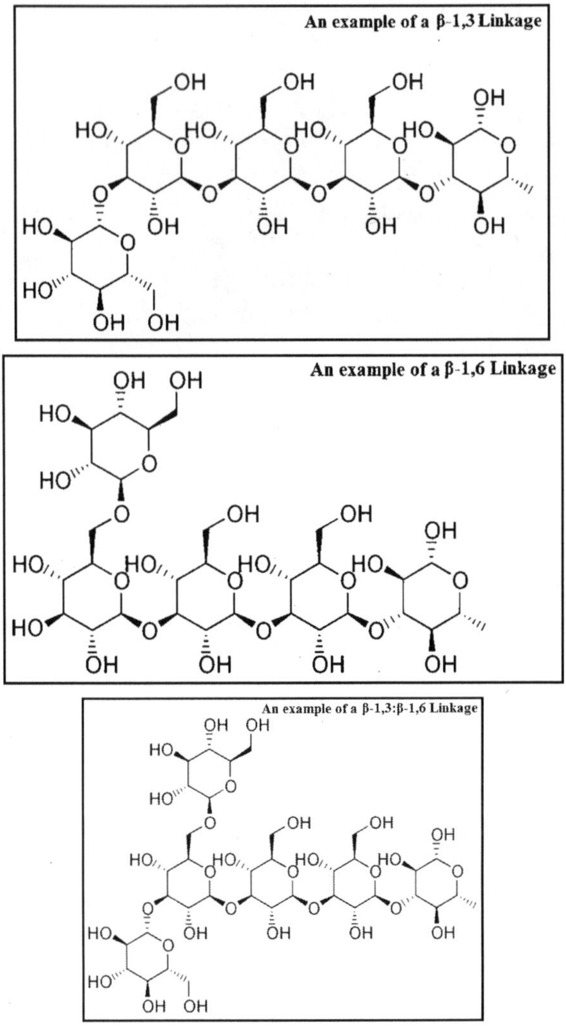

Fig. Examples of Various β-glucan glycosidic linkages.

β-Glucans are polysaccharides that contain only glucose as structural components, and are linked with β-glycosidic bonds. Glycosidic bonds are etheric oxygen bridges which link the monosaccharide units in a polysaccharide, and they are designated by a pair of numbers to indicate which carbons in each of the monosaccharide units are linked (according to the standard numbering for simple monosaccharides). When a glycosidic bond involves the number 1 carbon of an aldose monosaccharide (or the number 2 carbon of a ketose monosaccharide), a second designation is needed to indicate the spatial orientation of the etheric oxygen linkage at that carbon (the anomeric carbon). An "α-" (alpha) indicates the etheric oxygen linkage attaches to the anomeric carbon below the ring, and a "β-" (beta) indicates the etheric oxygen linkage attaches to the anomeric carbon above the ring (in the standard Haworth projection). The orientation of the etheric linkage at any of the other carbons is predetermined and fixed by the identity of the monosaccharide and the position, and no further descriptor is required for the glycosidic bond.

Thus, the designation of β(1-3) for a glycosidic linkage indicates that the etheric oxygen bridge between two consecutive monosaccharide units of the polysaccharide connects the number 1 carbon of the first unit to the number 3 carbon of the second unit, and that etheric oxygen bridge attaches to carbon 1 of the first unit from above the ring.

Likewise, the designation of β(1-6) for a glycosidic linkage indicates that the etheric oxygen bridge between two consecutive monosaccharide units of the polysaccharide connects the number 1 carbon of the first unit to the number 6 carbon of the second unit, and that etheric oxygen bridge attaches to carbon 1 of the first unit from above the ring.

Beta-glucan Chemistry

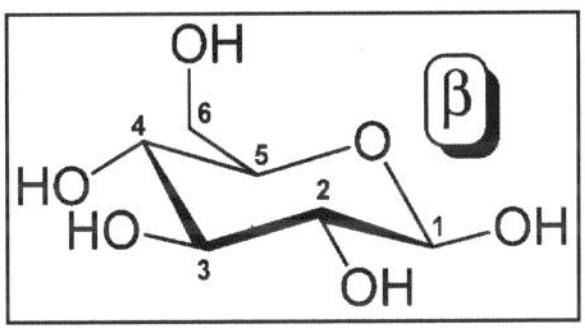

Fig. Glucose Molecule, Showing Carbon Numbering Notation and β Orientation.

By definition, beta-glucans are chains of D-glucose polysaccharides, linked by beta type glycosidic bonds. These six-sided D-glucose rings can be connected to one another, on a variety of positions on the D-glucose ring structure. Some β-glucan compounds are continual repeats of D-glucose attached at a specific position.

However, β-glucans can be more diverse than molecules like starch. For instance, a β-glucan molecule can be composed of repeating D-glucose units linked with β-glycosidic bonds at one position like starch, but have branching glucose side-chains attached to other positions on the main D-glucose chain. These smaller side-chains can branch off the β-glucan "backbone" (in the case of starch, the backbone would be D-glucose chains linked at the (1,4) position) at other positions like that of the 3 and 6 position. In addition, these side-chains can be attached to other types of molecules, like proteins. An example of a β-glucan that has proteins attached to it is Polysaccharide-K.

The most active forms of β-glucans are those comprising D-glucose units with (1,3) links and with side-chains of D-glucose attached at the (1,6) position. These are referred to as β-1,3/1,6 glucan. Some researchers have suggested that it is the frequency, location, and length of the side-chains rather than the backbone of β-glucans that determine their immune system activity. Another variable is the fact that some of these compounds exist as single strand chains, while the backbones of other β(1,3)-glucans exist as double or triple stranded helix chains. In some cases, proteins linked to the β(1,3)-glucan backbone may also be involved in providing therapeutic activity. Although these compounds can potentially enhance immune function, it must be emphasized that this research is in its infancy. In addition, there are differing opinions on which molecular weight, shape, structure, and source of β(1,3)-glucans provide the greatest biological activity.

β-glucan Sources in Nature

Fig. The Shiitake Mushroom Contains Beta-glucans.

One of the most common sources of β(1,3)D-glucan for supplement use is derived from the cell wall of baker's yeast (*Saccharomyces cerevisiae*). However, β(1,3)(1,4)-glucans are also extracted from the bran of some grains, such as oats and barley, and to a much lesser degree in rye and wheat. The β(1,3)D-glucans from yeast are often insoluble.

Those extracted from grains tend to be both soluble and insoluble. Other sources include some types of seaweed, and various species of mushrooms, such as reishi, shiitake, and maitake.

β-Glucan and the Immune System

β-Glucans are known as "biological response modifiers" because of their ability to activate the immune system. Immunologists have discovered that receptors on the surface of innate immune cells called dectin-1 and complement receptor 3 (CR3 or CD11b/CD18) are responsible for binding to β-glucans, allowing the immune cells to recognise them as "non-self".

β-glucan and Blood Cholesterol

Several health claims requests were submitted to the EFSA NDA Panel (Dietetic Products, Nutrition and Allergies), related to the role of â-glucans in "*maintenance of normal blood cholesterol concentrations and maintenance or achievement of a normal body weight*". In July 2009, the Scientific Committee issued the following statements:

- On the basis of the data available, the Panel concludes that a cause and effect relationship has been established between the consumption of beta-glucans and the reduction of blood cholesterol concentrations.
- The following wording reflects the scientific evidence: "Regular consumption of beta-glucans contributes to maintenance of normal blood cholesterol concentrations". In order to bear the claim, foods should provide at least 3 g/d of beta-glucans from oats, oat bran, barley, barley bran, or from mixtures of non-processed or minimally processed beta-glucans in one or more servings. The target population is adults with normal or mildly elevated blood cholesterol concentrations.
- On the basis of the data available the Panel concludes that a cause and effect relationship has not been established between the consumption of beta-glucans and the maintenance or achievement of a normal body weight.

In November 2011, the EU Commission published its decision in favour of oat beta-glucans in respect of Article 14 of the EC Regulation on the labelling of foodstuffs with nutrition and health claim statements permitting oat beta-glucan to be described as beneficial to health. Following the opinion of the Panel on Dietetic Products, Nutrition and Allergies (NDA) the EFSA and the Regulation (EU) no. 1160/2011 of the Commission, foodstuffs through which 3g/day of oat beta-glucan are consumed (1g of oat beta-glucan per portion) are allowed to display the following health claim: «Oat beta-glucan reduces the cholesterol level in the blood. The lowering of the blood cholesterol level can reduce the risk of coronary heart disease.»

Research

Tumoricidal Effects

The tumoricidal properties of β-glucans have been studied in several *in vitro* and *in vivo* animal models. In a mouse model study, β-1,3 glucan in conjunction with interferon gamma inhibited tumors and liver metastasis. In some studies, β-1,3 glucans enhanced the effects of chemotherapy. In a mouse carcinoma model, β-1,3 glucans did not reduce tumor incidence, but were associated with reduced mortality in combination with cyclophosphamide. In human patients with advanced gastric or colourectal cancer, the administration of β-1,3 glucans derived from shiitake mushrooms, in conjunction with chemotherapy, resulted in prolonged survival times.

Prevention of Infection

Alpha-Beta Technologies conducted a series of human clinical trials in the 1990s to evaluate the impact of β-glucan therapy for controlling infections in high-risk surgical patients. In the initial trial, 34 patients were randomly (double-blind, placebo-controlled) assigned to treatment or placebo groups. Patients who received the PGG-glucan had significantly fewer infectious complications than the placebo group (1.4 infections per infected patient for PGG-glucan group vs. 3.4 infections per infected patient for the placebo group). Additional data from the clinical trial revealed intravenous antibiotic use was decreased, and stays in the intensive care unit were shorter for the patients receiving PGG-glucan vs. patients receiving the placebo. A subsequent human clinical trial studied the effect of â-glucan on the incidence of infection in high-risk surgical patients. A total of 67 patients were randomised to treatment with a placebo or a dose of 0.1, 0.5, 1.0 or 2.0 mg PGG-glucan per kilogram of body weight. Serious infections occurred in four patients who received the placebo, three patients who received the low dose (0.1 mg/kg) of PGG-glucan, and one patient who received the highest dose of 2.0 mg/kg.

The results of a phase III human clinical trial showed that PGG-glucan therapy reduced serious postoperative infections by 39 per cent after high-risk noncolourectal operations. This study was conducted in patients who were already at high risk because of the type of surgery and were more susceptible to infections and other complications.

A study conducted by the Canadian Department of Defence showed that orally administered yeast β-glucan given with or without antibiotics protected mice against anthrax infection. A dose of antibiotics along with oral whole glucan particles (2 mg/kg body weight or 20 mg/kg body weight) for eight days prior to infection with *Bacillus anthracis* protected mice against anthrax infection over the 10-day postexposure test period. Mice treated with the antibiotic alone did not survive.

A second experiment was conducted to investigate the effect of yeast β-glucan orally consumed after exposure of mice to *B. anthracis*. The results were similar to the previous experiment, with an 80-90 per cent survival rate for mice treated with β-glucan, but only 30 per cent for the control group after 10 days of exposure.

Early research by Onderdonk *et al.* investigated the ability of yeast b-glucan to reduce septic infections using *in vivo* models. They found that mice challenged with *E. coli* or *S. aureus* bacteria were protected against septic infections when they were injected with PGG-glucan 4–6 hr prior to infection. Preventative dosing of yeast β-glucan prior to infection with *S. aureus* prevented sepsis in a guinea pig model. Research has been conducted in animals on the use of yeast β-glucans for the treatment and prophylaxis of bacterial sepsis and protection against oxidative organ injury.

In a prospective, randomised, double-blind study, 38 trauma patients received a soluble, yeast-derived glucan intravenously for seven days or placebo. The total mortality rate was significantly less in the glucan group (0 per cent vs. 29 per cent), with also a decrease in septic morbidity (9.5 per cent vs. 49 per cent). Yeast-derived beta-glucan significantly enhanced phagocytic activity in an experimental mouse model of sepsis induced by *Candida albicans* and midline laparotomy. Mice not operated on, on glucan, had a 100 per cent survival vs. 73 per cent in the surgical group.

Radiation Exposure

Specific hematopoietic activity was first demonstrated with β-glucan in the mid-1980s in an analogous manner as granulocyte monocyte–colony stimulating factor (GM-CSF). Research was carried out initially with particulate β-glucan and later with soluble β-glucans, all of which were administered intravenously (IV) to mice. Mice exposed to 500-900 cGy (500-900 mrads) of gamma radiation exhibited a significantly enhanced recovery of blood leukocyte, platelet and red blood cell counts when given IV β-glucan. Other reports showed that â-glucan could reverse the myelosuppression produced with chemotherapeutic drugs such as fluorouracil, carboplatinum, or cyclophosphamide. Moreover, the anti-infective activity of β-glucan combined with its hematopoiesis-stimulating activity resulted in enhanced survival of mice receiving a lethal dose of 900-1200 cGy of radiation. *In vitro* studies showed β-glucan could enhance granulocyte and megakaryocyte colony formation by hematopoietic stem progenitor cells when used in combination with GM-CSF and interleukin-3, respectively.

Original studies delivered glucan almost entirely by injection. Later, numerous studies tried to evaluate the possibility that glucan can be delivered orally without compromising its biological activities, opening the oral route of administration as a more pleasant alternative. Oral beta-glucan had hematopoietic effects analogous to beta-glucan administered by IV, and orally administered whole glucan particulate functions to accelerate hematopoiesis following irradiation in an analogous manner as IV-administered β-glucan. The mechanisms of the glucan transfer through the gastrointestinal tract were reported. Oral β-glucan stimulates hematopoiesis in radiation-treated mice.

Allergic Rhinitis

This disease is caused by an IgE-mediated allergic inflammation of the nasal mucosa. Orally administered yeast-glucan decreased levels of IL-4 and IL-5 cytokines responsible for the clinical manifestation of this disease, while increased the levels of IL-12.

Arthritis

A study employing paramagnetic resonance spectroscopy indicated that yeast-derived beta-glucan could reduce free-radical formation *in vitro* and reduce thc plasma levels of protein carbonyls in a rodent arthritis model.

Additional Applications

Consuming certain cereals (barley, oats) and edible mushrooms decreases the levels of serum cholesterol and liver low-density lipoproteins, leading to lowering of atherosclerosis and cardiovascular disease hazards; this is also mediated by β-glucans. These cereals, mushrooms, and yeast facilitate bowel motility and can be used in amelioration of intestinal problems, particularly obstipation. Indigestible β-glucans, forming a remarkable portion of these materials, are also able to modulate mucosal immunity of the intestinal tract.

β-Glucan Absorption

For maximal absorption, oral β(1,3)-D-glucan should be taken on an empty stomach. Enterocytes reportedly facilitate the transportation of β(1,3)-glucans and similar compounds across the intestinal cell wall into the lymph, where they begin to interact with macrophages to activate immune function. Radiolabeled studies have verified that both small and large fragments of β-glucans are found in the serum, which indicates they are absorbed from the intestinal tract. M cells within the Peyer's patches physically transport the insoluble whole glucan particles into the gut-associated lymphoid tissue.

Role in Diagnostics

β-D-Glucan forms part of the cell wall of certain medically important fungi, especially *Aspergillus* and *Agaricus* species. An assay to detect its presence in blood is marketed as a means of diagnosing invasive fungal infection in patients. False positives may occur because of fungal contaminants in the antibiotics amoxicillin-clavulanate, and piperacillin/tazobactam. False positives can also occur with contamination of clinical specimens with the bacteria *Streptococcus pneumoniae*, *Pseudomonas aeruginosa*, and *Alcaligenes faecalis* which also produce (1→3)β-D-glucan.

INONOTUS OBLIQUUS

Inonotus obliquus, commonly known as chaga mushroom (a Latinisation of the Russian term 'ψaΓa'), is a fungus in Hymenochaetaceae family. It is parasitic on birch and other trees. The sterile conk is irregularly formed and has the appearance of burnt charcoal. It is not the fruiting body of the fungus, but a mass of mycelium, mostly black due to the presence of massive amounts of melanin.

The fertile fruiting body can be found very rarely as a resupinate (crustose) fungus on or near the clinker, usually appearing after the host tree is dead. I. obliquus grows in birch forests of Russia, Korea, Eastern and Northern Europe, northern areas of the United States, in the North Carolina mountains and in Canada. The chaga mushroom is considered a medicinal mushroom in Russian and Eastern European folk medicine.

Name

The name "chaga" (pronounced "tsjaa-ga") comes from the Russian word of the mushroom (anglicised from *czaga*), which in turn is purportedly derived from the word for the fungus in Komi-Permyak, the language of the indigenous peoples in the Kama River Basin, west of the Ural Mountains. It is also known as the clinker polypore, cinder conk, black mass and birch canker polypore. In Norwegian, the name is *kreftkjuke'* which literally translates as "cancer polypore", referring to the fungus' appearance or to its alleged medicinal properties.

In England and Canada, it is known as the sterile conk trunk rot of birch, which refers to the fruiting bodies growing under the outer layers of wood surrounding the sterile conk once the tree is dead, to spread the spores. In France, it is called the *carie blanche spongieuse de bouleau* (spongy white birch tree rot), and in Germany it is known as *Schiefer Schillerporling* (slate *Inonotus*). The Dutch name is *berkenweerschijnzwam* (birch mushroom glow). Within the wilderness survival community, it is called tinder fungus (this name is also claimed for *fomes fomentarius*), because it can catch a weak spark and form a coal that can be used to kindle a fire.

Medicinal Use

Since the 16th century, chaga mushrooms were recorded as being used in folk medicine and the botanical medicine of the Eastern European countries as a remedy for cancer, gastritis, ulcers, and tuberculosis of the bones. A review from 2010 stated, "*As early as in the 16th century, chaga was used as an effective folk medicine in Russia and Northern Europe to treat several human malicious tumors and other diseases in the absence of any unacceptable toxic side effects.*"

Chemical analysis shows that *I. obliquus* produces during its development and growth a range of secondary metabolites, including phenolic compounds such as melanins, and lanostane-type triterpenes, which include a small percentage of betulinic acid. Among these metabolites are biologically active components which have been researched and tested for their potential antioxidant, antitumoral, and antiviral activities. Both betulin and betulinic acid are being studied for use as chemotherapeutic agents and are already used as anti-HIV agents). In an animal study, researchers found betulin from birch bark lowered cholesterol, obesity and improved insulin resistance. In 1958, scientific studies in Finland and Russia found chaga provided an epochal effect in breast cancer, liver cancer, uterine cancer, and gastric cancer, as well as in hypertension and diabetes.

Cultivation

Geographically this fungus is mostly found in very cold habitats. It grows very slowly, suggesting it is not a reliable source of bioactive compounds in the long run. Attempts at cultivating this fungus all resulted in a reduced and markedly different production of bioactive metabolites. Secondary metabolites were either absent or present in very different ratios, and in general showed significantly less potency in cultivated Chaga. Cultivated Chaga furthermore results in a reduced diversity of phytosterols, particularly lanosterol, an intermediate in the synthesis of ergosterol and lanostane-type triterpenes. This effect was partially reversed by the addition of silver ion, an inhibitor of ergosterol biosynthesis.

Additionally, the bioactive triterpene betulinic acid is completely absent in cultivated Chaga. In nature Chaga grows pre-dominantly on birches, and birch bark contains up to 22 per cent of betulin. Betulin is poorly absorbed by humans, even when taken intravenously; its bioavailability is very limited. However, the Chaga mushroom converts betulin into betulinic acid, and many internet sources state Chaga's betulinic acid is bioavailable, even when taken orally. Unfortunately there is no research that confirms this claims.

Chaga Dietary Supplements

In China, Japan and South Korea, extracts of chaga and other mushrooms from the family Hymenochaetaceae are being produced, sold and exported as anticancer medicinal supplements. The main bioactive ingredient in these extracts are usually (1>3)(1>6) Beta-D-glucans, a type of water-soluble polysaccharide. The biologic properties of crude preparations of these specific β-D-glucans have been subject of research since the 1960s.

Although these macro-molecules exhibit a wide range of biologic functions, including antitumor activity, their ability to prevent a range of infectious diseases (by triggering and supporting the immune function) has been studied in the greatest detail.

Recent scientific research in Japan and China has been focused more on the anticancer potential and showed the effects of these specific beta-glucans to be comparable to chemotherapy and radiation, but without the side effects. Further research indicated these polysaccharides have strong anti-inflammatory and immune balancing properties, stimulating the body to produce natural killer (NK) cells to battle infections and tumor growth, instead of showing a direct toxicity against pathogens.

This property makes well-prepared medicinal mushroom extracts stand out from standard pharmaceuticals - no side effects will occur or develop; the body is healing itself, triggered into action by the BRM effect of the chaga extract. Herbalist David Winston maintains it is the strongest anticancer medicinal mushroom.

Russian literature Nobel Prize laureate Alexandr Solzhenitsyn wrote two pages on the medicinal use and value of chaga in his autobiographical novel, based on his experiences in a hospital in Tashkent, *Cancer Ward* (1968).

Clinical Trials

The majority of research has been performed *in vitro* and *in vivo* in animals; few human clinical trials have been conducted. In a 48-patient human clinical trial in Poland in 1957, 10 patients treated with chaga showed a reduction of tumor size, a decrease in pain, a decrease in the intensity and the frequency of hemorrhaging, and a recovery accompanied with better sleep, appetite and feelings of improvement. Most of these patients were females treated with chaga for cancer of the genital organs or breast cancer. In 1973 a case study including 50 patients about the effect of a Chaga extract on psoriasis was published in the Russian journal Vestnik Dermatologii i Venerologii. The outcome was almost 100 per cent successful.

Preparation

Chaga is traditionally grated into a fine powder and used to brew a beverage resembling coffee or tea. For medicinal use, an extraction process is needed to make at least some of the bio-active components bioavailable. These bio-actives are found in the mostly indigestible chitin cell walls of the chaga. Humans lack the enzyme chitinase, so cannot fully digest raw mushrooms or their derivatives, and the digestive process works too fast for the stomach acid to take effect. Scientific studies and research are in general also based on highly concentrated extracts, and traditional Russian usage is also based on a form of hot-water extraction (by preparing zavarka).

Currently, three extraction processes are used, each with a different outcome:

- Hot water extraction is the most common and the cheapest method. It can be compared to the traditional tea-making process. All water-soluble components will be present in the resulting extract. Water-insoluble components, such as phytosterols, betulinic acid and betulin, will be absent. Several extraction rounds combined with modern pharmaceutical techniques can result in high levels of polysaccharides, up to almost 60 per cent. The β-D-glucans, the bio-active part of these polysaccharides, might add up to ±20 per cent. Polyphenolic components are water-solubles and will also be present.
- Ethanol or methanol extraction isolates the water-insoluble components, betulinic acid, betulin and the phytosterols. This extraction process is in general used as a second step after hot-water extraction, since ethanol alone will not break down chitin effectively - heat is essential.
- Fermentation is the most time-consuming, so is the most expensive; this method is not used very often. Because fermentation methods are not standardised (many types of bacteria and fungi can be used in the process), the outcome is also not standardised.

Extracts with a therapeutic value usually combine two methods, usually hot water and ethanol extraction. This will result in all bioactive components being present. Cheap, mass-produced extracts are in general hot water, low percentage (4-20 per cent) polysaccharide extracts with limited therapeutic value. The information on the supplements' label will usually reveal inclusion or exclusion of components. However, the majority of mushroom dietary supplements that are sold are non-extracted, being the cheapest option. To achieve at least some therapeutic effects the consumer has to make tea from it.

Research

A 1998 study in Poland demonstrated chaga's inhibiting effects on tumor growth. Noda and colleagues found betulin seems to work highly selectively on tumor cells because the interior pH of tumor tissues is generally lower than that of normal tissues, and betulinic acid is only active at those lower levels. Fulda *et al.* found, in 1997, once inside the cells, betulinic acid induces apoptosis (programmed cell death) in the tumors. In 2005, *I. obliquus* was evaluated for its potential for protecting against oxidative damage to DNA in a human keratinocyte cell line. The study found the polyphenolic extract protected these cells against hydrogen peroxide-induced oxidative stress.

Another study that year found the endopolysaccharide of chaga produced indirect anticancer effects via immunostimulation. The mycelial endopolysaccharide of *I. obliquus* was identified as a candidate for use as an immune response modifier and indicated the anticancer effect of endopolysaccharide is not directly tumoricidal, but rather is immunostimulation. It also has anti-inflammatory properties. Saitoh Akiko published on the antimutagenic effects of chaga in 1996. Mizuno *et al.* published on the antitumor and hypoglycemic activities of the polysaccharides from the sclerotia and mycelia of chaga. Due to the serum glucose-lowering activity of polysaccharides, caution should be taken by those with hypoglycemia.

7

Diseases Resistance of Animals

GENETICS OF SCRAPIE RESISTANCE IN SHEEP

Scrapie is an infectious disease of sheep that affects the central nervous system and is always fatal. Upon necropsy, infected animals will have holes or vacuoles in the tissue of the brain. The disease is classified as a transmissible spongiform encephalopathy (TSE). Some TSE's in other species include bovine spongiform encephalopathy (BSE, Mad Cow Disease), Creutzfeldt-Jakob Disease (CJD) in man, new variant CJD (vCJD) in man thought to result from eating meat from BSE-infected cattle, transmissible mink encephalopathy (TSE), chronic wasting disease (CWD) in some U.S., populations of deer and elk, and feline spongiform ecephalopathy (FSE). Federal and State animal health authorities have considered scrapie a high priority disease for elimination for many years, but the relatively recent evidence that ties vCJD in humans to BSE in cattle has resulted in increased attention on scrapie in sheep.

GENETICS OF SCRAPIE

Scrapie is a 100 per cent fatal, degenerative disease affecting the central nervous system of sheep and goats. The disease has been reported in countries throughout the world with few notable exceptions (Australia and New Zealand). The first case of scrapie was discovered in the United States in 1947. The current incidence is 2/10 of 1 per cent or 1 in 500 U.S., sheep. The incidence of scrapie in the U.S., goat population is not currently known.

While bovine spongiform encephalopathy is believed to be caused by the consumption of contaminated feed (meat and bone meal), scrapie is transmitted during lambing when lambs come into contact with infected placenta and birth fluids from infected ewes. Rams can get scrapie, but are not known to transmit scrapie.

Scrapie is NOT caused by genetics, but the genetic make-up (DNA) of an animal determines whether it will get scrapie if it is exposed to the infective agent. In other words, if a genetically susceptible lamb is exposed to a scrapie-infected placenta, it will develop scrapie. It takes from 2 to 5 years after exposure for an animal to show clinical signs of scrapie. If a genetically resistant lamb is exposed to a scrapie-infected placenta, it will not develop scrapie. No resistant genotypes have been identified in goats.

Incidence

Scrapie was first diagnosed in the U.S., in 1947 in a flock of Suffolk sheep in Michigan that had imported sheep from Canada of U.K., origin for several years. From 1947 through July 2001, scrapie had been diagnosed in over 1,000 flocks, and approximately 1,600 individual cases of natural sheep scrapie and 7 individual cases of natural goat scrapie have been found. Through 1992, the breed distribution was as follows: 87 per cent Suffolk, 6 per cent Hampshire, and 7 per cent other breeds and crosses. While 1,600 appears to be a large number of cases of sheep scrapie, it actually is not since it represents an average of only about 30 cases per year. Relative to most other sheep diseases, the incidence of scrapie is quite low.

Clinical Signs

Early signs of scrapie include subtle changes in behaviour and temperament. These changes may be followed by scratching and rubbing against fixed objects. Other signs are loss of coordination, weight loss despite a good appetite, biting of feet and limbs, and lip smacking. Gait abnormalities such as high-stepping of the forelegs, hopping like a rabbit, and swaying of the back end may be seen. Signs or effects of scrapie usually do not appear until 2 to 5 years after an animal is infected. Sheep may live 1 to 6 months (sometimes longer) after the onset of clinical signs.

Infected animals show no immune response to the infective agent as is the case with viral and bacterial diseases. There is currently no live-animal test to determine if an animal is infected. This is a major area of research, and some promising live-animal tests are in the development and evaluation stages. The scrapie agent is thought to be spread most often from ewe to offspring and to other lambs in contemporary lambing groups through contact of lambs with the placenta and placental fluids of infected ewes. Lateral transmission from infected rams to ewes and lambs, from infected ewes to other ewes, and from an infected environment to adult animals are not thought to be major routes of transmission; although there are documented cases of scrapie where such routes of transmission are thought to have occurred.

There is not agreement among scientists on the nature of the infectious agent. However, infected animals have large amounts of an abnormal type of prion protein (denoted as PrP^{Sc}), especially in the central nervous system tissues of the brain and spinal cord. Scrapie-free animals have normal prion protein (PrP^{C}) and no PrP^{Sc}. Some feel that PrP^{Sc} itself is the infectious agent while others feel that PrP^{Sc} is the result of another infectious agent. However, scientific evidence increasingly supports the theory of PrP^{Sc} itself as the infectious agent.

Prion Protein

Prion protein is a constituent of normal mammalian cells. Sheep prion protein is a protein of 256 amino acids in length. There is some variation among sheep in the particular amino acids that comprise their prion protein. However, the PrP^{Sc} found in a scrapie-infected sheep is of the same amino acid sequence as the normal prion protein (PrP^{C}) found in that sheep. The difference between PrP^{Sc} and PrP^{C} is in its molecular shape. The prion infectious agent theory contends that PrP^{Sc} infects a normal sheep and serves as a template to change the normal shape of PrP^{C} to the infective shape of PrP^{Sc}.

Of the 256 locations for amino acids along the sheep prion protein molecule, amino acid changes in at least three of these locations (amino acids number 136, 154, and 171) have been shown to confer increased or decreased susceptibility to scrapie. Each particular amino acid is specified in the prion protein gene by a group of three nucleotide bases known as a codon. Therefore, these three important locations are referred to as codon 136, codon 154, and codon 171 on the prion protein gene. The most common amino acids coded for at these three codons are: alanine (A) or valine (V) at codon 136, arginine (R) or histidine (H) at codon 154, and glutamine (Q) or Arginine (R) at codon 171. Since genes occur in pairs, sheep have two prion protein genes. Each gene can be the same in which case the sheep will produce only one type of prion protein. If the genes are different, then two types of prion protein are produced. Just considering these three codons, a population of sheep can contain individuals with one of 27 different prion protein genotypes.

Table. Possible Genotypes for the Sheep Prion Protein at Codons 136, 154, and 171 and a Scrapie Resistance Index for each Genotype

Codons			Scrapie Resistance Index
136	154	171	for U.S. Sheep[a]
AA	RR	QQ	3
AA	RR	QR	9
AA	RR	RR	13
AA	RH	QQ	3
AA	RH	QR	9
AA	RH	RR	13
AA	HH	QQ	3
AA	HH	QR	9
AA	HH	RR	13
AV	RR	QQ	1
AV	RR	QR	7
AV	RR	RR	11
AV	RH	QQ	1
AV	RH	QR	7
AV	RH	RR	11

AV	HH	QQ	1
AV	HH	QR	7
AV	HH	RR	11
VV	RR	QQ	0
VV	RR	QR	6
VV	RR	RR	10
VV	RH	QQ	0
VV	RH	QR	6
VV	RH	RR	10
VV	HH	QQ	0
VV	HH	QR	6
VV	HH	RR	10

Note: [a]Index is sum of numeric value assigned for each individual codon genotype: codon 136: AA=3, AV=1, VV=0; codon 154: RR=RH=HH=0; codon 171: RR=10, RQ=6, QQ=0. Higher values indicate greater expected resistance.

GENETIC MODIFICATION: SCIENCE, MICE AND PATENTS

The most commonly genetically modified animal is almost certainly the mouse. It is small, short-lived and sufficiently similar to humans to be an almost ideal laboratory animal. As a result, mice have not only been cloned and modified, they have led to an actual industry in the production of "knockout mice," that is, mice with a particular gene or set of genes inactivated for research purposes. Trans Genic Inc asserts, "Currently, we are able to produce almost 1,000 strains of Knockout mouse in a year." "Pharming". Cattle, sheep, goats, chickens, rabbits and pigs have been genetically modified with the aim of producing human proteins that are useful, generally as medicines. The gene transfer process is typically very inefficient, and cloning is seen as another way of propagating the GM animal. A 1999 USDA report cited estimates that there was a $24 billion market for human proteins, and theoretically 600 transgenic cows could supply the worldwide demands for some drugs. In practice, however, several companies that have pursued this line have gone bust, and the profit potential seems less than it once did.

Genetic modification of animals in order to improve the prospects of organ transplants is also being investigated.

Genetically Modified Fish as Pets

A tropical fish genetically modified to glow in the dark went on sale in Taiwan in 2003 for about $17 each. A different variety of zebrafish, called "GloFish," which were created in Singapore, reached the United States market in January 2004.

The distributor says that GloFish were originally developed to fluoresce only in the presence of pollutants, but that is not the form in which they are being sold. They cost about $5 each, and are intended to live in aquariums, but can breed and, in the right conditions, live in the wild.

The Food and Drug Administration (FDA) approved the sale without ceremony. A coalition led by the Center for Food Safety filed suit against the decision, but sales went ahead. In California, the Fish and Game Commission initially banned the fish but later agreed to hold hearings at the request of the distributor.

Allergy-free Cats

A company called Transgenic Pets, in Syracuse, NY, was widely reported in 2001 to be working with scientists at the University of Connecticut to "remove the allergen gene" from cats. The company hoped to raise $2 million and sell the modified animals for $1,000 each. Funding problems ended the project.

Genetic Modification for Aquaculture

Most discussion of bioengineered food focuses on plants, but work on animals and fish is well under way. Genetically modified salmon have already been created, though FDA approval is not expected before at least 2006 and the target date has repeatedly been delayed. The Biotechnology Industry Organization (BIO) and the developers, Aqua Bounty Farms, claim that the GE salmon grow faster but not larger than ordinary salmon, but this is strongly disputed. Opponents also cite studies showing dramatic population crashes and unpredictable environmental impacts. Salmon are not the only fish species being modified. Acting FDA Commissioner Lester M. Crawford summarized the position in a speech given in March 2004: "Less well known is that catfish and tilapia have been also genetically modified to grow faster and more efficiently than their non-transgenic counterparts. Rainbow trout has been engineered to increase its contents of omega 3 fatty acids, and shellfish is being modified to reduce its allergenicity and make it grow faster."

The matter-of-fact, if not approving, tone of these comments is disturbing, as it comes from the head of the agency that is supposed to be regulating these technologies. Other countries, including Canada, have moved much faster than the United States to ban or at least place a moratorium on the genetic modification of fish. Malcolm Grant, Chair of the UK Agriculture and Environment Biotechnology Commission, points out that

"Once the fish has escaped, there's virtually nothing that can be done to recall it... Genetic biotechnology has opened a new chapter in the relationship between man and animals. We must therefore prepare now for developments that may be many years away." He also argues that pet cloning is "trivial, distasteful and could be sold to gullible owners."

Genetically Modified Farm Animals

Acting Commissioner Crawford continued: "Cows can be bioengineered to produce several varieties of milk: milk with a lower level of a protein that's allergenic to some infants; milk that is more easily digested by people who are lactose intolerant; milk that has more naturally occurring antimicrobial enzyme, and therefore has longer shelf life; and milk that makes better cheese because it has altered distribution of caseins or less fat."

Genetic Modification for Livestock

Research is also underway to use genetic modification to improve the health of cows and pigs. Advocates hope to produce cattle that would be resistant to Mad Cow disease, for example. There have also been reports of cows being genetically modified to "produce milk with higher than normal levels of protein, which would speed the process of making cheese."

CHIMERAS

Some researchers have created embryos with genetic material from both a human and an animal, also known as chimeras. Most claim this is for research purposes only, and such embryos would never be implanted or brought to term. But some bioethicists are unwilling to draw a line that would prevent it.

For example, Jason Scott Robert and Francoise Baylis assert that they take "no stance at all" on whether "interspecies hybrids or chimeras from human materials should be forbidden or embraced." But much of their article is devoted to their contention that "the arguments against... creating novel part-human beings... are largely unsatisfactory."

The prospect of such human-animal chimeras being born raises a number of troubling questions. Does such an organism have human rights? What if it were 99.9 per cent human and 0.1 per cent chimpanzee? What of the reverse situation? The mixing of species can occur at three different levels. First, some DNA from one organism can be inserted into another organism's genome. Most often, however, these are described as transgenic animals rather than chimeras. For example, scientists have produced transgenic mice which contain some human genes.

Another type of chimera involves a blastocyst containing components from multiple species. James Grifo has used this technique for fertility research. After the US federal government informed him he could no longer continue his experiments, his team moved to China. There, they removed the nucleus from a fertilized rabbit egg and inserted a nucleus from a human somatic cell. The resulting embryos, which still contained rabbit mitochondrial DNA, were allowed to grow for 14 days before destruction.

Finally, scientists have placed cells from one species, typically human stem cells, into a multicellular embryo of another species. In 2004, researchers were surprised to discover that injecting human stem cells into a pig embryo resulted in a fetus whose cellular components were intermingled down to the genetic level. Many cells contained chromosomal DNA of both pig and human origin.

Opposition

Several environmental and animal rights organizations have expressed opposition to pet cloning, including Friends of the Earth and the Humane Society of the United States, the largest animal welfare organization in the country. Jeremy Rifkin from the Foundation on Economic Trends and Stuart Newman of the New York Medical College and the Council for Responsible Genetics have pursued a novel path of opposition. They filed a patent application with the US Patent and Trademark Office for chimeric embryos and animals containing human cells. They did not intent to create a "humouse." Instead, they tried to force the PTO to take a position and, if the patent is granted, to prevent chimeras from being developed.

CLONING AND GENETICALLY ALTERED ANIMALS

At Eurogroup for Animals, we believe that humans have a duty to safeguard animals' welfare and health and we oppose all human activity that causes distress or suffering. In particular in the area of animal breeding, we are of the opinion that animals genetic constitution should not be changed if doing so causes suffering to themselves or to future generations.

Genetic selection must always seek to maximise the wellbeing of the resultant animal. Animal breeding should support a diverse gene pool in order that the genetic resources to maximise wellbeing are sustained for future generations. Techniques and technologies used in animal reproduction must not be used if they cause distress or suffering. For these reasons Eurogroup is opposed to the cloning of animals and would urge the EU to introduce an immediate ban on the cloning of animals for food production, and on the sale of imported food products from cloned animals and their offspring. Eurogroup is concerned that delaying a decision will mean that food from cloned animals will enter the supply chain uncontrolled and without consideration of the opposition of EU citizens and consumers. Under Article 13 of the Lisbon Treaty, in the EU it is a legal obligation to pay full regard to the welfare requirements of animals when formulating and implementing the Union's agriculture, fisheries, transport, internal market, research and technological development and space policies.

Our concerns are based on the fact that some of the developments and applications of animal cloning and other genetic alteration techniques:

- Involve procedures that may cause animals pain, suffering or distress;
- Use a very large number of animals;
- Encourage a wider variety of applications leading to increased animal use;
- Increase the perception of animals as commodities for human use and/or gain, such as research tools or units of production;
- Are progressing at a rate that is outstripping public understanding and ethical and public debate.

CONCERNS RELATING TO HUMAN AND ANIMAL HEALTH

The main applications of cloned and other genetically altered (GA) animals (and their offspring) that are of concern at the present time are:

- The creation of genetically manipulated (GA) and cloned animals;
- The subsequent use of these GA animals (and their offspring):
  - As livestock on farms (used to standardize/maximise quantity and reduce the costs of processing/producing animal derived food products *e.g.* milk, meat, eggs);
  - As disease models;
  - In fundamental research *e.g.* to understand gene function;
  - In toxicity testing;
  - As bioreactors to produce biologically active compounds for experimental and/or medical purposes;
  - As sources of cells, tissues and organs for xenotransplantation;
  - In the creation of cloned pets, sports animals and 'living art'.

However, any new biotechnology, or new application of existing technologies, that is developed and/or tested in animal models, or that causes the animals pain, suffering, distress or lasting harm is of serious concern to Eurogroup. Examples include tissue engineering and the development of nanotechnology generally. Production of cloned and other GA animals

GENERAL ETHICAL AND ANIMAL WELFARE CONCERNS

Numbers of Animals Used

Use of animals in experiments is a matter of serious public concern and pressure to reduce numbers has contributed to a downward trend in some countries. Creation and use of GA animals is reversing this trend and there has been an exponential rise in the number of GA animals used in scientific procedures each year in some EU member states. For example, in the UK, the number of GA animals has risen from around 50,000 in 1990 to 900,000 in 2004. In Germany, Ireland, Finland, Sweden and the Netherlands, there is also increasing creation and use of GA animals.

GA animals can be produced by a number of different techniques, all of which are inherently wasteful with respect to the number of embryos that are manipulated relative to the number of GA offspring subsequently produced. Current estimates suggest only a 3-5 per cent success rate when generating new GA animals. The remaining offspring are surplus to requirements and are killed.

A large number of animals are therefore used to provide sufficient eggs or embryos for genetic manipulation, and to act as recipients and foster mothers for the manipulated embryos. These animals will usually be killed, either before harvesting the eggs or embryos, or, in the case of recipients, once their young are weaned. Thus GA technology is wasteful of animal life and the ethical implications of such wastage are important and need to be acknowledged.

Potential for Pain, Suffering or Distress During Creation of GA or Cloned Animals

The procedures used to produce GA and/or nuclear transfer cloned animals involve hormonal and surgical interventions that can cause pain, suffering and distress. These include; superovulation, vasectomy, semen and embryo collection and embryo transfer. Many of the manipulated embryos die during gestation. This may also cause suffering, although this depends on the stage of development at which death occurs and the species involved.

Adverse Effects as a Consequence of Genetic Modification and Cloning

Where GA animals are created as models of specific diseases, they can experience a range of adverse effects associated with the condition in question, which can have profound effects on their health and welfare. However, being genetically altered does not necessarily compromise the welfare of individual animals, and indeed some GA animals are indistinguishable from their non-GA siblings.

It also needs to be acknowledged that adverse effects may only become apparent when animals are subsequently maintained in a less well defined or controlled, environment than that of the laboratory or experimental farm. Nevertheless, in many cases, genetic alteration can have a deleterious effect on animal welfare and the harms caused depend on a number of factors.

Studies have shown that some *in vitro* culture procedures carried out during genetic manipulation or cloning protocols can lead to unpredictable complications in the animals subsequently produced. An example of this is Large Offspring Syndrome (LOS), which affects sheep, mice and cows following nuclear transfer.

As well as causing complications for the mother during birth, LOS encompasses a range of debilitating pathologies for the offspring including malformations in the liver, brain and urogenital tract, immune dysfunction, placental abnormalities, stillbirth, fetal overgrowth, respiratory failure and circulatory problems.

A further concern is that there is currently no mechanism to ensure consistency in the training of personnel in specific GA technologies and procedures, or to ensure that experimental and other refinements are disseminated throughout the biotechnology community.

This can have a profound influence on the levels of pain, suffering and distress experienced by animals undergoing these procedures. It can also have a negative impact on the success rates achieved and levels of animal wastage.

Concerns Regarding Mutagenesis Programmes

Mutagenesis (using chemicals such as ENU, or physical mutagens such as radiation, to increase the natural mutation rate of DNA) is the quickest method for producing large numbers of GA mice. This technique is being widely applied both in Europe and beyond, with mutagenesis programmes aiming to explore the function of every gene in the mouse genome.

This project presents a number of animal welfare concerns:

- The number of animals involved in mutagenesis programmes is vast; *e.g.* around 35-36,000 for a 3 year project, or around 50 animals per mutant line established;
- The process of mutagenising animals has significant animal welfare implications for the animals involved, for example males require 12-14 weeks to recover their fertility following treatment with a mutagen. Furthermore, only around 50 per cent of mutagenised males are able to go on to sire offspring with the remainder being culled;
- The mutations induced are by their very nature unpredictable and the scientific usefulness of mutagenised animals cannot therefore be predicted, nor can the effect of any mutation on the health and welfare of the animals. This makes the justification for producing GA mice in this way highly questionable.

CONCERNS REGARDING KNOCK-IN AND KNOCK-OUT TECHNOLOGY

A more targeted approach to the production of GA animals uses DNA constructs inserted into the animals genome to either completely remove a gene of interest (knock-out), or to replace a given gene with an altered version (knock-in). This is done either in the whole animal, or within specific tissues of an animal (conditional).

The animal welfare concerns include:

- At the current level of efficiency, the numbers of animals used is high - at least 200 animals will be used in the production of a single GA animal;
- The site of insertion into the host genome and the number of copies of the DNA construct inserted cannot always be controlled unless ES cell manipulation is used. This can be a problem because the incorporation of the construct DNA at an incorrect location can result in the random inactivation of other genes or alterations in the expression of surrounding genes, both of which can impact on the health and welfare of the animals produced;
- When a knockout animal is generated the level, or pattern of expression observed for other genes can be altered to compensate for the lost gene. Thus any effects observed in the GA animal may

not only reflect the loss of the gene of interest, which means that investigation of the effects is not always straightforward. The use of knock-out as an alternative to knock-in or conditional technology, therefore needs careful consideration and justification.

Special Concerns for Non-human Primates

This submission relates to all animals used in biotechnologies, but Eurogroup has particular concerns about the application of such technologies to non-human primates. Macaques have already been produced by nuclear transfer cloning in the USA and recent technical advances show that the production of GA primates is currently increasing. Eurogroup firmly believes that the genetic modification of non-human primates should not be allowed for any purpose, given the ethical issues raised by such developments, the large number of animals required to produce each GA animal, and the associated potential for harms.

ANIMAL MODELS OF DISEASE

A common justification for the creation of GA animals is that they will provide, or contribute to, 'improved', more predictive models of disease. This is an oversimplification because a) this is not necessarily always true, and b) even where a new GA model is more appropriate, it may be used alongside other, older models by different researchers. There is no mechanism for ensuring only the most relevant are used and that newer models are available to all researchers.

In such cases the GA animal is not the new *definitive* model but just an *additional* one. The motivation for the research may merely be an interest in the model for its own sake, but a medical application may be used to justify the work since this is likely to be more acceptable publicly and politically. Basic, fundamental research carried out within academic research establishments is not regulated under Directive 86/609, so it may not undergo an ethical review with appropriate assessment and weighing of harms and benefits. This directive is under review and is expected to be adopted end 2010. The revised directive may incorporate basic, fundamental research with proper review, however only if implemented and enforced properly.

There is also currently no requirement for GA animals to be cryopreserved and stored within central archive facilities or depositories. Such methods can reduce repetition and duplication of work by providing a central resource for use by the wider scientific community and protect against adverse events such as environmental disasters or genetic drift.

They also reduce the need for live transportation of GA animals, with associated welfare problems, because frozen gametes or embryos could be sent instead.

GA ANIMALS IN TOXICITY TESTING

GA mice and rats are increasingly used in genotoxicity and carcinogenicity testing, within studies that are done to fulfil regulatory requirements for the marketing of chemicals and pharmaceuticals. Such animals are likely to suffer equivalent (or even more severe) levels of pain and distress to those experienced by animals in traditional tests, but it has been claimed that fewer animals will be needed.

Eurogroup believes that reducing the numbers of animals used in research and testing is an important goal, provided that this can be done without increasing the level of suffering experienced by individual animals. However, relatively severe adverse effects have been reported in some GA strains used in toxicity testing. For example, mortality rates are higher in *c-neu* and *c-myc* mice used in carcinogenicity testing than in conventional mice used in the same type of test. Many GA mice used in carcinogenicity tests are more susceptible to developing cancer and so will develop tumours more rapidly, which could make it more difficult to implement humane endpoints. Reducing numbers is thus not automatically a positive outcome for animals - the impact on individuals must be taken into account and it may be justifiable to use more animals who will suffer less.

In any case, the contribution that biotechnologies can make to reducing animal numbers in carcinogenicity and genotoxicity is not consistent, largely due to actual or potential regulatory requirements. In carcinogenicity testing, GA mice were introduced to try to reduce animal numbers and the time taken to obtain test results. However, there are proposals to add additional control groups (a positive control and treated and untreated controls using non-GA mice) that would decrease the magnitude of the reduction if they are adopted. It is also uncertain whether testing on one GA strain will be regarded as sufficient by regulators, in which case the reductions will be further diminished by requirements for results obtained using other strains.

For genotoxicity testing, GA models have distinct scientific advantages over existing *in vivo* assays when used as second tier tests for *in vitro* genotoxins. The reduction in numbers of animals used would be fairly substantial, perhaps from 50 to 20 per substance tested, but again it is not certain that a test on one GA model would be regarded as sufficient. There are potentially much greater savings in animals if GA tests can be used in place of very large tests of heritable mutation which are currently used, albeit rarely, for chemicals of high concern.

Conversely, using GA animal models in toxicity testing could increase the use of animals by making some tests more practical, or by increasing the amount of information they produce. For example, the use of GA animals could make investigations feasible that would otherwise require very large and impractical numbers of non-GA animals. Regulators might be inclined to ask for the GA test in cases where the conventional test would not have been requested (and could therefore presumably have been done without) because it was regarded as too cumbersome and possibly uninformative. Eurogroup believes that, in toxicity testing as in the other research fields discussed in this document, the focus should be on developing *in vitro* alternatives to replace animals and not on developing different animal models.

Animals as 'Bioreactors'

This category of GA animal use includes:

- "Pharmed" animals who produce therapeutic substances *e.g.* pigs who express the blood protein Factor IX in milk, and goats that express the anti-clotting agent Atryn in their milk;
- Animals producing specialist materials *e.g.* goats who produce spiders' silk proteins in their milk for use to make 'biosteel'. This has both medical and non-medical applications (*e.g.* in sutures and bullet-proof vests respectively) (Nexia Biotechnologies, Quebec);

The substance produced may have an adverse effect on the animal, either at the point of expression, or if it can enter the animal's bloodstream. For example, a strain of rabbits genetically engineered to express human erythropoietin (EPO) in the mammary glands also expresses the protein at low levels in other organs, resulting in greatly elevated numbers of red blood cells, infertility and premature death. Using animals in this way, and referring to them as 'bioreactors', reinforces the perception of animals as units of production and/or biological tools, rather than as sentient beings with the ability to experience pain, suffering and distress.

Xenotransplantation

The use of GA animals to supply organs, tissues or cells for transplantation into humans is a highly controversial issue which has been the subject of a great deal of debate. There are many legal, scientific, human health, animal welfare and ethical concerns, which have been described in a number of documents. Only the ethical and welfare issues relating to animals are addressed here.

These include:

- The ethics of genetically modifying animals of any species as a source of cells, tissues and organs for human transplantation;
- The harms associated with the initial creation of GA animals as source animals;
- The suffering and/or distress associated with production and maintenance systems for high health status source herds. This includes hysterotomy-derivation, early weaning practices, and barren husbandry environments, which have a serious negative impact on

animal welfare because they prevent animals from satisfying their physical, social and behavioural needs.

Furthermore, development of xeno technology to a point where it can be used still requires a great deal of pre-clinical research. To date, such research has included studies of efficacy, physiology, immunology, and infection risks in a range of species including primates, goats and dogs. This research, by its very nature, causes considerable suffering. Experiments involving organ transplantation require major surgery, which in itself causes suffering that is exacerbated by tissue rejection and immunosuppressive treatment.

Xenotransplantation is also an example of a biotechnology where over-optimistic claims are made to justify the approach, the funding and the use of animals. For example, in September 1995, the UK company Imutran *"envisaged the first xenotransplants of transgenic pig hearts into human patients taking place in 1996"*. Yet despite some progress, particularly with cell transplants, the transplant of whole organs is no closer and xenografts still rarely survive for more than a few months.

Cloning Companion Animals

Eurogroup believes that, without doubt, some applications of modern biotechnology are trivial, scientifically unnecessary and ethically unjustifiable.

Examples include;

- The cloning of champion racing and show jumping horses for sport;
- The generation of a green fluorescent rabbit 'GFP Bunny' as transgenic "art";
- The cloning of companion animals for example cats, purely to satisfy humans' emotional requirements.

CONCERNS RELATING TO AGRICULTURAL PRODUCTION

Gene mapping was/is the most widely used technique in agriculture to enhance selective breeding schemes, however as more and more 'super' animals exist this is being overtaken by the import/export of embryos and/or sperm from cloned 'super' animals. Examples of agricultural applications of gene mapping and other genetic altering techniques that have given rise to ethical and animal welfare concerns include:

Increasing Productivity or Changing Body Composition

Gene mapping has been used primarily to select individuals for breeding with the aim of increasing productivity *i.e.* growth rates, litter sizes and production traits such as egg laying, milk volume and meat quantity and quality (including the proportion of lean meat to fat). Whilst still being practiced to create 'super' individuals for subsequent breeding, gene mapping is being/has been replaced by the import/export of embryos and/or sperm from cloned 'super' livestock (primarily from the US). Farmed species have also been genetically manipulated to alter the composition of meat and milk. For example, pigs and cows respectively have been genetically altered to have higher levels of Omega 3 in their muscle, and to express higher levels of casein in their milk.

Increasing Disease Resistance

Animals are genetically manipulated to be resistant to disease, for example, cattle have been engineered to express the antibiotic lysostaphin in their milk, which results in increased resistance to mastitis.

Animals (including pigs, sheep, mice and rabbits) have also been modified to express antibodies providing immunity to specific diseases, for example mice have been generated with protection against prion disease.

Making Animals More 'environmentally Friendly'

An example of this application is the 'Enviropig', which has been engineered to contain the enzyme phytase in the pigs' saliva so that they can digest sources of dietary phosphorus. This results in faeces with a lower phosphorus content, which in turn reduces the pollution of surface and ground water with phosphorus.

ETHICAL AND ANIMAL WELFARE CONCERNS

The selective breeding of farm animals has been conducted for thousands of years, and has given rise to a number of welfare concerns. However, Eurogroup believes that the use of cloning or other GA technologies to speed this process, or to introduce genes that could never be incorporated into the genomes of farm animals by any natural process, is a serious ethical and welfare issue. Directly altering an animal's genome is viewed by many as an unacceptable assault on the integrity of the animal that is incompatible with the concept of respecting farmed animals, and minimising the harms that are caused to them for human benefit. These views are important and should be respected as a legitimate part of the debate on biotechnology and farmed animal welfare.

Eurogroup also questions the necessity of further increasing production in farm animals. In many cases, productivity is already pushing animals to their physical and metabolic limits, so with any further increase there is an enhanced likelihood of animal welfare problems. Enhancing selective breeding, by gene mapping or genetic modification, can also cause suffering if the trait that is selected for has a negative impact on the rest of the animals' physiology. For example, hens who produce high numbers of eggs suffer from osteoporosis because the majority of the calcium they ingest is used in eggshell production.

There can also be less direct effects on welfare, in that some GA animals may receive lower standards of husbandry than conventional animals. For example, clinical mastitis is a major welfare problem in dairy systems with suboptimal standards of hygiene, and early detection is reliant on routine inspections by parlour staff at milking. The creation of cattle resistant to mastitis may encourage the perception that mastitis is no longer a problem. This could not only compromise standards of parlour hygiene, but may reduce the level of attention paid to each animal at milking. This would increase the potential for other clinical or welfare problems to go undetected.

"High productivity" animals may also be at risk if their husbandry is not appropriate. It may be possible for them to be properly managed and cared for in the controlled environment of a breeding company or experimental farm, but there are serious concerns regarding the welfare of such animals once they are released into commercial agriculture.

Concluding Remarks

Modern biotechnologies have had, and will continue to have, a serious adverse impact on animals, particularly with regard to their use in scientific research and agriculture. This adverse impact relates to the numbers of animals used and the nature of the harms caused to them. In addition, directly altering an animal's genome, as occurs in many applications of biotechnology, is viewed by many as altering the integrity of the animal in a way that is incompatible with the concept of respecting animals, and minimising the harms that are caused to them for human benefit. Lastly, the technology is progressing at a rate that is outstripping public understanding and ethical and public debate.

The development and application of novel biotechnologies therefore poses new challenges for existing regulatory regimes in a number of fields of science, medicine, agriculture and the environment. Eurogroup believes that the broader issues surrounding the ethical and social acceptability of such uses of animals, as set out in this submission, cannot be effectively addressed within the current regulatory systems. The following principles are fundamental to ensuring that the lives and welfare of animals involved in all modern and future biotechnologies are awarded due priority.

- It is critically important that all relevant regulatory systems are updated to take into account the animal welfare, ethical and social implications and societal concerns of the development and intended use of all modern biotechnologies.
- Biotechnology is applied in many different fields so there needs to be effective liaison, co-ordination and definition of responsibilities within and between all the relevant legislative and regulatory bodies concerned with a particular issue. This includes, for example, the different bodies regulating use of animals in experiments and those setting requirements for product regulation.
- The regulatory framework for each technology must encompass a process which enables a critical scrutiny of the potential harms to

animals and the intended benefits, and a careful and fair weighing of these. This applies to broad research directions as well as individual projects and the further application of new technologies that result from these. Critical assessment of justification needs to be done prior to the development and/or application of a technology and must then be reviewed regularly to check whether the harms and benefits are as expected so that appropriate action can be taken if necessary.

- A mechanism should be set in place to ensure that the justification and clinical relevance of all research involving the production and use of GA animals is critically scrutinised, such that animals are not used simply because the technology is available. This also needs to ensure that animal models of disease are regularly reviewed, so that redundant models are no longer routinely used for research purposes.
- There needs to be greater transparency with regard to the use of animals in biotechnology throughout Europe. Clearer information on the numbers of animals, and nature and level of any suffering that they experience, is essential in order to be able to identify issues of concern and assess trends. It is also vital that the public is well informed and therefore able to engage in constructive debate on the associated ethical issues.
- Restrictions should be placed on the species of animal that it is permissible to genetically modify. Non-human primates should not be genetically modified or cloned for any purpose, nor used as source animals for cells, tissues or organs.
- The production of GA livestock where the intention is to modify traits such as increased lean to fat ratio, growth rate, or litter size should not be allowed, as levels of productivity are already causing serious welfare problems.
- There should be far greater effort devoted to developing and validating alternatives to animal use in all fields. There should be greater commitment to, and endorsement of, the principles of the Three Rs of reduction, refinement and replacement in animal experiments.

  Examples especially relevant to modern biotechnology are:
  - The development of GA germ cells for *in vitro* testing;
  - The production of drugs, proteins, or material by bacteria rather than 'bioreactor' animals;
  - Generating cells, tissues and organs for treatment or transplantation using a patients own cells, eg human bladders.

CURRENT CONTEXT OF GENETICALLY ENGINEERED ANIMALS

Genetic engineering technology has numerous applications involving companion, wild, and farm animals, and animal models used in scientific research. The majority of genetically engineered animals are still in the research phase, rather than actually in use for their intended applications, or commercially available.

COMPANION ANIMALS

By inserting genes from sea anemone and jellyfish, zebrafish have been genetically engineered to express fluorescent proteins — hence the commonly termed "GloFish." GloFish began to be marketed in the United States in 2003 as ornamental pet fish; however, their sale sparked controversial ethical debates in California — the only US state to prohibit the sale of GloFish as pets. In addition to the insertion of foreign genes, gene knock-out techniques are also being used to create designer companion animals. For example, in the creation of hypoallergenic cats some companies use genetic engineering techniques to remove the gene that codes for the major cat allergen.

Companion species have also been derived by cloning. The first cloned cat, "CC," was created in 2002. At the time, the ability to clone mammals was a coveted prize, and after just a few years scientists created the first cloned dog, "Snuppy". With the exception of a couple of isolated cases, the genetically engineered pet industry is yet to move forward. However, it remains feasible that genetically engineered pets could become part of day-to-day life for practicing veterinarians, and there is evidence that clients have started to enquire about genetic engineering services, in particular the cloning of deceased pets.

Wild Animals

The primary application of genetic engineering to wild species involves cloning. This technology could be applied to either extinct or endangered species; for example, there have been plans to clone the extinct thylacine and the woolly mammoth. Holt et al point out that, "As many conservationists are still suspicious of reproductive technologies, it is unlikely that cloning techniques would be easily accepted. Individuals involved in field conservation often harbour suspicions that hi-tech approaches, backed by high profile publicity would divert funding away from their own efforts." However, cloning may prove to be an important tool to be used alongside other forms of assisted reproduction to help retain genetic diversity in small populations of endangered species.

Farm Animals

As reviewed by Laible, there is "an assorted range of agricultural livestock applications [for genetic engineering] aimed at improving animal productivity; food quality and disease resistance; and environmental sustainability." Productivity of farm animal species can be increased using genetic engineering. Examples include transgenic pigs and sheep that have been genetically altered to express higher levels of growth hormone. Genetically engineered farm animals can be created to enhance food quality. For example, pigs have been genetically engineered to express the Ä12 fatty acid desaturase gene (from spinach) for higher levels of omega-3, and goats have been genetically engineered to express human lysozyme in their milk. Such advances may add to the nutritional value of animal-based products.

Farm species may be genetically engineered to create disease-resistant animals. Specific examples include conferring immunity to offspring via antibody expression in the milk of the mother; disruption of the virus entry mechanism (which is applicable to diseases such as pseudorabies); resistance to prion diseases; parasite control (especially in sheep); and mastitis resistance (particularly in cattle).

Genetic engineering has also been applied with the aim of reducing agricultural pollution. The best-known example is the Enviropig; a pig that is genetically engineered to produce an enzyme that breaks down dietary phosphorus (phytase), thus limiting the amount of phosphorus released in its manure.

Despite resistance to the commercialization of genetically engineered animals for food production, primarily due to lack of support from the public, a recent debate over genetically engineered AquAdvantage Atlantic salmon may result in these animals being introduced into commercial production.

Effort has also been made to generate genetically engineered farm species such as cows, goats, and sheep that express medically important proteins in their milk. According to Dyck et al, "transgenic animal bioreactors represent a powerful tool to address the growing need for therapeutic recombinant proteins."

In 2006, ATryn® became the first therapeutic protein produced by genetically engineered animals to be approved by the Food and Drug Administration (FDA) of the United States. This product is used as a prophylactic treatment for patients that have hereditary antithrombin deficiency and are undergoing surgical procedures.

Research Animals

Biomedical applications of genetically engineered animals are numerous, and include understanding of gene function, modeling of human disease to either understand disease mechanisms or to aid drug development, and xenotransplantation.

Through the addition, removal, or alteration of genes, scientists can pinpoint what a gene does by observing the biological systems that are affected. While some genetic alterations have no obvious effect, others may produce different phenotypes that can be used by researchers to understand the function of the affected genes. Genetic engineering has enabled the creation of human disease models that were previously unavailable. Animal models of human disease are valuable resources for understanding how and why a particular disease develops, and what can be done to halt or reverse the process. As a result, efforts have focused on developing new genetically engineered animal models of conditions such as Alzheimer's disease, amyotrophic lateral sclerosis (ALS), Parkinson's disease, and cancer. However, as Wells (13) points out: "these [genetically engineered animal] models do not always accurately reflect the human condition, and care must be taken to understand the limitation of such models."

The use of genetically engineered animals has also become routine within the pharmaceutical industry, for drug discovery, drug development, and risk assessment.

As discussed by Rudmann and Durham: "Transgenic and knock out mouse models are extremely useful in drug discovery, especially when defining potential therapeutic targets for modifying immune and inflammatory responses...Specific areas for which [genetically engineered animal models] may be useful are in screening for drug induced immunotoxicity, genotoxicity, and carcinogenicity, and in understanding toxicity related drug metabolizing enzyme systems."

Perhaps the most controversial use of genetically engineered animals in science is to develop the basic research on xenotrans-plantation — that is, the transplant of cells, tissues, or whole organs from animal donors into human recipients.

In relation to organ transplants, scientists have developed a genetically engineered pig with the aim of reducing rejection of pig organs by human recipients. This particular application of genetic engineering is currently at the basic research stage, but it shows great promise in alleviating the long waiting lists for organ transplants, as the number of people needing transplants currently far outweighs the number of donated organs.

However, as a direct result of public consultation, a moratorium is currently in place preventing pig organ transplantation from entering a clinical trial phase until the public is assured that the potential disease transfer from pigs to humans can be satisfactorily managed. According to Health Canada, "xenotransplantation is currently not prohibited in Canada. However, the live cells and organs from animal sources are considered to be therapeutic products (drugs or medical devices)...No clinical trial involving xenotransplantation has yet been approved by Health Canada".

ETHICAL ISSUES OF GENETIC ENGINEERING

Ethical issues, including concerns for animal welfare, can arise at all stages in the generation and life span of an individual genetically engineered animal. The following sections detail some of the issues that have arisen during the peer-driven guidelines development process and associated impact analysis consultations carried out by the CCAC. The CCAC works to an accepted ethic of animal use in science, which includes the principles of the Three Rs (Reduction of animal numbers, Refinement of practices and husbandry to minimize pain and distress, and Replacement of animals with non-animal alternatives wherever possible).

Together the Three Rs aim to minimize any pain and distress experienced by the animals used, and as such, they are considered the principles of humane experimental technique. However, despite the steps taken to minimize pain and distress, there is evidence of public concerns that go beyond the Three Rs and animal welfare regarding the creation and use of genetically engineered animals.

GENETIC VULNERABILITY OR RESISTANCE

Genetic resistance to scrapie depends not only on the prion genotype of the sheep but also on the strain of scrapie present. Genotypes found to be resistant to one strain of scrapie have been shown to be susceptible to another strain. It appears that there are differences between the U.S., and Europe in the strains of scrapie that are present. The fact that the U.S., is free of certain scrapie strains is good justification for the stringent regulations for importation of sheep and goats into the U.S., from countries with scrapie.

Codon 136: Sheep homozygous for alanine (AA) have been shown to be more resistant to scrapie than sheep homozygous for valine (VV) or heterozygous (AV) in European studies. While a lower incidence of scrapie is anticipated in AA sheep in Europe than in the other genotypes, some AA sheep still have been diagnosed with scrapie. AA sheep are not 100 per cent resistant to scrapie. In the U.S., amino acid changes at codon 136 appear to be less important to scrapie susceptibility than in Europe. This may be due to the different strains of scrapie found in the two regions.

Codon 154: Of the three important codons, amino acid changes at 154 appear to have a slightly less dramatic effect on scrapie susceptibility than do the other two, and the susceptible genotypes are not consistent across studies. In some studies, sheep with the arginine (R) allele have a lowered incidence of scrapie, and in other studies, sheep with the histidine (H) allele have a lowered incidence. At the present time, it appears that screening sheep on the basis of codon 154 genotype has little value in increasing resistance to scrapie.

Codon 171: Amino acid changes at codon 171 have a large effect on scrapie susceptibility in sheep in both Europe and the U.S. Virtually no sheep homozygous for arginine (RR) have been identified with scrapie. The one exception is a single RR Suffolk in Japan that was diagnosed with scrapie. The frequency of heterozygous (QR) sheep also is very low among scrapie-infected sheep. However, the frequency of sheep homozygous for glutamine (QQ) is very high among scrapie-infected sheep. A scrapie resistance index for each genotype in U.S., sheep. The higher the number, the greater the expected resistance to scrapie. The index is weighted heavily in favour of RR and QR genotypes at codon 171, less heavily in favour of the AA genotype at codon 136, and neutral with respect to the genotypes at codon 154.

SUSCEPTIBILITY VERSUS SCRAPIE-INFECTED

It must be remembered that a susceptible genotype like QQ at codon 171 does not imply that the animal is scrapie-infected. Scrapie is not a genetic disease. Scrapie is not caused by a particular genotype. An infectious agent causes scrapie. In order to have scrapie, both an infectious agent and a susceptible genotype need to be present. For example, Australia and New Zealand are scrapie-free, but Suffolk, Cheviot, Merino, and Poll Dorset sheep of susceptible genotypes are found in these countries. There is no scrapie in these countries because the scrapie agent is not present.

Likewise, susceptible genotypes are found in many breeds of sheep in the U.S., but almost 90 per cent of the scrapie cases are in the Suffolk breed. This may be due to the major mode of transmission of the disease from an infected ewe to her newborn lambs. Scrapie was brought into the U.S., in Suffolk sheep. A possible scenario that followed was that infected Suffolk ewes infected their lambs, and infected females from infected ewes continued to pass the disease to their lambs. However, infected Suffolk rams would not tend to infect the ewes they were mated to or their lambs, so infected Suffolk rams mated to ewes of a different breed, as is common in crossbreeding systems in the U.S., would not tend to pass scrapie onto that breed.

While scrapie is primarily a disease of Suffolk sheep in the U.S., the disease is found in the Suffolk, Cheviot, Swaledale, Bleu du Maine, Herdwick, Poll Dorset, Shetland, and Soay breeds in the U.K., Texel breed in The Netherlands, Suffolk and Corriedale breeds in Japan, Romanov and Lacaune breeds in France, Rygja breed in Norway, and Icelandic breed in Iceland. This is not an exhaustive list of sheep breeds or countries with scrapie, but a sample to indicate that scrapie is found in several breeds. A discussion of theories on how scrapie initially becomes established in a breed of sheep would be interesting.

Breeding for Scrapie Resistance in the U.S.

Selection for resistant prion protein genotypes need only be conducted in breeds where scrapie is found, *i.e.* Suffolks and, perhaps, other blackfaced breeds that have been "improved" by illicit crossing with the Suffolk. With the current strain(s) of scrapie present in the U.S., selection for the arginine (R) allele at codon 171 appears to be sufficient.

A flock where every ewe is of the susceptible genotype QQ can be converted to a flock of QR and RR resistant ewes fairly quickly by simply purchasing and using only RR rams. With a normal ewe replacement rate of 20 per cent per year, over 2/3 of the ewe flock will be of the QR or RR genotypes after five years of RR ram use; even with no selection of ewes or ewe lambs on prion protein genotype. Conversion of the flock to QR and RR ewes will be even faster if, in addition to using RR rams, replacement ewe lambs are preferentially selected from QR or RR dams and if QQ ewes are preferentially culled.

DISEASES IN NON- SALMONID FISH

Considering that 70 per cent of the international fishing grounds are overexploited and that the current volume of captures by fishing efforts has practically reached a maximum level, the increased consumption of fishery related products must be sustained by aquaculture. Indeed, it has recently been estimated that approximately 30 per cent of the fisheries products that are consumed in the world are provided by aquaculture and this aquatic production will increase in the near future.

However, mortalities due to disease can occasionally be high in fish farms. Although there are several commercial vaccines that help in the prevention of numerous diseases, further studies are needed for the improvement of existing vaccines and for the development of new vaccines and vaccination strategies. Aquaculture can greatly benefit from the use of molecular and biotechnological tools to identify and characterize genes and regulatory genetic networks of potential use. In this sense, molecular biology can be applied in aquaculture in the diagnosis of diseases, in the design of new vaccines, in the control of growth and reproduction, and in increasing disease resistance of cultivated species. Moreover the availability of genetic maps of the appropriate density is of high interest for the identification of genomic regions with characters of interest in production by means of assisted marker selection programmes.

This knowledge acquired in fish can be used not only to improve aquaculture production but can also be applied in other industries. In fact, for a long time now, fish have had an impact in Genomics development. Approximately 30 years ago, a popular tropical fish named *Danio rerio* (zebrafish) was chosen as a great candidate for the genetic analysis due to numerous interesting features such as its short life cycle (3 months to reach sexual maturity), relatively high egg clutches throughout all the year, easy culture and external development with transparent embryos.

All this combined with large scale mutagenesis experiments conducted at the beginning of the 90s allowed zebrafish studies to fill a gap in the vertebrate development biology because of the ability to help in the study of genes through mutant phenotypes, as in *Drosophila melanogaster* o *Caenorrhabditis elegans*.

Since then great advances on the knowledge and understanding of vertebrate development and human disease have been accrued using this model fish. Fish became an important actor in genomic studies (DNA sequencing and data mining) in 1993, when Sydney Brenner suggested the "pufferfish", *Takifugu rubripes* (fugu) as a genome model. On top of its attraction to gourmets in Japan and China fugu has one of the smallest vertebrate genomes. This characteristic already known from its freshwater relative *Tetraodon nigroviridis* in 1968, constitutes a great advantage to access a gene catalogue for a modest cost compared to higher vertebrates. It is important to point out that in recent years available fish genomic resources have increased dramatically, especially for salmonids.

CLONING OF IMMUNE RELATED GENES

A hypothesis that is becoming quite popular among comparative immunologists is that innate immunity of lower vertebrates constitutes an important protection against pathogens for this group of animals. The reason is the relative inefficiency of their acquired immune response due to the evolutionary status and their poikilothermic nature. In fact, innate parameters are considered as relatively independent of temperature and more active at the lower range temperatures at which fish live. However, acquired immunity parameters such as antibody production and lymphocytic activity are more effective at higher temperatures. This translates into a limited antibody repertoire and lower memory and a very slow lymphocyte proliferation in contrast with the almost instantaneous innate response.

Despite of all this, a few non- specific immune genes have been cloned in fish, although this group has grown significantly in the last years. What follows is a review of the "innate and specific" genes characterized up to now in fish (model and commercial species). Most of these studies are based on the cloning and characterization of individual genes, complete sequencing, phylogeny, expression analysis, etc., and recently also the profile expression analysis.

Salmonid Fish

Salmonids, especially rainbow trout (*Oncorhynchus mykiss*), are the group in which the molecular basis of the immune response has been more extensively studied. Among other genes, there have been sequencing and expression studies on interleukins (IL): IL-1β, IL-6, IL-11 and IL-18 and Tumor Necrosis Factor-alpha (TNF-α). Interferon (IFN) and related genes such as Mx have been also studied because of the importance of viral aetiology in diseases and the lack of treatments. In rainbow trout the following genes have been characterized: two Toll Like Receptors (TLRs); IFN regulatory factors (IRFs); Mx proteins; a macrophage activating factor (MAF) similar to IFN-g, expression and gene structure of Inducible Nitric Oxide Synthase (iNOS). Both in rainbow trout and Atlantic salmon (*Salmo salar*) the guanylate binding protein (GBP) induced by IFN-g in mammals has been characterized.

Several antimicrobial peptides (AMPs) have been sequenced and their expression characterized, the same was done with complement factors such as C4 and molecules type C1, anaphylatoxins C3a, C3a receptor and several chemokines. Transforming Growth Factor-beta (TGF-β) has also been characterized and an Immunoglobulin T (IgT) has been described as well. Concerning IFN-γ inducible chemokines a CXC, CXCL8 type and another with homology with the subgroup CXCL9, CXCL10, CXCL11, apparently with two forms and inducible by viral infections have been identified.

Two chemokines, CK1 and CK2 have been characterized. β1,3-glucan receptors in macrophages have been found in Atlantic salmon. A type C lectin receptor has been sequenced and its expression characterized and a lectin associated to pathogens recognition has been detected in serum. A protein associated to IL-1 receptor has been cloned. Two IgM isotypes and an IgD isotype have been cloned. IFN type I genes and some of the inducible proteins have been cloned and their expression studied, some IFN-g have been identified. A lectin has been partially characterized in Chinook salmon (*Oncorhynchus tshawytscha*) embryos.

Non- Salmonid European Fish

In comparison with those reported in salmonids there are not many characterized genes in non- salmonid fish species. The non- salmonid cloned genes up to date. In sea bream (*Sparus aurata*) several genes induced by interferon (α or γ) such as Mx or IRF-1 have been characterized. Viral infections are able to induce the expression of these factors suggesting an anti-viral function. Sunyer *et al.* have underlined the diversity of functional forms of complement factor 3, as already described in trout.

A type F lectin has been characterized. Several factors involved in the immune and inflammatory response have been studied such as TNF, IL-1 or its receptors or TGF. All of these factors are expressed constitutively. In contrary to what is expected with the mammalian perspective, TNF does not increase its expression level after a lipopolysaccharide (LPS) treatment, whilst IL-1 does.

However, TNF shows pro-inflammatory and proliferatory functions when administered *in vivo*, suggesting a similar role as in mammals. There is also evidence of other roles not directly immune such as the regulation of fat tissue in this species. TLR-9 has high expression levels in the immune organs (spleen, head kidney) and in the mucosal and epithelial barriers.

However, its expression did not change in the spleen when fish were infected with bacteria. A second form was found, generated by alternative splicing, although its biological meaning has to be further studied. Recently Major Histocompatibility Complex (MHC) class II alpha chain has been characterized and is constitutively expressed in several tissues and the expression was increased in kidney cells when incubated with bacteria and yeast cells. In addition the CD8a co-receptor has recently been cloned and characterized in sea bream. As in sea bream several genes modulated by interferon α, β and γ have been characterized in turbot (*Psetta maxima*) such as Mx and IRF-1.

Recently, class II α and β MHC and TNF have been cloned. The expression profile in the life cycle has also been studied for hepcidin and T cell receptors. In sole (*Solea senegalensis*), only the Mx gene has been characterized so far. In sea bass (*Dicentrarchus labrax*) there has been a growing effort in the characterization of immune genes. Caspase genes such as caspase 3 and 9, cyclooxygenase-2 (COX-2), T cell receptor-β, MHC class II, CD8α, hepcidin and interleukins such as IL-1, IL-10 and IL-12 have all been characterized recently. The light chain of IgM has also been cloned. It is interesting to point out the study on the biological function of IL-1 that has been proposed as an immunostimulant or adjuvant to be used in aquaculture.

Non- Salmonid Fish Out of Europe

β1,3-glucan receptors have been characterized in catfish (*Ictalurus punctatus*) neutrophils. Also in this species an IgD, Mx and IFN-γ similar sequences have been recently described. In the carp (*Cyprinus carpio*) the α1-anti-protease has been purified and characterized. Two Mannose Binding Protein (MBP) homologues and an MBP associated serine protease have been characterized. Among the different characterized molecules we can quote the following: a chimerical immunoglobulin IgM-IgZ; the protein Nitric Oxide Synthase-2 NOS2; a factor similar to TGF-β-2; a chemokine CXC similar to subgroup CXCL9, CXCL10, CXCL11, a chemokine CC similar to C4-CC, a chemokine receptor CXC and two IL-11 paralogues.

In the herbivorous carp (*Ctenopharyngodon idellus*) the α2-macroglobulin has been purified and characterized. In the Japanese flounder (*Paralichthys olivaceous*) the following immune genes have been cloned: an IgD cDNA; IFN receptor factors; Mx and the chemokines CXC type CXCL8 and two CC chemokines, C6-CC and C4-CC, with a high homology with mammalian CCL11. Three isoforms with homology with mammalian TGF-b-1/4/5, β-2 and β-3 have been described in plaice (*Pleuronectes platessa*). In halibut (*Hippoglossus hippoglossus*) the Mx proteins have also been characterized. In cod (*Gadus morhua*) a soluble and membrane associated IgM form with an unusual alternative splicing and an IgD have been described. In the stripped sea bass (*Morone saxatilis*)a type F lectin has been characterized.

In the cobia (*Rachycentron canadum*), the most important cultured fish in Taiwan, a mannose specific lectin has been cloned and characterized that exhibited a bactericidal and mitogenic activity. The primary structure of a Japanese eel (*Anguilla japonica*) lectin from the epithelial mucus has been characterized. The gene and promoter of viperine (an antiviral protein) from Chinese perch (*Siniperca chuatsi*) have been cloned.

Model Species of Primitive Fish

The lamprey (*Lampetra fluviatilis*), belonging to the most primitive fish, the agnathans, have been used as a model to study the evolutionary relations with other model fish: fugu, *Tetraodon*, zebrafish, goldfish (*Carassius auratus*) and other species. Although the studies of the immune system are not very abundant, the first fish chemokine is a lamprey CXC that codes a mammal type CXCL8 peptide. A MBP has also been cloned and characterized. Although structurally similar sequences can be found between groups this does not necessarily indicates a functional similarity but rather how conserved the defence related mechanisms can be throughout evolution. An immunoglobulin isotype has been identified in fugu (*Takifugu rubripes*, fugu), whose structure and expression analysis suggest differences from the IgH locus isotypes previously described; IFN-γ, TLRs, IRFs phagocytic oxidase subunits (oxidase NADPH) and Macrophage Colony-Stimulating Factor (M-CSF) receptor sequences have been also identified. In *Tetraodon nigroviridis*, freshwater fugu, comparative genomics shows an independent expansion in vertebrates of type I IFN receptors and their ligands. Janus kinases (JAKs)-Signal Transducers and Activators of Transcription (STATs) pathway (JAK-STAT pathway) genes have also been characterized.

Several IgZ isotypes have been recently described in zebrafish. A molecular and functional analysis of the following genes have been conducted in zebrafish: IFN, TLRs, several JAK-STAT genes and Mx proteins and their promoters. Sequences of the M-CSF and two chemokines, a CXC similar to mammalian CXCL14 and CXCL12 type have been identified. In goldfish studies have been mostly focused in interferon and associated proteins. TLRs, whose expression increases in activated macrophages, IFN regulatory factors, JAK-STAT genes and antiviral proteins such as Mx, Interferon Stimulated Gene (ISG) and dsRNA Protein Kinase (PKR) have been characterized.

EXPRESSED SEQUENCE TAGS (ESTS)

Immune response understanding is crucial for minimising losses in aquaculture. The use of immunostimulants, vaccines and genetic selection programmes has helped to attain this goal. The use of genomic tools such as ESTs libraries, subtractive hybridizations (SSH) and microarrays have given essential information to identify homologous genes, new genetic functions, expression profiles, routes, candidate genes. EST production, directly from organs related with the function under study or by SSH, has generated multiple databases with hundreds or thousand of genes, which constitute the platform needed for microarray construction. The developing and application of bioinformatics tools in parallel has permitted an accurate management of a large number of sequences for expression profiles and gene annotation. The developing of microarrays for the simultaneous analysis of the expression of hundreds to thousands of genes is a basic tool for the identification of signatures and candidate genes. The main advantage of the modern techniques in genomics is the large quantity of results produced but this also implies two different risks: the massive production of data in few individuals leads to the choice of models with too high parameters (overparameterization) and obtaining of false positives.

Obtaining consistent results by this strategy demands an appropriate experimental design, a correct application of statistics in data analysis, and recently, the convenience of results modelling using genetic nets has been demonstrated in order to correctly interpret gene expression patterns.

The increase of EST databases has been dramatic in recent years. Salmonids are the group in which this is more apparent. More than 175 cDNA libraries have been constructed from a wide variety of tissues and different developmental stages and more than 300.000 salmonid cDNA sequence reads have been combined from a consortium comprising groups from Canada (Ben Koop and Willie Davidson and the Genomics Research on Atlantic Salmon Project, GRASP; Susan Douglas *et al.* and the Institute for Marine Biosciences, IMB-NRC); France (Yann Guiguen *et al.* and INRA-SCRIBE); Norway (Bjorn Hoyheim *et al.* and the Norwegian School of Veterinary Science, NSVS) and the U.S.A. (Caird Rexroad III and the USDA/ARS National Center for Cool and Cold Water Aquaculture).

These sequences have been assembled into over 40.000 unique contigs. A preliminary microarray of 3.557 cDNAs has been constructed and assessed for its ability to provide new data in the study of cellular and tissue responses to pollutants, diseases and stress, as well as for reproduction and development. On the basis of these results, a larger array of 16.006 genes has been constructed and initial results have shown sensitivity of gene expression patterns to disease challenge, and to small environmental and physiological changes.

BASSMAP, an EU funded project, is focused on sea bass genomics. A Bacterial Artificial Chromosome (BAC) library was constructed with an insert mean of 164 kb and a final cover equivalent to 7 times of total genome. The BAC clones are being used to generate ESTs in order to detect genetic markers such as Single Nucleotide Polymorphisms (SNPs) and microsatellites. BRIDGEMAP is another EU funded project focused on sea bream genomics, where 1.500 ESTs have been generated with the aim of improving the knowledge of several aspects of the biology of this species. EU has funded and is supporting several research projects that have the common goal of increasing the knowledge of genomics in aspects such as health (AVINSI, AQUAFIRST, IMAQUANIM), welfare (WEALTH, STRESSGENE), reproduction (PROBASS, CRYOCITE, PUBERTIMING), growth and nutrition (FISHCAL, fPPARS). Moreover the network "Marine Genomics Europe" (MGE) has a major goal for the development, use and dissemination of genomic approaches to the study of marine organism biology and marine ecosystems.

One of the goals of the fish and shellfish node has been the production of genomic tools, including cDNA libraries, molecular markers (macro and microsatellites, SNPs), microarrays, etc., integrated in a bioinformatics platform. These tools will dramatically improve the available genomic resources that will help solving numerous scientific questions. Despite of all these efforts there is a long way in front of us. As an objective data, while the GeneBank number of sequences for rainbow trout is 244.984 and 236.009 for Atlantic salmon, there are only 200 for turbot, 2.448 for sea bream and 24.452 for sea bass. As a comparison there are 833.880 zebrafish ESTs.

IMMUNE RESPONSE AND DISEASE RESISTANCE

Another approach that is being used to characterize genes related with immune response and disease resistance if the study of differential expression of genes using Suppression Subtractive Hybridization (SSH). SSH is a powerful technique that allows the comparison of two mRNA populations, cloning genes that express in one but not in the other. Two pioneer SSH examples are the identification of T cell receptors and the identification of activated genes in *Xenopus laevis* gastrulation. Since then several modifications of the technique have been described.

Alonso and Leong used SSH Chinook salmon and staghorn sculpin (*Hemilepdotus spinosus*) to determine differentially expressed genes against Poly I:C, a potent IFN inducer, finding that the overexpressed genes were quite similar to those described as IFN inducible in mammals. Bayne *et al.* identified rainbow trout immune genes constructing a subtracted library with livers from fish injected with *Vibrio* bacterine.

The resulting library consisted in 300-600 bp fragments with 25 genes related with the immune system, of which 15 had not been previously described in salmonid fish and 12 not in any fish at all.

The library included acute phase genes such as amyloid A protein, transferrine and precerebeline type protein, complement proteins and lectins, a putative antimicrobial peptide, several membrane receptors, such as TLRs and IL-1 receptor.

O'Farrell *et al.* also used SSH to identify changes in host cells against viral infections (rainbow trout/Viral Hemorrhagic Septicemia, VHS). 24 genes were identified as induced by the virus, most of them were IFN inducible. Zhang *et al.* used SSH to study the model grass carp/Grass Carp Hemorrhagic Virus (GCHV) to conclude that IFN and IFN induced genes are the main tool against virus. Also, with a viral infection model using Chinese perch and Infectious Spleen and Kidney Necrosis Virus (ISKNV), He *et al.* found genes related with the immune response and apoptosis and with the proteasome degradation pathway of ubiquinated protein that was found also in the SSH library generated by O'Farrell *et al.* and also by Dios *et al.*

Another approach was cDNA libraries generation from primed organs and randomly sequencing of a group of clones. These were successful approaches but with high cost and low efficacy. The same approach was used by Fernandez-Gonzalez *et al.* to detect immune genes in carp experimentally infected with *Icthyophthirius multifilis* sampled at 3 and 72 hours post-infection. In a total of 3.500 ESTs, 82 ortologues with immune relevance, previously described in other organisms, were found. 61 of them have not been previously described in carp. Among them the complete Prostaglandin D2 Synthetase (PGDS), the chemokine SCYA103 and a second βm2 macroglobulin gene were found.

The expression modulation by this parasite was determined by RealTime-PCR (RT-PCR). Tsoi *et al.* used SSH to determine the differential expression of Atlantic salmon genes against the furunculosis etiological agent *Aeromonas salmonicida*. Subtracted cDNA libraries were prepared from three immune organs at two times after infection. A forward and a reverse library were constructed and 200 clones of each library were sequenced, giving a total of 1.778 ESTs that were annotated according functional categories and deposited in GenBank. Numerous genes involved in signalling, innate immunity and other processes were identified.

They include genes that are involved in acute phase and other more novel genes such as tachylectin, hepcidin, genes similar to precerebeline or methyltransferase, a putative protein that binds saxitoxine, etc. A subgroup of genes was studied with more detail by virtual Northern or RT-PCR to verify the differential expression as a result of the infection. Matejusova *et al.* used SSH to determine *Gyrodactylus salaris* infected Atlantic salmon gene expression profiles. Dios *et al.* used this technique to determine gene profile expression in sea bream brain after nodavirus infection that has the CNS as target. Forward and reverse libraries were generated one day post-infection and the ESTs expressed in infected tissues were catalogued as belonging to stress an immune response. In the reverse library (ESTs expressed in controls when compared with infected tissues) the most expressed genes were ribosomal and mitochondrial.

PHYSICAL PLATFORMS OF MICROARRAYS

Microarrays are physical platforms or chips (usually glass slides) with genes or gene sequences printed, which allow for the simultaneous analysis of the expression of hundred or thousands of genes. The study of these expression profiles in tissues or cell pools constitute the basis for functional genomics. Microarrays can be prepared by printing ESTs, generated from cDNA libraries (subtracted or not) or by short oligos (25-70 bp), designed from the ESTs specific regions such as 3' Untranslated Regions (3'UTRs) or code regions, if alternative splicing studies are the goal. Sometimes cDNA microarrays are prone to the ambiguous EST identification, as they can produce cross-hybridizations among homologous genes and they are not able to distinguish different products of the same gene resulting from differential splicing. In contrast, oligo microarrays avoid some of these problems, but are more expensive. Thus, its application in projects of species involving an important research using arrays could be justified.

Comparative studies between different platforms do not permit a definitive conclusion in this matter. In fact, controversial results between oligo and cDNA platforms have been described. The use of microarrays allows the detection of gene expression profile changes under different experimental conditions such as pathogens, probiotics, developmental stages, contaminants, etc..

The use of microarrays in fish diseases studies was based on human cDNA microarrays. Commercially available human cDNA microarrays were used to compare differential expression in the livers of Atlantic salmon infected with *A. salmonicida* compared to healthy fish. cDNA probes were prepared from total RNA isolated from livers of control salmon and infected salmon by reverse transcription in the presence of 33P-dCTP and independently hybridized to human GENEFILTERS GF211 microarrays. Of the 4.131 known genes on the microarray, 241 spots gave clearly detectable signals using labelled RNA from the control salmon liver.

Of these, 4 spots were consistently found to have a greater than 2-fold increase in infected salmon compared with controls when using the same pair of filters to generate hybridization data from triplicates. These up-regulated genes were ADP/ATP Translocase (AAT2), Na+/K+ ATPase, Acyloxyacyl Hydrolase (AOAH), and Platelet-Derived Growth Factor (PDGF-A). A BlastN search revealed an AAT2 homolog from Atlantic salmon, and a reverse transcriptase polymerase chain reaction assay using primers based on this sequence confirmed its up-regulation (approx. 1.8-fold) during early infection. This work demonstrates the feasibility of using human microarrays to facilitate the discovery of differentially expressed genes in Atlantic salmon.

This was corroborated by Renn *et al.*, which used a microarray prepared with 4.500 genes from a specific cichlid *Astalotilapia burtoni* brain cDNA library. This microarray was used with different fish species and the degree of concordance in expression profiles (number of genes and changes in the levels of expression) was consistent. Rise *et al.* used microarrays constructed with ESTs previously described to study macrophage and haematopoietic kidney gene expression modulated by *Piscirickettsia salmonis* infection. Results were validated by RT-PCR. In infected salmon macrophages, 71 different transcripts were up-regulated and 31 different transcripts were down-regulated. In infected haematopoietic kidney, 30 different transcripts were up-regulated and 39 different transcripts were down-regulated.

Ten antioxidant genes, including glutathione *S*-transferase, glutathione reductase, glutathione peroxidase, and cytochrome b558 and subunits, were up-regulated in infected macrophages but not in infected haematopoietic kidney. Changes in redox status of infected macrophages may allow these cells to tolerate *P. salmonis* infection raising the possibility that the treatment with antioxidants may reduce haematopoietic tissue damage caused by this rickettsial infection. The down-regulation of transcripts involved in adaptive immune responses (*e.g.*, T cell receptor chain and C-C chemokine receptor 7) in infected haematopoietic kidney but not in infected macrophages may contribute to infection-induced kidney tissue damage.

Molecular biomarkers of *P. salmonis* infection, characterized by immune-relevant functional annotations and high fold differences in expression between infected and non- infected samples, may aid in the development of anti-piscirickettsial vaccines and therapeutics. Purcell *et al.* also used this microarray to study Atlantic salmon gene profile against a DNA vector or an Infectious Haematopoietic Necrosis Virus (IHNV) DNA vaccine. Eighty different genes were significantly modulated in the DNA vector group while 910 genes were modulated in the IHNV DNA vaccinated group relative to control group. RT-PCR was used to examine expression of selected immune genes at the Intra Muscular (I.M.) site and in other secondary tissues. In the localized response (I.M. site), the magnitude of gene expression changes was much greater in the vaccinate group relative to the DNA vector group for the majority of genes analyzed. At secondary systemic sites (*e.g.* gill, kidney and spleen), type I IFN-related genes were up-regulated in only the IHNV DNA vaccinated group. The results presented here suggest that the IHNV DNA vaccine induces up-regulation of the type I IFN system across multiple tissues, which is the functional basis of early anti-viral immunity.

In a similar way Martin *et al.* used this microarray to study the response of Atlantic salmon against an *A. salmonicida* vaccine and found that the greatest increase in expression identified in the array analysis was a liver antibacterial peptide, hepcidin, which was increased 11-fold following the challenge. Roberge *et al.* also used this microarray to determine the levels for gene expression in *Saprolegnia* infected salmon, confirming the importance of non- specific immune response in the resistance against this disease. MacKenzie *et al.* used another microarray developed by Krasnov *et al.* including 1.380 genes printed in six replicates. They studied the ability of cortisol to directly modulate the transcriptional response of rainbow trout macrophages to the cellular activator LPS. The results indicate that cortisol significantly inhibits the well-described LPS-dependent induction of the expression of TNF-alpha2, a pro-inflammatory cytokine through a complex network of interactions. Gerwick *et al.* used an oligo microarray to study gene profile in the inflammatory process of de *Listonella* (*Vibrio*) *anguillarum* bacterine injected to rainbow trout livers with

Freund adjuvant. Microarray analysis determined that individual variability was high probably due to variable resistance to the disease. Li and Waldbieser have also reported the construction and use of a microarray for catfish aimed to the study of innate immune response.

The high density microarray was prepared from oligos based on catfish ESTs. This platform was used to study the gene profile in catfish spleens 2, 4, 8 and 24 hours post-LPS injection. 38 genes were modulated by LPS treatment. The expression of 9 genes determined by RT-PCR was positively correlated with microarray data. Byon *et al.* determined the antiviral response in Japanese flounder using a microarray. Non-specific immune response genes such as Macrophage Inflammatory Protein 1-a (MIP 1-a) receptor of Natural Killer (NK) and Kupffer cells and Mx1 protein gene were observed to be up-regulated by the VHSV G-protein DNA vaccine at 1 and 3 days post-immunization.

Also, specific immune-related genes including the CD20 receptor, CD8 alpha chain, CD40 and B lymphocyte cell adhesion molecule were also up-regulated during that time. In a later work Byon *et al.* determined that humoral defence-related genes such as complement component C3, complement regulatory plasma proteins, IgM, IgD, MHC class II-associated invariant chain and CD20 receptor were observed to be up-regulated by the VHSg recombinant protein vaccine at 1 or 21 days post-vaccination.

On the other hand, cellular defence-related genes such as CD8 alpha chain, T-cell immune regulator, MIP 1-a and apoptosis-associated protein were not detected. With a second version of this microarray Matsuyama *et al.* determined the immune response of Japanese flounder against *Edwardsiella tarda*. Among the 1.187 analyzed genes, 42 genes were up-regulated during the course of infection either in vaccinated or non-vaccinated fish.

These genes included immune-related genes, such as Matrix Metalloproteinases (MMP) MMP-9, MMP-13, CXC chemokine, CD20 receptor and hepcidin. Some immune-related genes were down-regulated after the *E. tarda* challenge, *i.e.* interferon inducible Mx protein, MHC class II-associated invariant chain, MHC class II alpha and MHC class II beta encoding genes, immunoglobulin light chain precursor, immunoglobulin light chain and IgM. Kurobe *et al.* using cDNA microarray containing 871 unique cDNAs including 91 putative immune-related genes studied the gene expression of *in vitro* grown kidney cells stimulated with mitogens such as Concanavalin A (ConA), Phorbol Myristate Acetate (PMA), LPS or infected with hirame rhabdovirus (HRV). The number of genes whose expression was increased or decreased by these factors was: 17 by Con A, 139 by PMA, 76 by LPS and 182 by HRV infection. The treatment of ConA for 1 and 6 h affected the expression of only a few of the immune-related genes. PMA down-regulated much more genes from the ones it up-regulated. Apoptosis-related factors such as c-fos, NGF induced protein IB and NR13 genes were among the genes whose expression was induced by PMA. LPS induced the expression of inflammation-related genes, such as IL-1β, monocyte chemotactic protein 1 and collagenase. The expression of many genes was induced after 3 h HRV infection but some of them were decreased to the basal level after 6 h HRV infection. The expression of some genes of unknown function was induced or reduced by Con A, PMA or LPS or by HRV infection in different time periods. Dios *et al.* studied gene profile expression of nodavirus infected sea bream. cDNA SSH libraries were generated 1 day post-infection. Some of the genes were included in functional categories of stress and immune system. To check the modulation of expression profiles, samples were taken 1, 3 and 7 days post-infection and hybridized with a macroarray with 765 genes constructed from PCR products with more than 300 bp from the SSH forward (385 genes) and reverse (380) libraries. Significant changes of antiviral genes were detected.

CLONED AND GENETICALLY MODIFIED ANIMALS

Several animal species have already been genetically modified, and at least eleven have been cloned, though some scientists doubt the health of those clones that survived to birth. Some of these efforts are commercial, either for agribusiness or for sale directly to consumers as pets. Others are scientific experiments, usually defended as advancing, directly or indirectly, the cause of medicine for humans. The cloning or genetic modification of pets serves no justifiable purpose. These efforts serve only to play upon one set of emotions-our affection for our companion pets-in order to desensitize another set of emotions-our repugnance at the idea of treating them as artifacts that can be "designed" and manufactured. The modification of livestock and the possibility of cloned meat entering the food chain are very controversial. So is the possible use of genetically modified animals to "grow" either pharmaceutical products or organs for transplant into humans.

CLONING

There have been published reports of the following species being cloned: carp, sheep, mice, cattle, goats, pigs, cats, rabbits, mules, horses, rats, and a deer. Some closely related species have also been cloned (a banteng, a wild cow, and a mouflon, a kind of sheep). A gaur, a wild ox, was cloned but died within two days. Attempts have also been made, without success, to clone monkeys, dogs, pandas, chickens, and at least two extinct species: the Tasmanian tiger and the woolly mammoth. The mammoth experiment used an elephant surrogate and tissue found in permafrost. Three major British institutions announced in July 2004 that they were setting up a tissue bank to preserve the DNA of endangered species, even after their extinction. Future cloning is seen as a possibility. The Roslin Institute, where the first mammal was cloned, maintains records of all published mammalian cloning experiments up to July 2002 - 50 papers detailing 68 experiments, with 386 surviving clones. They conclude that the "overall efficiency of cloning is typically between 0 and 3 per cent (number of live offspring as a percentage of the number of nuclear transfer embryos), irrespective of the species, the donor cell type or technique." There is no evidence that efficiency has significantly improved since.

Cloning for Livestock

About 300 bulls have been cloned, with the aim of improving the quality of breeding stock. Cloning is too expensive, at around $20,000 per bull, to clone directly for meat, but breeding bulls are worth much more than that. The meat and milk of cloned cattle is presently kept off the market pending FDA approval. Nevertheless, BIO is actively promoting the use of cloned and transgenic animals for human food, as well as for pharmaceuticals. The FDA is investigating the possibility of selling meat from cloned pigs as well as cattle, and in 2003 released a draft executive summary well in advance of the report it was supposed to summarize. "The FDA wants to gauge public reaction to the prospect of food from cloned farm animals before it decides whether to require government approval of the products before they are sold. That decision is expected to take another year."

Pet Cloning

In 1998 Arizona billionaire John Sperling gave $3.7 million to Texas A&M University to clone his pet dog, Missy (the "Missyplicity Project"). Sperling is controversial. He became wealthy as founder of the University of Phoenix, a distance learning university that has been accused by many of being a "diploma mill." He has since used his assets to support a number of idiosyncratic endeavors. The dog-cloning effort failed, and the team, reportedly against Sperling's wishes, began parallel efforts to clone a cat. The birth of the first cloned domestic cat was announced by the Texas A&M research team in February 2002. Born on December 22, 2001, "CopyCat" or "CC" was produced by fusing an ovarian tissue cell from an adult cat with a cat egg, and implanting the clonal embryo into an adult cat. This cloning "success" occurred after 188 failed attempts. Sperling retained Lou Hawthorne, a publicist from Marin County, California, to handle public relations for his dog-cloning effort. Hawthorne subsequently established "Genetic Savings and Clone" as a profit-making venture to provide pet cloning services.

The company is offering to clone cats for $50,000. They also offer "gene banking" services for up to $1395, plus up to $150/year, all of which can be credited against the future cost of cloning. They announced in April 2004 that five customers had paid the fee, work had already begun on three other cats for staff members and one more slot was available, to make "nine lives." They hope to increase production to several thousand a year, and to reduce the cost substantially.

Bibliography

A.C. Shuttleworth and R.H. Smythe : *Clinical Veterinary Surgery*, Greenworld Publication, Delhi, 2000.

Ajay Kumar Upadhyaya : *Text Book of Preventive Veterinary Medicine*, International Book, Delhi, 2005.

Anuj Sharma : *Encyclopaedia of Veterinary Science*, Anmol Publication, Delhi, 2006.

Ashis Kumar Ghosh : *Ethnomedicine for Human and Veterinary Development*, Daya Publication, Delhi, 2009.

Debasis Jana and Nilotpal Ghosh : *Essentials of Veterinary Practice*, Daya Publication, Delhi, 2011.

Dinesh Arora : *Biotech's Dictionary of Veterinary*, Biotech, Delhi, 2004.

F W Nicholas : *Introduction to Veterinary Genetics*, Blackwell Science, 2004.

G.D. Bagchi and Sushil Kumar : *Indian Traditional Veterinary Medicinal Plants*, Central Institute of Medicinal and Aromatic

Plants, 2000.

Gary R. Mullen and Lance A. Durden : *Medical and Veterinary Entomology*, Academic Press an imprint of Elsevier, Delhi, 2013.

Geo F. Boddie : *Diagnostic Methods in Veterinary Medicine*, Greenworld Publication, Delhi, 2005.

Geoffrey Lapage : *Monnig's Veterinary Helminthology and Entomology*, Greenworld Publication, Delhi,l 2000.

Gwilym O. Davies : *Gaiger and Davies' Veterinary Pathology and Bacteriology*, Greenworld, 2000.

Gyanendra Pandey : *Pasvayurveda : Study on Fauna and Veterinary Medicine in Ayurveda*, Sri Satguru Publication, Delhi, 2010.

Har Pal Singh and S.N. Maurya : *Instant Veterinary Drug Index: Biologics and Pharmaceutics*, International Book Distributing Co, 1998.

Indranil Samanta : *Veterinary Bacteriology*, New India Publishing Agency, 2013.

J L Vegad : *A Textbook of Veterinary General Pathology*, International, Delhi, 2012.

J.F. Craig : *Flemings's Veterinary Obstetrics*, Greenworld, Delhi, 2000.

J.L. Vegad and A.K. Katiyar : *A Textbook of Veterinary Special Pathology*, IBDCO Publication, Delhi, 2005.

J.L. Vegad and Madhu Swamy : *A Textbook of Veterinary Systemic Pathology*, IBDC Publishers, 2010.

John T. Abrams : *Linton's Animal Nutrition and Veterinary Dietetics*, Greenworld, Delhi, 2000.

Jules J. Haberman : *The Farmer's Veterinary Handbook* , Greenworld, Delhi, 2001.

M.K. Shukla : *Applied Veterinary Andrology and Frozen Semen Technology*, New India Publication, Delhi, 2011.

M.V. Thrusfield and E.A.M. Graat : *Application of Quantitative Methods in Veterinary Epidemiology*, International Publication,

Delhi, 2003.

Mahesh Kumar and R.D. Sharma : *Textbook of Clinical Veterinary Medicine*, Indian Council of Agri Res, 2009.

Mohd. Amanullah : *Veterinary Biochemistry and Biotechnology*,

Neelesh Sharma and S.R. Upadhyay : *Clinical Veterinary Medicine - II*, New India Publication, Delhi, 2009.

P. Kinjavdekar, H.P. Aithal and A.M. Pawde : *Anaesthesia and Analgesia for Veterinary Graduates*, Satish Serial Publishing

House, Delhi, 2013.

P. Srinivasan : *Veterinary Anatomy Of the Ox*, Bio-Green Books, Delhi, 2012.

Pawan Kumar Yadav and Mamta Kumari : *Veterinary and Human Health*, Ravi Kant Verma, Ved Prakash Saini, Agrotech Publishing Academy, 2011.

R.S. Chauhan : *Text Book of Veterinary Pathology*, IBDC Publication, Delhi, 2010.

S.K. Gupta : *Medical, Veterinary and Public Health Important Mites and Ticks*, Nature Books India, 2010.

Sohrab Ali : *Elements of Veterinary Genetics*, Arise Publication, Delhi, 2008.

Vivek M. Patil : *Information Technology in Veterinary Science*, New India Publication, Delhi, 2009.

Index